THE UNHEEDED CRY

THE Unheeded CRY

Animal Consciousness, Animal Pain, and Science

Expanded Edition

BERNARD E. ROLLIN

BERNARD E. ROLLIN is University Distinguished Professor of Philosophy, Biomedical Sciences, and Animal Sciences at Colorado State University in Fort Collins. He is the author of fourteen books, including *The Frankenstein Syndrome: Ethical and Social Issues in the Genetic Engineering of Animals*, *Farm Animal Welfare*, and *Animal rights and Human Morality.*

First edition, Oxford University Press, 1989
Expanded edition, John Wiley & Sons, 1998

LIBRARY OF CONGRESS CATALOGING-IN-PUBLICATION DATA
Rollin, Bernard E.
The unheeded cry: animal consciousness, animal pain, and science/Bernard E. Rollin.—Expanded ed.
p. cm.
Includes bibliographical references and index.
ISBN 978-0-557-57821-4
1. Animal experimentation. 2. Animal welfare. 3. Animal rights. 4. Animal psychology. I. Title.
HV4915.R65 1998

179/.3—dc21 98-3110

To my mother,
my wife, and
Michael David Hume,
with thanks

FOREWORD

When I read Bernard Rollin's book *Animal Rights and Human Morality* (New York, 1981) I realized that the animals had gained a powerful champion. Thus I was delighted when I found that another book on the subject was in preparation and considerably honoured when I was asked to write a Foreword. *The Unheeded Cry* is a book that I can recommend enthusiastically to all serious-minded people who care about animals and the ways in which they are used—or abused—in our society. It will be of equal interest to those whose interest lies in the field of human morality and ethics. Rollin writes in a clear, crisp style that compels attention. His argument is logical, rational, and persuasive.

Over the past ten years the general public has become increasingly aware of human exploitation of non-human animals. In particular, as a result of the efforts of animal activists, the public is learning more about the atrocities that may be perpetrated behind the closed doors of underground animal research laboratories around the world. How is it that a scientist, who appears to be quite fond of his dog at home can, when he dons his white coat, become so seemingly insensitive to animal suffering? How can science teach us that animals feel no pain when we 'know', from our common-sense perspective of the world, that they do feel pain? This is one of the main issues addressed in this book.

With considerable authority Rollin sets the present growth of international concern about animal welfare issues into historical perspective. He traces the changes in social mores and economic pressures that have influenced the ideology of science and the scientist's attitude towards animal nature. In addition to his own lucid commentary on these changes, Rollin allows many of the key figures to speak for themselves. He quotes long passages from their writings that convey vividly the flavour of their personalities and the scientific atmosphere of their times.

As Rollin points out, the common-sense view of most people has always been that animals, like pre-verbal children and the mentally retarded, experience a variety of human-like feelings including pain. It was from this perspective that Darwin, for example, argued that there was continuity in the evolution of mind as well as structure. Subsequently, however, American psychology introduced the concept of behaviourism and it became fashionable for scientists (at least during working hours) to view animals as animated, mindless machines—walking bundles of stimulus-responses. Not until quite recently has mainstream scientific ideology begun to change again with the admission that perhaps, after all, mind and consciousness in animals (including human animals) may be real enough to justify investigation.

But *The Unheeded Cry* offers a great deal more than an overview of changing scientific attitudes towards animals. It is not overtly an animal rights/welfare book: there are few graphic descriptions of the actual cruelties that human beings inflict on non-human beings. Yet despite this—perhaps because of it—Rollin conveys poignantly the extent to which animals suffer at our hands. And this is the more shocking because so much of their pain has been for nothing.

The book is enlivened by many anecdotes and Rollin's sense of humour bubbles throughout. Sometimes he pokes deliberate fun at the posturings of mere humans who, once they have donned their white coats, become (only too often in their *own* eyes) demi-gods. At other times we can almost hear him chuckling in the background as he allows them to speak to us themselves.

This is not an anti-science book—far from it. But it does illustrate clearly that scientific ideology is susceptible to social, political, and economic pressures, and that changes in scientific attitudes have by no means always been for the good. Rollin's plea is for the integration of human morality and ethics into scientific ideology and methodology. His plea is backed by such compelling logic that it will, I suspect, gain the support of many open-minded readers, scientists and non-scientists alike, who have been sitting on the fence, concerned about the plight of animals but afraid of being considered sentimental cranks if they express their feelings. It is particularly important for students to read the arguments presented here.

Rollin is by no means an armchair philosopher pontificating from behind the closed doors of a book-lined academician's study. He first became involved in animal welfare issues when he learned, with incredulity and dismay, something of the lack of true compassion for animals then pervading veterinary education. And, having made this discovery, he rolled up his sleeves and started to try to do something about it. He developed a course in veterinary ethics and eventually got the go-ahead from the administration of his school to present his material to the students. The project was a success—the course became one of the best attended and was integrated into the mainstream teaching.

One reason why I welcome this book is because of its implicit message of hope. Darwin, despite his belief that animals experience human-like emotions and feelings, was no more concerned with the moral implications of animal abuse than was the average man in the street at that time—and cruelty to animals, usually through ignorance, was pervasive. But today, as Rollin stresses, it is the growing moral concern for animals and their welfare among the general public that is putting pressure on scientists to investigate animal consciousness and suffering. This book, especially if it becomes required reading in science courses (as it surely must), will give added credibility, and an additional and powerful incentive, to the non-invasive, scientific investigation of animal intelligence and awareness. And surely, once there is sufficient scientific 'proof' to provide substance to the concerns of those who care, the 'unheeded cry' of millions of suffering creatures will be addressed responsibly at last.

Jane Goodall

Dar es Salaam
Tanzania
December 1988

The tale is told of a man who purchased a camel from a savant. The savant informed him that as the camel matures, it is likely to become bad-tempered and irascible, and that he would do well to castrate it at the first sign of such behaviour.

'And how do I accomplish this, O wise one?' asked the man.

'You must take two large stones, hold one in each hand, place the testicles between them, and bring the stones sharply together,' replied the sage.

'But surely this will cause severe pain,' said the astonished man.

'Not if you take great care to keep your thumbs out of the way,' responded the savant.

PREFACE

For the past decade, I have been concerned in my teaching, lecturing, and writing with raising questions about the morality of animal use in the sciences and attempting to effect significant change in this area. Unlike most philosophers, I enjoy the opportunity of working directly with scientists in many fields on a daily basis, and engaging them in extended dialogue on a wide array of moral and conceptual topics. During most of this time, my activities have been predicated on the assumption that what is most needed in this area is an ideal theory concerning the moral status of animals, which could be used to provide a criterion for assessing current practice and as a guide to change. After all, as Aristotle pointed out long ago, just as archers can sharpen their skill only if they have a target to aim at, so too, we need a measure by which to guide our moral activities in any given area.

Until very recently, Western society in general and philosophers in particular have neglected to develop such an ideal ethic for the treatment of animals; indeed, more has been written on this question during the past ten years than in the previous three thousand. In approaching this issue, I have taken the tack that such a theory might best be constructed by following through the logical implications of the ethical theory about humans implicit in our socal practice and our laws, as rationally extended to animals.[1] Further, I have become convinced on the basis of my own activities that one could elicit acquiescence from scientists to such a theory through rational dialogue, which would help them lay bare their own moral assumptions and what follows from these, something typically unrecognized by most of us, scientists and non-scientists alike, even including philosophers. Thus I have seen my task as Socratic; in Plato's judicious metaphor, as helping people recollect and appropriate in conscious fashion what they already carry within them. In my own, much cruder metaphor, I see the battle as one which could be won only by means of a form of intellectual judo—using opponents' own force to move them—rather than by intellectual *sumo*—attempting to muscle their ideas out of the arena and overpower them with mine. I believe that such a Socratic strategy is the correct one for a moral philosopher; Plato was right in his view that we cannot teach, but only remind.

While I have found this strategy quite successful, I failed to realize initially the extent to which what I call in this book the ideology, or common sense, of science overrides what to non-scientists is ordinary common sense. I soon learned that I could expect the same sorts of initial responses from scientists of all varieties the world over to my raising of ethical questions: 'Science is value-free.' 'Ethical questions are outside the purview of science proper.' And, most astonishingly, the claim that one cannot know 'scientifically' that what we do to animals matters to them. In short, built into the common sense of science is the idea that we cannot know that animals experience pain, fear, suffering, distress, anxiety, and all the other subjective states of consciousness which are so essential to our moral concern for and deliberations about our moral obligations to other people. Increasingly, I found myself in the position of being forced to 'prove' that animals were conscious, and to provide good, 'scientifically acceptable' grounds even for claiming that animals feel pain. Thus, it became clear to me that an integral part of becoming a scientist is learning to abandon ordinary common sense in a number of areas, including that of ascribing mentation and subjective experience to animals. Even more surprisingly, I found that very often such a stance was donned along with a laboratory coat, and that in their non-scientific garb, most scientists used the same mentalistic locutions in talking about animals as the rest of us. Thus, my attention gradually turned to trying to confute this morally pernicious, ideologically based scepticism about animal subjective experience.

The issue of animal consciousness, particularly subjective states like pain which are directly relevant to moral thinking about animals, forms the main subject of this book. I hope to show that denial of subjective states in animals is not an essential feature of a scientific stance, but rather a contingent, historical aberration which can be changed—and indeed must be changed—to make science both coherent and morally responsible.

My hope is that this book will help scientists break the ideological bonds which keep them from ascribing subjective mental states to animals. Additionally, I hope that it will help non-scientists with an interest in the moral status of animals to persevere in their attempt to penetrate the fortress of scientific ideology and practice, and effect change therein. Such changes are already taking place, and it is socially, morally, and scientifically necessary that they proceed rationally, rather than haphazardly.

As in my previous book, I make extensive use of anecdote in the course of developing my argument, in part, for the sake of readability, and in part, and more important, because, like Hegel, I am a believer in the existence and value of the concrete universal—that is, particular cases which vividly instantiate and communicate a general truth. So, as one reader of the manuscript remarked, it is no wonder that I am so sympathetic to George Romanes's anecdotal method!

I am grateful to the hundreds of scientists in a variety of disciplines who have treated me as a colleague and helped me to understand their thinking on many of the issues discussed in this book. Though I have encountered isolated pockets of hostility and occasionally what can only be termed 'dementia' as a reaction to my ideas and activities, the vast majority of scientists have been unusually generous in giving me a hearing and an education, and have been extraordinarily open to new ideas. I trust that they will take this book in the spirit in which it was written—not as 'anti-science', but as a constructive critique of questionable philosophical assumptions which underlie much current scientific activity, whose abandonment can only enhance the intellectual and moral validity of scientific efforts to understand the world.

In particular, I should like to thank my scientific colleagues at Colorado State University, whose responsiveness to my ideas has helped make the institution an acknowledged world leader in moral regard for research animals,[2] and who have allowed and encouraged me to teach courses under the aegis of biology, physiology, veterinary medicine, and animal science. I must especially acknowledge Dr Lynne Kesel, laboratory animal veterinarian, with whom I work every day; Dr David Robertshaw, physiologist; Dr Murray Nabors, botanist; Dr Phil Lehner, ethologist; Dr Wayne Viney, psychologist; and the following veterinary scientists and surgeons: Drs James Voss, Harry Gorman, and Dennis McCurnin. Thanks go, too, to Dr Roy Henrickson of Berkeley, who first forced me to examine the issue of animal pain, and Dr Andrew Rowan of Tufts University, who has helped catalyze my thinking.

I am also grateful to the following people, who critically read, discussed, and argued about all or part of the manuscript as it developed: Linda Rollin, Steve Sapontzis, Ron Williams, Peter Singer, and Stephen Clark. Irene Lewus, Judi Day, and Carol Monthei did not merely type the manuscript; they somehow found the Rosetta stone which enabled them to decipher my handwriting. My editors at Oxford University Press made many invaluable suggestions to improve the text. And anyone who has compiled an index knows how grateful I am to Emily Zanori for preparing this one.

In the class by itself is the debt I owe to my colleague and friend Professor Richard Kitchener, philosopher and psychologist, with whom I have discussed philosophy and science for fifteen years, and who read and commented incisively on the entire manuscript. His encyclopaedic knowledge of both philosophy and psychology has been invaluable, his patience limitless, and his friendship and encouragement unbounded.

CONTENTS

1. Science, common sense, and the common sense of science 1

Science and the compartmentalization of common sense 1
The common sense of science 8
Can common sense ever correct science? 13
Conclusion 20

2. Animal consciousness as an object of study 23

Compartmentalization and animal consciousness in scientific ideology 23
Is anthropomorphism a fallacy? 24
Common sense and animal consciousness 29
Animal consciousness and Darwinian science 32
The work of George Romanes 33
The anecdotal method 35
Anecdote versus laboratory experimentation 39
Romanes's principles 42
Romanes's method applied 46

3. Aspects of change in science and philosophy 52

How does science change?—the orthodox view 52
An alternative approach to scientific change 54
The philosophical and valuational bases for scientific change 57
Relevance of this view to the science of animal consciousness: some preliminary remarks 64
The rise of positivism 66
Positivism, behaviourism, and consciousness 69

4. The tortuous path from Romanes to Watson 74

Lloyd Morgan and his Canon 74

Loeb's mechanism	79
H. S. Jennings	83
Titchener's introspective psychology	90
Thorndike	92
Watson and behaviourism	97
Why did behaviourism triumph?	100
5. Animal pain: the ideology cashed out, 1	107
The common sense of science and the rejection of value questions	107
Ideology and incoherence	113
The case of pain	114
Moral consequences	118
Avoiding incoherence—the physicalization of pain and stress	123
Reprise: philosophy, value, and the neglect of consciousness	129
Ideology and human pain	131
6. Animal pain: the ideology cashed out, 2	135
Arguments for the insignificance of animal pain	135
The claim that animals lack concepts	136
The Wittgenstein version	140
Pain and the intellectual limitations of animals	144
Anthropomorphism and animal pain	145
Why do such arguments thrive?	147
Logical behaviourism and the ascription of pain to animals	147
Variability of pain experience in humans and animals	150
The scientific incoherence of denying pain in animals	153
The alleged unobservability of mental states	155
Mental states as a perceptual category	159
Morality and the perception of mental states	161
Application of this theory to animals	163

7. Morality and animal pain: the reappropriation of common sense 167

The rise of social concern about the morality of animal use 167
Moral concern and its impact on science 170
Law and local project review as a factor in moral change in the United States 177
British legislation 181
Law and policy in Canada and Australia 185
The new moral attitudes and the study of pain 187
Ideology crumbling: the subjective acknowledged 191
Subjective mental states as explanatory: the case of stress 196
How ignoring mental states may jeopardize research 201

8. Consciousness lost 206

Consciousness under behaviourism 206
European approaches to animal psychology: ethology 209
Lorenz and consciousness 213
Vestiges of consciousness: America and Britain 216
Tolman's purposive behaviourism 217
McDougall's unabashed mentalism 218
Vestiges of consciousness: Europe 221
Köhler and *Gestalt* psychology 221
The idealistic tradition—Buytendijk and von Uexküll 225

9. Consciousness regained: psychology 231

Internal critiques of behaviourism 231
Attacks on learning theory 235
Chomsky's critique of behaviourism 236
Cyril Burt's defence of consciousness 238
The work of R. B. Joynson 241
Ernest Hilgard and the return of consciousness 242
The cognitive turn 244

Animal consciousness regained: the study of primate communication 246
Animal consciousness in the psychological literature 247

10. Consciousness regained: ethology and beyond 251

The work of Donald Griffin 251
Social concern with farm animals as a factor in reinstating consciousness 254
The work of Marian Dawkins and other farm-animal ethologists 256
Other ethological advances in the farm-animal area 259
New methods for studying animal consciousness: Herrnstein and Gallup 261
A new approach to studying animal consciousness 264
The study of animal consciousness in mainstream science 266
Convincing the sceptics: a thought experiment 268
Conclusion 269

11. *The Unheeded Cry* revisited 273

I 273
II 277
III 283
IV 289
V 291

Notes 293

Bibliography 309

Index 323

1 Science, common sense, and the common sense of science

Science and the compartmentalization of common sense

In his philosophical writings, David Hume draws a refreshing distinction, which has undoubtedly endeared him to generations of fledgling philosophers, between what he can deduce and ponder as a philosopher, and what he can take seriously as a man. His philosophical musings had led him to a position where he could not rationally defend his belief in causation, physical objects, other minds, or for that matter his own mind. Yet, Hume informs us, when he hears the sound of laughter and the tinkle of wineglasses, he must set aside such dry abstruse philsophical reflections and be drawn into a world where the scepticism which they generate is irrelevant and ought to have no place. Philosophical reasoning, however elegant and brilliant, he says, cannot survive long if it has no point of contact with ordinary existence. Thus the same Hume who professes scepticism about the existence of other people writes volumes on how they ought to be treated. 'Be a philosopher,' he tells us, 'but amidst all your philosophy, be still a man.' Putting this edict into the mouth of nature personified, he has her warn the abstract thinker: 'Abstruse thought and profound researches I prohibit and will severely punish, by the pensive melancholy which they introduce, by the endless uncertainty in which they involve you, and by the cold reception which your pretended discoveries shall meet with, when communicated.'[1]

Here, as elsewhere, Hume points to a profound truth. Abstract reason and common sense must perennially engage one another in dialogue, and check each other's excesses. When common sense predominates to the exclusion of speculation and 'outrageous' argument which strikes at its very heart, we have sanctified conservatism by reifying the status quo and glorifying the 'we have always done it this way so why change' mentality. Had there never been frontal attacks on common sense, we would still live in a Ptolemaic, geocentric universe;

we would still own slaves; and we would still be looking at fossils as species of plants and animals which happen to be made of stone.

On the other hand, it is eminently possible to err in the other direction and to give too little credit to the sagacity of ordinary experience. Common sense is, after all, practical wisdom, which to a large extent endures because it has proved its worth in the experience of the race or the individual. The danger is very clearly present in philosophy, in which we have a tendency to soar to stratospheric heights of absurdity, losing all touch with common sense—witness the excesses of Cratylus, who, going one better than his teacher Heraclitus, who had taught that one cannot step into the same river twice, resoundingly declared that we cannot step into the same river once, and who so despaired of the possibility of communication that he ceased to attempt it, and would only wiggle his finger to indicate the futility of linguistic expression in a world of flux.

It is manifest that some degree of separation—what psychologists have called 'compartmentalization'—between one's theoretical and one's common-sense stances is inevitable; this much is implicit in our first quote from Hume. Indeed, one would be hard pressed not to separate one's career as a research scientist from one's religious commitment to fundamentalism, or one's profession as a gynaecologist from one's extracurricular activities as a lover. What is pernicious is complete, systematic separation of theory, in philosophy *or* science, from common sense, as if they were two hats, so that when one wears one hat, one feels it necessary to totally discard the other. (This is especially the case when science omits from its purview the ethical questions and their implications which its activity raises.) Yet the desirability of such compartmentalization is a major theme in the ideology of modern science, at least in so far as modern science is an institutionalized social phenomenon.

Thus, for example, at the American Association for the Advancement of Science meetings in 1985, a scientist announced the exciting discovery that one's facial expressions can determine one's moods, a fact recognized since antiquity in the writings of Aristotle, Thomas Reid, and other philosophers, in common sense, among actors, and in popular music—witness 'Whistle a Happy Tune'. He had made it a 'scientific' fact, however, by testing it in the laboratory. We have all laughed at cases in which scientists felt the need to prove scientifically what we all knew anyway. A very prominent physiologist spoke derisively to me of the flurry of recent activity to 'prove' that pets can make elderly and other lonely persons feel better about their lives. 'There are lots of things we don't need to prove or explain scientifically,' he said. 'For example, I know very well that when someone drives with the emergency brake on, the gas mileage will decrease. I can't explain it scientifically; I've seen no literature in the area, and I don't need to!'

But in becoming a scientist, too often one must foreswear what has not been arrived at by the 'scientific method' (whatever that may be), and commit oneself totally to the pursuit of 'objectivity' (whatever that may be), a calling which is generally construed as leaving no place for the accumulated wisdom of common sense unless it happens to have been transmuted into legitimate scientific knowledge by experiment or jargon. As a scientist, one learns that values, common sense, 'being a man', must be banished; their call is the song of Circe, beckoning one to intellectual ruin. They are epiphenomena, shadows, secondary qualities, as the data of the senses were to Lockean Newtonians in the eighteenth century. But when they are banished, there is nothing to check the excesses of scientism. If it violates common sense and ordinary sensibility for the physician to regard the human person as a biochemical machine, then so much the worse for common sense—and, all too often, for the common man. 'I was not trained for fifteen years in arcane lore to mouth common sense,' says the doctor to himself. Besides, did not Newtonian and Einsteinian physics both violate common sense? And what about quantum theory? It is a very small step from noting the undeniable fact that much good science violates common sense to unconsciously making the less defensible claim that all good science must or ought to do so, or that such violation is the very essence of science.

A major concern of this book is the rejection of common sense in twentieth-century science, most especially in psychology, as regards the existence and knowability of consciousness in animals. Ever since the advent of behaviourism, psychology deliberately distanced itself from common sense, in precisely the manner outlined above. In his 1939 presidential address to the American Psychological Association, Gordon Allport cited a widespread belief among psychologists 'that the very claim made by some psychologists that their work remains true to life, close to untrammeled common sense *is the very thing that disqualifies this work from being scientific*'.[2]

The Aristotelian world-view held sway for two millennia, in part because of its compatibility with common sense. ('If the earth moves, we ought to feel it move'; 'Elephants fall faster than feathers'; 'We do not see new kinds of animals appear in nature.') Overturning it was in large measure a matter of discrediting the prejudices of common sense (and, as we shall see, introducing new prejudices)—witness Descartes's polemic against what we all take for granted. Indeed, it often seems that each step forward in the science of physics is accompanied by a routing of common sense. For most people, one need not invoke anything beyond Newtonian physics to make this point; common sense is still pre-Newtonian, as is common language. In common language, the sun still 'rises' and 'sets'; and to many college students, it is still a mystery that Australians do not fall off the earth. The mathematization of nature, the movement from a closed world to an infinite universe, in Alexander Koyré's judicious locution, the relation of secondary qualities to fictions, the symmetry between motion and rest—all render classical physics and its implications as initially mystifying to the student or layperson today as they were to philosophers of the eighteenth century. And there is no question that common sense has lost its virginal status after its rape by physics. Could Hume have seriously denied the reality of mind, body, and causality except to an audience already shaken by a denial of the reality of colour, taste, and smell and by the postulation of a universal, totally inexplicable secular force called gravitation? And could Berkeley have seriously attempted to defend the traditional qualitative universe of common sense by appealing to the notion of physical objects as nothing but bundles of sensations at a time when common sense had not been seriously weakened by repeated assaults from mathematical physics? (A physics and mathematics which, incidentally, Berkeley found inconceivable!)

The remarkable successes of physics, conceptually and practically, from Galileo to the present time, have unquestionably weakened the hold of ordinary experience on the psyche of our culture: scholars of art and literature are forever attributing artistic crises to the unsettling influence of physics. And in so far as physics has had a commanding influence on the development of the other sciences, in form as well as content, they too have moved away from common sense. Only in a world in which the importance of qualitative differences was diminished could the traditional chemical elements—earth, air, fire, and water—be dethroned, and not until about 1800! Only in a world in which the earth was just one small cog in a great celestial machine could man be seen as just one small step in the inexorable—and mechanical—evolution of species.

The retreat from common sense which has characterized the development of the sciences since Newton is replicated in the education of each generation of scientists. Here ontogeny recapitulates phylogeny. Who among us who has studied physics can forget the numbing yet exhilarating shock of grasping concepts in classical physics like inertia, instantaneous velocity, and additive velocities; and then, having mastered those, experiencing the second-order shock engendered by their revision or overthrow by modern physics? So too in chemistry and biology—science overcomes and transcends common sense, which cannot even begin to deal with the ultimate unreality of species, the probabilistic basis of knowledge of quanta, thermodynamics, and populations, or even the connection between genotype and phenotype.

So in becoming educated as scientists, we often abandon common sense and the categories which govern our interpretation of ordinary experience. The study of science becomes in a remarkable way like immersion in a religious tradition such as the Talmud or in scholastic philosophy—that is, in something self-contained, complete in itself, inexplicable in categories appropriate to other domains, and protected by its own ideology and exclusions. Indeed, as the sciences become increasingly specialized and subdivided, one need not even have an overview, or conceptual map, of the whole scientific terrain; it is enough to have a so-called working knowledge of some small area. Most biologists, for example, cannot even understand their own particular area of study in terms of current evolutionary theory, though everyone would freely admit that all modern biological thought is played out on an evolutionary stage. It is as if each individual biologist is a cartographer drafting an enormously detailed map of one small part of one small country, knowing on some level that what he or she is doing is part of a greater geographical whole, but having no real need ever to refer to anything beyond his or her own region, either in space or in time—which brings us to another important point—namely, that scientific knowledge and its ideology are almost invariably devoid of historical perspective.

Thus science becomes increasingly remote from common sense and ordinary experience, for a variety of reasons, all of which pull inexorably in the same direction. In the first place, the basic concepts of science are often at odds with common sense. Secondly, the concepts of science are often not even interpretable or explicable in the language of common sense, especially in the case of physics, the exemplar for all the other sciences. Thirdly, the details of a given area of science are generally irrelevant to common sense, at least in a direct way. Fourthly, many of the rules governing scientific behaviour are radically different from the rules governing ordinary behaviour. Data, evidence, and explanations which are adequate for one are often deemed totally inadequate or irrelevant for the other.

Thus the tendency to compartmentalization flagged by Hume for philosophers is seen as a virtual necessity for scientists. 'Be a scientist, and be a man—but not at the same time' becomes their maxim. The common sense of science decrees that a wedge be driven between the truths of science and the truths of ordinary life, and that each set of truths be supported by its own metaphysics, epistemology, ideology, and valuational base, though few scientists are able to articulate these in any detail. The paediatrician who in his research sees illnesses as mechanical defects in a biological machine cannot so view his own sick child (though, unfortunately, he may so view the sick children of others). The pool-shooting microphysicist foreswears in this activity the principle of indeterminacy. And the erotically infatuated neuro-endocrinologist experiences the same highs and lows as the rest of us, totally untouched by his greater understanding of their biochemical underpinnings. The experimental physiologist would never see his own pet dog as an 'animal model', and few behaviourists can avoid mentalistic attributions in their daily dealings with friends and family.

It is as if each scientist feels that he or she must live in two worlds: the common-sense world in which we all live and the world delineated by the science he or she pursues. At some time or other a scientist may make an attempt to bring the two together, but this is usually short-lived and sometimes conceptually disastrous. Illustrative of the problems involved is the famous example of Eddington, who earnestly misapplied the categories of microphysics to the experiences of life, arguing that floors were not really solid, but were really like swarms of flies, and that it was truly a miracle that they held us up at all.[3]

The point we have just made is echoed by Allport in his 1939 speech about the state of behaviouristic psychology. In discussing psychology's retreat from common sense, Allport describes the typical psychologist who has been indoctrinated in the common sense of science as follows:

> Outside the laboratory he lives a cultured and varied life of a free agent and useful citizen. Yet his methodical work in the laboratory overspreads very little of his daily experience and prevents integration in his life. Though he

generally repudiates a dualism of mind and body, he welcomes the equally stultifying dualism of laboratory and life.[4]

Sometimes the slippage across compartmental barriers is hilarious. Recently, I was having dinner with a group of senior veterinary scientists, and the conversation turned to the subject of this book: namely, scientific ideology's disavowal of our ability to talk meaningfully about animal consciousness, thought, and awareness. One man, a famous dairy scientist, became quite heated. 'It's absurd to deny animal consciousness,' he exclaimed loudly. 'My dog thinks, makes decisions and plans, etc., etc.' All of which he proceeded to exemplify with the kind of anecdotes we all invoke in such common-sense discussions. When he finally stopped, I turned to him and asked, 'How about your dairy cows?' 'Beg pardon?' he said. 'Your dairy cows,' I repeated; 'do they have conscious awareness and thought?' 'Of course not,' he snapped, then proceeded to redden as he realized the clash between ideology and common sense, and what a strange universe this would be if the only conscious beings were humans and dogs, perhaps humans and *his* dog.

In general, then, science and common sense coexist like multiple personalities or Siamese twins, in the same body. And science has inevitably developed its own philosophical base—its own common sense, as it were—which is shared among its practitioners in much the same way as ordinary common sense is shared, and by scientists in their non-scientific moments. The Humean dichotomy is fully realized—far too fully, for the dialogue between the two common senses either falls prey to absurdities like Eddington's or, far more often, is simply never initiated. Science may continue to erode the province of common sense, but, in a scientistic age, common sense has little leavening effect on science.

This compartmentalization is inevitable and understandable, but is not necessarily beneficial; for neither way of knowing is absolute unless one makes the dubious assumption that some tradition is beyond criticism. Science has convincingly shown that common sense can be wrong. Is it then inconceivable that common sense can correct science? Certainly not, for as we have seen, science rests on its own 'common sense', its own philosophical, valuational, ideological, and metaphysical presuppositions, which are by no means necessarily true or even self-correcting, and which are rarely examined. Furthermore, criticism from within a tradition is inevitably far less trenchant than criticism from outside; and this is as true of science as it is of religion.

Moreover, it is most dramatically true where the unexamined presuppositions of science include valuational commitments with moral implications which are unrecognized. It is thus desirable to reinstate the dialogue and to examine the common sense, or ideology, of science in critical fashion; but before this can be done, its fundamental commitments must be delineated. As we shall see, many of the key assumptions of the current common sense of science arose virtually by historical happenstance in the early twentieth century, are extremely implausible, and do not stand up well to close scrutiny.

The common sense of science

By the time a student has completed an undergraduate degree, and certainly by the time he or she has earned a doctorate in one of the 'hard sciences', or even in one of the softer ones with aspirations to hardness (note that this distinction is itself part of science's ideology), he or she will pretty much have developed the following philosophical commitments. First, the student will very likely have come to believe that science deals only with *facts*, based, of course, on 'observations', and may even affirm, as the head of the National Society for Medical Research did in a public debate with me, that science is just the gathering of facts. Above all, he or she will believe that science is value-free, devoid of any basis in valuational judgements of any sort. If values have anything whatever to do with science, it is with the *applications* of the results of science decided on by politicians, as in the use of atomic theory for weaponry; but such decisions are not the concern of the scientist *qua* scientist, only *qua* citizen. (This claim was in fact made by many scientists who worked on the atomic bomb.) If science itself is rooted in values of any sort, it cannot be empirical or objective, it is thought, and it then loses its right to claim superiority as a way of knowing about the world. Scientists are generally quite adamant on this point, and seem to believe that allowing the existence of values in science would make it little better than theology or literary criticism. The full extent of this repudiation of values may be gleaned from a remark made to me by a colleague in the biological sciences after attending a lecture of mine on values in science. 'There are no values in science,' he proclaimed. 'We scientists have no truck with values. All we care about is knowing, and we do whatever it takes to know.' He was dumbfounded when I pointed out to him that his own statement itself reflected a resounding valuational commitment.

The belief that science has nothing to do with values is closely connected with the notion that only what can be tied to 'experience' is a legitimate part of science. Any concepts, valuational or otherwise, which are not 'empirical' do not belong in science. Thus scientists are also quick to distance themselves from philosophical concerns—be they ethical, metaphysical, or epistemological—in their discipline. This primitive positivism is one of the most formidable obstacles to getting scientists to look at valuational questions. At some stage in their education, they have been infused with a smattering of philosophy, taught as gospel, specifically with something like the logical positivism and operationism which had its heyday in the 1920s and 1930s, which insisted that anything not observable or reducible to observations or operations was mystical dry rot which would undermine the foundation and practice of science. This sort of positivistic discussion is often found in what one of my colleagues calls the 'throat-clearing' introductions to basic science textbooks.[5] Ironically, they typically see themselves not as adopting a philoso-phical position (and a crude one at that) in their beliefs, but merely as adhering to the rules which are constitutive of genuine science. Yet clearly, to say that philosophical assumptions and questions are irrelevant to science is precisely to make a highly questionable philosophical assumption.

Indeed, one of my most difficult tasks in trying to teach bioethics to scientists is getting them to realize that value-notions and philosophical issues in general are part and parcel of the cake of science, not merely frosting which may or may not be added later. Most are quite shaken, for example, when, in dealing with the subject of our moral obligations to animals, I point out that every invasive use of animals in scientific research involves a tacit moral judgement: namely, that the data gained in such research are of greater value or importance than the pain or suffering or death or fear engendered in the animals in the course of gathering such data. I then go on to point out that if, as biological scientists themselves believe, modern biology is inseparable from and could not exist without the invasive use of animals in research, then surely modern biology rests on a debatable valuational presupposition. What is shocking is not so much that many researchers are unprepared to defend such invasive use when it is questioned (though this is often the case), but rather that they are totally unaware of the fact that such use involves value-judgements and commitments.

Most abhorrent to scientific ideology is the suggestion that science is *fundamentally* involved with moral values. Echoing the positivism which breathed life into and informed the common sense of science, this ideology tends to see moral notions as 'emotive', as matters of mere taste and opinion. In fact, I have found it far harder to get scientists to admit that the issue of animal use is a rationally adjudicable moral one than to get them to admit that a given practice is wrong once they have accepted the relevance of moral categories to the discussion in the first place. It is perhaps because they do not see moral issues as rational questions that their responses to questions in this area often tend to be emotive and irrational, and thus a perfect complement to the irrationalism endemic among some extreme antivivisectionists.

Typical of this reaction was the response I received at a formal dinner from a prominent medical researcher who asked me to outline my concerns about animal research. I indicated that my chief concern was that many scientists do not even admit that there are any moral issues in this area. He then accused me and 'all others like you' of attempting to 'lay their trip on everyone else'. 'Morality', he told me, was 'a matter of taste in a free country.' To attempt to pass legislation restricting animal research was 'fascistic' and 'totalitarian'. I was entitled to my opinion, he admitted, but ought not to try to impose it on him. He continued in that vein at length, and wound down only when I pointed out to him that absence of constraints on the use of animals in research meant that *he* was 'imposing his trip on me', which was equally unacceptable, by his reasoning, in a democracy.

I wish I could say that such a response was atypical. But in fact, since the ideology of science tends to exclude consideration of moral and other valuational questions, such questions are rarely dealt with in science education or in scientific forums and journals, so scientists tend to be naïve and simplistic in their approaches to them, and very emotive in their responses. It is as if there were no onus on them to be rational in this area, since science, which excludes these questions, is coextensive with rationality, and as if a non-rational outburst is somehow appropriate to what they believe to be a non-rational issue. So it was that I was essentially labelled a supporter of 'Nazis' and 'lab trashers' in a review of my book on animal rights in the *New England Journal of Medicine*,[6] which furthermore refused to print my letter of response, as well as the letters of a number of major figures in the biomedical community who were disturbed less by the attack on me than by the apparent willingness of the world's leading medical journal to lay reason aside in matters dealing with ethics.

The notion that science is the activity of gathering 'real' data out there in the world, devoid of valuational commitments and arguable philosophical assumptions is but one aspect of the standard world-view of many scientists. In addition, there is the belief in a distinct scientific method by which one can inexorably arrive at true knowledge. Though most scientists are hard pressed to articulate the method, or to show precisely how they are following it in their own work, they are convinced that it is there and, more significant, that progress and conceptual change in science, when they have occurred, have always been engendered by adherence to this method. It is thus an implicit article of faith that what comes later in the history of science is better, since when notions have been abandoned, it has been because they were false. And thus the vast majority of scientists are totally ahistorical in their view of science; the history of science is merely a chronicling of past errors and their subsequent rectification. As a result, students of the sciences generally have no idea where ideas came from, or how anyone could have thought of them. Formulas in physics, for example, are to be mastered, not necessarily understood in terms of what real questions they were historically developed to answer.

So the common sense of science dictates that science is an empirical, fact-gathering, value-free activity, which makes progress in accordance with a specific method and whose history is essentially irrelevant to scientists, though perhaps of interest to historians or literary critics. In addition, as we have indicated, science is believed by most practising scientists to be clearly demarcated from philosophy. The notion that science rests on certain philosophical presuppositions is rejected or ignored by most scientists, unless they are in theoretical physics. To the average working scientist, philosophy exists in some Never Never Land, along with theology and poetry. If it is relevant to science at all, its relevance lies somewhere in the distant past, when science was in large part speculation, in the days before the so-called scientific method was articulated by somebody or other, perhaps Bacon or Galileo—it doesn't matter.

So far we have dealt with presuppositions that are easy to detect as constitutive of the common sense of scientists *qua* scientists. But there are also less obvious notions informing the world-view of working scientists. One such notion is the idea mentioned earlier that physics is the master science, the 'real' science, with the consequence that the validity of the various sciences is determined by the extent to which they approximate physics. For Aristotle, however, biology was the master science, and the categories of the other sciences—even physics—were derived from the functional categories of biology. But in the twentieth century, biologists have grown increasingly uncomfortable with organismic biology, seeing it as nothing more than natural history, and tending to feel that biology is scientific only when it approximates physico-chemistry. Thus, most modern biologists feel that molecular biology is 'where it's at', not only in practical terms of funding, but in terms of laying bare the way nature really is. And in general, the more reductionist a science is, the closer it is to explaining phenomena in terms of chemistry and physics, the more it is seen as legitimate. It is ironical that the physics which is the rallying point for biologists and others is nineteenth-century mechanistic physics, twentieth-century physics, especially quantum physics, having left these notions far behind.

This reductionist credo also has a mathematical component. Ever since the scientific revolution, which changed what counted as a fact from that which was presented to the senses to that which was expressible mathematically, the progress of a science has tended to be measured by the extent to which its results can be expressed in the language of mathematics. What is expressible mathematically is seen as value-free, universally valid, devoid of subjective taint, and 'more true' than what can be expressed only in qualitative terms. It is doubtless for this reason that much of the work of the 'softer' sciences, most notably sociology, often appears in obfuscatory quantitative jargon, which, upon careful examination, reveals itself to be little more than the pseudo-mathematization of the obvious. It is also in part for this reason that much contemporary ethology distrusts simple qualitative natural-historical observations, preferring to substitute quantifiable observations. (The works of researchers like Jane Goodall stand as shining, but rare, exceptions to this generalization.)

The mathematization of reality which characterizes the scientific revolution itself provides us with a powerful example of the essential valuational and philosophical basis of science and scientific change, to which we shall return later. Suffice it to say here that the change in world-view from the commonsensical, experientially based physics and cosmology of Aristotle and Ptolemy to the rigorously geometrical and mathematizable physics and cosmology of Galileo, Descartes, and Newton was not occasioned by 'new facts'—how could it have been? All of what *counted* as facts at the time clearly supported the old world-view: after all, stones do fall faster than feathers, and if the earth moved, surely we would feel it, and the clouds would be left behind. What brought about the revolution was, rather, a change in what was *valued*, and hence counted, as a fact. As Galileo and Descartes said or intimated throughout their writings, God is a geometer, and surely a universe governed by one set of mathematical rules is more aesthetically desirable and elegant than one which is a qualitative hodgepodge of miscellaneous, qualitatively different sense-data, since reason is a better master than sensation, Plato than Aristotle. A change in valuation thus generated a change in philosophical assumptions, which ushered in a change in science. In the same way, as we shall see later, at the hands of J. B. Watson, the very notion of what counted as psychological science changed, not for logical or empirical reasons, but for valuational ones.

It is not difficult to see from this brief sketch how all the notions we have delineated as constitutive of the common sense of science would militate in favour of a distancing of science from ordinary common sense. For the latter is tainted by value-judgements, is not grounded in a stance of objectivity, has not been certified by a method which guarantees its validity, is rooted in the past and fraught with outmoded concepts (such as the sun's 'rising'), is not translatable into mathematical notions, and does not make progress. Thus arises among scientists *qua* scientists a distrust of common sense, and even more dramatically, a distrust of sciences which are too close to common sense. In general, the more commonsensical a science is and the more accessible its concepts and truths are to the uninitiated, the less it is felt to be a real science.

Can common sense ever correct science?

Though the above sketch is brief, incomplete, and subject to a great deal of qualification and exposition of details, I believe that it can stand as a précis of what most scientists take for granted. It explains many things about the daily activity of scientists and about scientific education: why science and scientific education tend to be ahistorical, aconceptual, and aphilosophical; why scientists, unless prompted, do not concern themselves with value questions or philosophical questions raised by their respective sciences; and, most important, why a major aspect of the training of scientists involves a systematic purging of ordinary common sense (and of the sentiments and sensitivities which go along with it), and its replacement by the common sense of science. Why else is such a premium placed on the cultivation of 'scientific objectivity', 'professionalism', 'detachment', and so on in the training of young scientists, doctors, veterinarians, and the rest? Simply because one set of commitments and values, those of ordinary common sense and experience, are being replaced by another set, those of science, and because the two sets are deemed to be in some measure incompatible.

There must, inevitably, coexist in each scientist two common senses: that pertaining to his or her ordinary life and that pertaining to his or her professional life. And since establishing a dialectical balance between them is so difficult as to be almost impossible without special effort and compulsion, it is easier to keep them separate. To effect this compartmentalization, science educators feel it necessary to totally immerse a student in the new ideology, and often to discredit the old.

The success of this strategy is unquestionable on many levels. It has certainly worked on the level of increasing knowledge. Physics and chemistry as we know them could only have progressed as rapidly as they did by ignoring common sense. It also works on the level of the individual psyche: it is much easier for a doctor to see illness as a breakdown in a biomedical machine which conforms to general laws than as the plight of an individual person; or for a researcher who is inflicting pain on an animal to see it as a *model*, not a sentient being. Yet a price has been paid for this total cleavage: namely, the virtual abandonment of any dialogue between common sense and science, and the possibility of ordinary common sense having any input into science. On the other hand, science inexorably effects changes in ordinary common sense. As Paul Feyerabend has provocatively argued, dialogue has been replaced by dictatorship, with science assuming the role of expert and ordinary common sense the role of passive recipient, even in political life.[7] Yet to assume that this lack of dialogue is desirable is to assume that science always knows best, that the common sense of science is necessarily superior to ordinary common sense, and that the former has nothing to learn from the latter. It may be difficult to contest this claim when it comes to physics and chemistry; but it is clearly less persuasive when it comes to other areas of science.

Let us take some examples of how science might benefit from common sense. In a number of earlier writings,[8] I argued that contemporary biomedical science has isolated itself from ordinary common sense, with pernicious results. Virtually every precept of the common sense of science has dominated biomedical theory and practice. In its zeal to be 'scientific', biomedical science has repudiated any notion of a valuational basis for its activity. Correlatively, with the success of molecular biology and the blossoming of biochemistry, biomedicine has become increasingly mechanistic, reductionistic, and atomistic in its approach to disease and treatment, as well as to research. Similarly—and this is something we have all experienced—medicine looks more and more at regularities in disease and less and less at individual patients as anything other than cases manifesting some general, scientifically comprehensible process. Hence doctors are prone to talk of patients as 'the kidney in room 306', for example.

Almost everyone in our society is familiar with the damaging aspects of this *Gestalt*, as well as with its obvious benefits. Medicine has increasingly come to see illness as a fact to be discovered, a breakdown or flaw in biochemical or physiological machinery, and has forgotten almost completely that it makes no sense to talk of illness or disease without reference to the related concept of health, and that health is surely a matter of value, not merely of fact. No amount of data forces the conclusion that a person or animal is healthy or sick: that judgement depends on the value system of the culture or individual in question. Thus, the World Health Organization defines health as 'a complete state of mental, physical, and social well-being', surely an unabashedly valuational claim.[9] So authoritative are doctors in our society, so powerful medicine's Aesculapian authority, that for a long time ordinary common sense was suppressed, and on the basis of mechanical tests, diagnoses of illness and consequent treatment were lavishly bestowed even on asymptomatic individuals (over-use of skin tests for allergy and glucose-tolerance tests for hypoglycaemia are two excellent examples). Furthermore, as medicine became increasingly reductionistic, doctors looked more and more at bodies and parts of bodies, and less and less at persons, to the point where a person was equated with a body. An inevitable consequence of this perspective was an ever greater proliferation of tunnel vision among doctors, who looked only at the immediate medical problem, with little regard for the effect of treatment on the patient as a whole. This in turn led to a large number of iatrogenic, or treatment-caused, problems, which were then referred to other specialists, and so on. Allergists gave steroids to patients with asthma, which led to an ulcerative condition, which was treated by an internist, who prescribed cimetadine, which in turn depressed sperm count, and so on. A point was reached at which doctors no longer felt responsible for the person, or even for the person's body as a whole. Furthermore, seeing themselves as scientific biologists, doctors rarely concerned themselves with the social or even the psychological ramifications of their recommendations, ignoring, for example, the devastating consequences which telling an allergic person to get rid of his dog might have on the family. In a marvellous piece of irony, general practice itself became a speciality, and a very-low-prestige one in the medical community at that. What resulted for the public was more and more of what is summed up in the old joke that 'the operation was a success but the patient died'.[10]

In the same vein, theoretical medicine eclipsed empirical medicine, and subordinated common sense to scientific dogma. On one occasion when I was being treated for chronic asthma, it was suggested to me by a friend that a sedative might alleviate the tension which chronic asthma engenders in its sufferers, which tension then leads to asthmatic attacks. I tried it, and it worked. Shortly thereafter I mentioned it to my allergist (a prominent immunologist who fully accepted the reductionistic approach), and told him I felt better after taking the sedative. 'Nonsense,' he snapped, 'sedatives are biochemically counter-indicated in asthma; you only thought you felt better!' Thus scientific ideology eclipsed not only common sense, but logic as well.

Also inimical to common sense is the tendency endemic to modern medicine to create new diseases. In 1985, the American Medical Association declared that obesity was a major illness in the United States: not a cause of illness, but an illness, echoing a view already found in medical textbooks. If obesity is an illness, so too are motorcycle riding, walking in the rain, and overeating, since they too can cause somatic problems. Connected with this is what I have elsewhere called the 'medicalization of evil'. Today, drug addiction and alcoholism are seen as illnesses, not evils or weaknesses, thereby absolving alcoholics and addicts of any blame for their condition. (Even if there are genetic predispositions to alcoholism, that doesn't *ipso facto* make it an illness. One may have genetic predispositions to clumsiness in athletic activities, which may lead to being harmed bodily; but that does not mean that one is sick.) Even more incredibly, one now hears public service radio advertisements from the medical community declaring wife-beating to be a disease! Thus, moral problems, which are not allowed in science, are converted into scientific ones.

Over the past decade, common sense has reacted dramatically to the overly scientistic state of affairs in medicine. Movements promoting 'wellness', preventive medicine (which always had a low status in scientific medicine), holistic medicine, diet and exercise, Eastern medicine, and so on are all clear signals that large portions of society are growing tired of reductionist, value-free, scientific medicine.

The same dialectic may be seen elsewhere in society. The scientific doctor's inability to deal with death and dying as a problem of a person, not merely a breakdown of a machine, has led to massive rejection of so-called heroic medical measures and hospitalization for the terminally ill. Most doctors regard the prolongation of life as good, seeing it not as a value, but rather as a self-evident truth. But in fact, it is a value, and one not shared by all members of the populace, including some doctors in their non-scientific moments, as well as nurses, who complain that doctors are interested only in 'cure', while they are trained also to 'care'. Hence there is increasing public concern with the 'right to die', so-called quality of life, the hospice movement, and so on. This too is symptomatic of the clash between the ideology of science—or at least one branch of science—and ordinary common sense, one that I have explored in detail elsewhere.[11]

Scientistic medicine's ability to divorce itself from moral questions has led medical scientists to engage in research without regard for the most fundamental rights of their research subjects. If science is value-free and human research subjects are simply essential instruments in the research process, their treatment should be dictated by scientific imperatives, not moral ones. And so the twentieth century has witnessed the horrors of a medical science unfettered by moral considerations: Auschwitz, Willowbrook,[12] and indigents, derelicts, prisoners, and the insane routinely used for experiments without their consent, and so on, until common sense called a halt to unchecked 'freedom of enquiry', and society demanded codified moral constraints on human research, and, increasingly, animal research, as we shall see.

Common sense has also grown tired of other aspects of the medicalization of moral and valuational problems and issues. American medicine has told us, for example, that any physically healthy person who wishes to commit suicide must be 'mentally ill' and has convinced the law that it should enforce this value-judgement. Common sense, on the other hand, recognizes and honours many cases in which a rational person chooses, or has chosen, to die, from Patrick Henry's 'Give me liberty or give me death' to the man who sacrifices himself by falling on a grenade to save his buddies. In the same vein, the public has grown increasingly impatient with insanity (= illness) as a legal defence. In essence, medicine has defined evil out of existence—not surprisingly, since evil cannot be expressed in physico-chemical terms. Thus, if one has done something sufficiently heinous, one must be 'sick', and ought therefore to be treated, not punished. As Thomas Szasz has argued in his extensive writings, social deviance came to be called 'mental illness', and people so diagnosed, even when admittedly harmless, were banished to institutions, often no better than prisons, without trial, for indefinite periods of time, and without recourse.[13] Moreover, doctors pass themselves off—along with prostitutes, clergymen, and psychologists—as arbiters of what counts as a good sex life, surely a matter of value, not of biology or biomedicine.

This brief discussion of biomedicine illustrates how science can benefit by interaction with common sense or suffer through being isolated from it. Other areas of science and technology—which is not all that neatly separable from science, despite the official dictum of scientific ideology which aserts that it is—could be invoked to illustrate the same point. According to the common sense of science, for example, laboratory animals are viewed as essential to the progress of biological and biomedical science, as we mentioned earlier. They are essential 'models' of biological processes. Yet traditionally, science has been very cavalier in their treatment, and few biomedical scientists know or care much about the animals they use, save in so far as they model some process or system under investigation, even though, moral considerations aside, cavalier treatment of research animals leads to undependable, non-reproducible results. There is a wealth of literature showing how not controlling for animal stress and discomfort can skew results so badly as to render them meaningless—all of which, in broad outline, is what would be expected by lay people, on the basis of common sense. I have been asked by auto mechanics, construction workers, and salesmen, among others, how scientists can possibly obtain decent results unless they treat animals uniformly and well. But such common sense seems to get submerged beneath scientific common sense, for which animal care is a low priority. A vice-president for research at a major university involved in a great deal of biomedical research informed me that he was amazed to discover that his university was doing major projects in cardio-pulmonary research, yet housing its animals in a facility whose air system was continuous with that of an underground parking garage. Some of the leading medical research universities in the United States are notorious for the shoddiness of their animal facilities and care. It is as if one were to keep delicate physics instruments in one's barn with pigs and chickens, letting them get knocked around and covered with manure, and then, after dragging them to the laboratory behind a tractor, attempt to make microphysical precision measurements with them.

Other examples of science deviating from common sense to the peril of both are easy to come by. Our hubristic attempts to manage nature, whether in the form of 'scientific' decisions about land use, damming the Nile in the Aswan Dam, or using herbicides and pesticides, do not sit well with common sense, and invariably result in environmental disasters, disasters which ordinary people, not blinded by the ideology of science, could have predicted more often than not. Scientific assurances regarding the safety of food additives, chemicals like asbestos, nuclear power, current air-pollution levels, and so on do not jibe with common sense. 'I don't care what the experts say,' is often the response of the ordinary person living in an atmosphere that 'smells funny'; 'breathing that stuff for years can't possibly do you any good'. Toxicity testing of the LD50 variety is rendered virtually meaningless by the marked differences between animals upon whom substances are tested and humans, by the variability of results even within animal populations, depending on a multitude of factors like temperature, crowding, noise, and so on, and, most notably, by the impossibility of controlling for possible long-term effects and for synergistic effects among groups of substances.

Psychological aptitude and personality testing and increasing reliance on them by government, industry, and educational institutions also clash with common sense, and do not emerge well from the confrontation. It strains credulity, for example, to weigh the results of a four-hour examination as heavily as assessments of four years of high school or university work when deciding on a candidate's suitability for admission to college, graduate school, or professional school. It is beyond belief that how well a person does on a short test can give any idea at all of what sort of doctor or lawyer he or she will be; life does not come in multiple-choice answers. Common sense indicates that a stomach-ache, a bad night's sleep, or a fight with a girl-friend or boy-friend can wreak havoc with a test score, and that a person's career decision ought not be at the mercy of one examination. Yet scientific ideology says that all these factors are irrelevant, and continues to plump for more such testing.

There is much in common experience and common sense which flatly refutes a good deal of what is accepted by various sciences. One noteworthy example of this is provided by B.F. Skinner's *Verbal Behaviour*, an attempt to explain children's acquisition of language by simple conditioning. Even before Chomsky's devastating critique of this approach,[14] anyone who ever raised a child knew that one cannot reinforce a child into speaking, that children speak when they are good and ready, and no quantity of lollipops or spankings will affect this one way or the other. Yet so powerful is scientific ideology that it has no qualms about ignoring common experience. After all, say scientists, Newton and Einstein refuted common sense and common experience. Yes, one might reply, but there is only one Newton and one Einstein.

Conclusion

The examples we have noted in which science ignores common sense at its peril point to an important conclusion. Unquestionably, common sense can in principle always engage science in dialogue, even as science can always engage common sense. In practice, however, some areas of science have become so esoteric, so far removed from ordinary life, and so successful both in their own terms and in terms which common sense respects, as to have been effectively removed from dialogue. The frontiers of theoretical microphysics certainly violate common sense, for example, but common sense does not engage the field directly, save through the vehicle of physicists like Einstein and Bohm, who are valuationally committed to some of the precepts of common sense. In areas like biology, medicine, psychology, natural resource management, and others, however, the existential need for dialogue between science and common sense is always there. When medicine claims to be value-free, it is simply ignoring ordinary values and slipping in values of its own, values which can be criticized coherently as we saw above. Whenever a science has direct or immediately foreseeable impact on ordinary life or moral questions, it behoves common sense to reopen the dialogue, rather than to accept the science uncritically. For, as we have seen, science rests on its own values, assumptions, and metaphysics, which are often unexamined and uncritiqued by scientists *qua* scientists. (When doctors criticize nuclear defence strategies, they are doing so as concerned citizens, not as experts, and are opposing ordinary common sense to scientific common sense.)

Science which directly affects human or animal life or nature or how we look at or treat any of these must be tested in the arena of common sense. The ideology of science, which stresses knowledge above all else, and which claims cognitive superiority and that it has no truck with values, especially moral values, is naturally disposed to ignore moral considerations, and to ride roughshod over objects of moral concern. Thus scientists will admit that invasive animal use is essential to biology and that animals can be harmed or killed in the course of that use, but have been reluctant to draw the conclusion that such use is fraught with morally problematic questions, since they insist that science is value-free and that only knowledge is important. In such an area, science will never criticize itself; common sense, common experience, and ordinary values structured in philosophical, Socratic dialectic must be brought to bear upon it.

In what follows, I shall try to examine a neglected area of scientific activity in terms of some of the notions developed above; in particular, I will attempt to evaluate its assumptions and values on the basis of a philosophically informed common sense. This is the area of psychology, zoology, and ethology which concerns itself with animal thought, consciousness, mentation, and subjective experience. We will look at the attitude which held absolute sway for a long period, and is still dominant today, that animal mentation is unknowable, and that concern with it is not only unscientific but scientifically impossible—that is, not compatible with the common sense of science. This attitude colours all science involving animals, including biomedical research. We will need to look at some of the history of the issue, at some of the philosophical and valuational presuppositions underlying the common sense of science as it applies to this issue, and beyond the issue. What I hope to demonstrate is that ordinary common sense and experience, philosophically shaped, still have something to say to science, and that science needs some mechanism for ensuring that it recognize the valuational presuppositions and implications of its theoretical stances, and is ready to modify them in the light of rational philosophical criticism, which is not a normal part of daily scientific routine.

We began our discussion by referring to Hume's injunction that a philosopher must allow his functioning as a man to take precedence over his functioning as a philosopher. Elsewhere in his writings, Hume informs us that whereas the mistakes of religion may be dangerous, those of philosophy are merely ridiculous;[15] in other words, that philosophy cannot be as harmful as religion. I have attempted to apply Hume's dictum to science, to discuss the relationship between living by the common sense of science and living by the common sense of humanity. Hume, in all his prescience, clearly did not anticipate the path that would be taken by the ideology of science with regard to animal consciousness, for in the *Treatise of Human Nature* he wrote the following: 'Next to the ridicule of denying an evident truth, is that of taking much pains to defend it; and no truth appears to me more evident, than that beasts are endow'd with thought and reason as well as men. The arguments are in this case so obvious, that they never escape the most stupid and ignorant.'[16] Yet just such a denial has been fundamental to twentieth-century science, and has turned out to be as dangerous as any error in religion—hence the need for our defence of an evident truth.

It is essential to realize that the mistakes of science—its philosophical mistakes, at least—are more like the mistakes of religion than the mistakes of the speculative philosophy which Hume had in mind; for they can be incalculably dangerous and can transgress with ease basic moral principles emblazoned in common sense. This alone is enough to legitimize the demand that science pause to engage common sense, at least in those areas where failure to do so may have dangerous moral consequences. The issue of animal consciousness is just such an area, for the scientific presupposition that animal experience is unknowable has great moral consequences, as we shall see. It suggests, first, that we need not worry about animal suffering; for how can we worry about it if we can't know it? It suggests also that those of us who believe that we have moral obligations to animals have set ourselves an impossible task, since we cannot even begin to enumerate those obligations if what matters to an animal must forever remain a closed book to us.

2 Animal consciousness as an object of study

Compartmentalization and animal consciousness in scientific ideology

There is, perhaps, no area in which compartmentalization on the part of scientists is more evident than that of animal consciousness, awareness, and mental states. Professionally, researchers in these areas are trained and committed to the notion that talking about animal consciousness is 'unscientific'. After all, most will tell you, we cannot experience animal minds; we cannot gather empirical data about them; we cannot even meaningfully apply mentalistic terms like 'angry', 'frightened', 'suffering', 'in pain', and so on without committing gross anthropomorphic fallacies. 'Talking about whether or not animals feel pain,' said one researcher to me, 'is like talking about the existence of God.' 'In any case,' he continued with extraordinary frankness, 'it makes my job as a researcher a hell of a lot easier if I just act as if animals have no awareness.' Quite often, I run across researchers, even veterinarians, who deny that animals feel pain, or who assert that animal pain, if there is such a thing, cannot be anything like human pain.

It is not hard to see how such an attitude might grow naturally out of the common sense of science. How can we know what an animal experiences? How can we know that animals have subjective experiences at all? The data of science are factual, observable, testable, objectively determinable; how can we possibly gather such data about something as vague and inaccessible as the minds of other creatures? After all, psychology made no progress in the eighteenth and nineteenth centuries when it attempted to study human awareness by introspection; how, therefore, can we hope to make any progress where we can neither introspect nor communicate? Furthermore, as soon as one even begins to raise questions about animal consciousness, one seems to be confronted inexorably by philosophical questions, and, as we mentioned earlier, the ideology of science is very suspicious of such questions, which are seen as futile wheel-spinning, matters on which closure can never be reached.

The question of animal consciousness and experience seems to be too closely tied to valuational issues to be a comfortable topic for most scientists. Moreover, concern regarding animal awareness is most often expressed by individuals interested in animal welfare: 'humaniacs', in the words of one of my colleagues, 'who are trying to take my research animals away'. If the existence of animal consciousness is taken as a legitimate fact, amenable to study, it is, as we shall see in our discussion of pain, but a small step to raising questions about details of that consciousness and to wondering about the extent to which research animals suffer, fear, or are hurt, questions which strike too close to home. Most scientists, like most non-scientists, are moral, conscientious people, and would have difficulty inflicting pain, injury, fear, and other unpleasant experiences on anything they believed could suffer as they themselves do. It is undoubtedly in part for this reason that the epistemological scepticism about animal minds built into the positivistic common sense of science is so appealing to those who must work with animals. Much of the appeal of Descartes's notion that animals are machines with no mental lives must have derived from the mollifying effect which this view of animals would have had on the psyches of researchers in the growing new science of physiology, who were forced to work invasively on animals without anaesthesia.

Perhaps the most definitive squelch which the common sense of science invokes for those who would consider applying mental predicates referring to subjective states to animals or would postulate a consciousness which we can meaningfully discuss is the dreaded appellation 'anthropomorphism'. The scientist guilty of this fallacy may kiss his or her credibility goodbye. Since the only consciousness to which we have access is human, specifically our own, how can we apply such notions beyond ourselves? How can we go from human experience to animal experience, given the manifest gap between the species? Any attempt to do so comes to be seen as involving a logical fallacy, that of anthropomorphism, attributing human characteristics to non-human creatures.

Is anthropomorphism a fallacy?

Though much of our subsequent discussion will be devoted to confronting the common sense of science with a philosophically expressed critical reaction based on ordinary common sense on a variety of issues pertaining to animal mentation, it would be good to pause here and consider the force of the dreaded charge of anthropomorphism. How bad is it to be guilty of attributing human traits to non-humans? Wherein, precisely, lies the fallacy?

In the first place, must any such attribution be *logically* fallacious? Clearly not, for a logical fallacy would arise only if one were attributing human traits to the sort of thing which could not conceivably bear them. So, for example, it would certainly be logically fallacious to attribute cleverness to Tuesday or randiness to a theorem. But are animals the sorts of things to which human characteristics could not possibly apply? Certainly no one objects to attributing traits which we normally attribute to human bodies or biological processes to animal bodies or biological processes. In fact, if we did not feel that there was some commonality of biological characteristics between animals and humans, there would be no point doing biological and biomedical research on animals and extrapolating the results to humans. Indeed, we extrapolate in this way in the presence of manifest disanalogies: for example, in toxicology, lifespan studies of the effect of radiation on animals, nutritional studies, disease studies, and so on. Indeed, built into animal research is a presupposition of anthropomorphism, the assumption that many human traits are portrayed in some relevant fashion in animals.

The response of scientific ideology to my point seems obvious: that anthropomorphism is a fallacy only when it attempts to apply human *mental* traits to animals, for it is here that a discontinuity exists, rendering extrapolation invalid. But how well does this response withstand scrutiny? Such extrapolation would indeed constitute a fallacy if we knew with total certainty that animals were not the sorts of things to which mental predicates could be applied. But we have no such certainty. No rational person would apply the term 'good-natured' to a day of the week, but many sane, rational, and intelligent people in most cultures and during most historical periods have applied mental-state terms to animals. It is possible that they have been wrong in so doing; but it is certainly not a fallacy unless one begs the question by assuming what one needs to prove: namely, that animals are the sorts of things to which such terms cannot sensibly be applied. And how can the common sense of science do this when whole segments of scientific research, while denying the tenability of mentalistic attribution to animals, must in fact presuppose it in order to do their work?

To take a simple example, which we shall later examine at length, consider pain research, including the testing of analgesics, all of which is routinely done on animals. Obviously such research presents a dilemma for the common sense of science: either animals are experiencing pain (unpleasant subjective sensations) or they are not. If they are, and they are valid 'models' for human pain, that commits one to the presence of a human (or human-like) mental state in the animal, and renders one liable to the conviction of having committed the fallacy of anthropomorphism. If they are not, animals have nothing like experienced pain, in which case there is no point to such research, since one cannot model the human situation on it.

Ironically enough, such a dilemma is surely built into psychological research on fear, emotion, anxiety, mothering, learned helplessness, the effects of reward and punishment on learning, and so on, despite the efforts of behaviourism to banish mentalistic predicates from its lexicon. The fact is that, ideological pronouncements to the contrary throughout the past fifty years notwithstanding, even behaviouristic psychology often tries to explain things like fear, anxiety, pain, and so forth, things which common sense *knows* have subjective components, which are what make these phenomena interesting in the first place.

We shall return to psychology and to a detailed discussion of the dilemma later; here let us merely state the resolution. It is, of course, that in actual research contexts, a blind eye is turned to the commitment to anthropomorphism, the accusation being trotted out only when someone tries to confront the question of animal mentation head on.

There are a great many physical disanalogies between humans and animals, many of them quite glaring. This does not stop science from seeing a continuity to biological processes, or from using animals to 'model' human biological processes. Moreover, the physical disanalogies seem to be far more glaring than many of the mental ones. I am no less comfortable with the conclusion that what seemed to control a rat's pain will control mine than with the conclusion that what caused a rat's tumour will cause one in me. In fact, I have *more* confidence in the predictive power of simple animal analgesia studies *vis-à-vis* what would kill my pain than I do in animal carcinogen studies regarding what might cause cancer in me, for in the latter, differences in lifespan, metabolism, environment, dosage, and a multitude of other variables cast grave doubts on the validity of extrapolation to humans.

The current ideology of biological science seems conveniently to forget one of its most fundamental tenets when dealing with animal consciousness: the dictum universally accepted among modern biologists that all biology must be structured within the framework of evolutionary theory. It readily embraces this principle in all studies of physical processes, and hence its readiness to study fundamental biological mechanisms and apply the results up and down the phylogenetic scale. If this is the case for physical processes, ought it not also to be the case for mental ones? To the unbiased observer not steeped in the common sense of science, we may pose the following question: Which of the following propositions is more plausible and provides a better explanation of the world of living things: (A) that many of the mental states which appear in humans and have subjective dimensions—certainly simple ones like hunger, taste preference, fear, anxiety, anger, sexual desire, pain, pleasure, and so on—have analogies in the consciousness or mental states of animals—certainly in those animals in which physiological, behavioural, and contextual similarities to humans are apparent; or (B) that despite analogous behavioural responses—for example, to noxious stimuli like burns—and despite analogous neurophysiological processes and structures, consciousness is not present anywhere on the phylogenetic scale but in humans?

I shall demonstrate shortly that though the choice between A and B may be problematic for the contemporary scientist steeped in today's ideology of science, it was as unproblematic for the nineteenth-century scientist consistently imbued with the theoretical commitment to evolutionary continuity as it is for ordinary common sense today. (Certainly ordinary common sense and ordinary language find talk of animal awareness and mentation inevitable. One has only to get anyone—even a scientist, as we saw earlier—talking on the subject of his or her own pet dog to appreciate this point.) I also hope to show that, at least as far as the study of animal mind is concerned, later is not better in the history of science, and that from a logical, scientific, methodological, philosophical, and valuational point of view, nineteenth-century attitudes were more coherent than those of today, for then the doctrine of evolutionary continuity was applied consistently and its consequences were faced forthrightly, rather than being defined out of existence by a simplistic, convenient ideological positivism, even if, like today, moral implications were side-stepped.

Interestingly enough, although scientific common sense disavows talking about animal mental states, few scientists are able to adhere to its precepts, which leads to the compartmentalization referred to at the beginning of the chapter. In their daily lives, the most hard-headed, positivistic, even behaviouristic scientists slip back to ordinary common sense, and describe the behaviour of animals in mentalistic terms. If one calls them on this lapse, as I often do, they will generally claim that mentalistic talk is a sloppy shorthand, which could be eliminated if one wished to be precise, a claim which invariably remains promissory and is never fulfilled.

In fact, there is little call for scientists to come through on this. In their research reports, terms such as 'fear' and 'pain' do not appear; they are operationally replaced by 'response' notions like 'withdrawal behaviour' for fear, number and severity of 'tail flicks' (when heat is applied to the tail of a rat in standard pain protocols) or number of writhes per minute in the standard hotplate writhing test for pain. Nowhere must the bench scientist come to grips with the notion that the animal is *feeling* hurt or is frightened, since this is disallowed as non-empirical. It is forgotten that we would not be observing tail flicks or writhing in rats if we weren't interested in the experience of pain and its control in humans, and have tacitly assumed that the tail flick or the writhe is a *sign* of pain (which is inherently subjective), rather than identical with pain. What the rat is feeling is supposed to be irrelevant to the conduct and reporting of the experiment, even though the experiment presupposes that the rat feels *something*, otherwise there could be no experiment.

The only times when the compartmentalized balance is disturbed in scientists is either when they are pressed on the logic of their position by philosophers (a rare occurrence), or when they are called to account on animal welfare issues by the public, as has been happening with increasing frequency. They are then forced to confront ordinary common sense, which of course sees animals as thinking, feeling beings, which cannot accept positivistic scepticism, and which has recently begun to *care* about how animals feel! Thus the scientist must defend what he or she does *qua* scientist in the idiom of common sense. The comfortable schism between what is believed as scientist and what is believed as ordinary person breaks down; he or she must now be both at once. The tension is generally resolved as it was at a scientific seminar on laboratory animal welfare, which I was invited to address. All the participants were top scientists attempting to address public concern regarding the welfare of laboratory animals. Interestingly enough, most of them made use of mentalistic predicates such as 'happy', 'bored', 'suffering', when talking about animals, but stressed that they were 'talking in quotes'—'When are the animals quote happy end quote?' Thus, they felt, they could steer a middle ground between ordinary and scientific common sense! I then pressed them and demonstrated that the quotes were not helpful; that in the final analysis, they had either to embrace the view that animals had mental experiences or to reject it; that at best the quotes served to underscore the obvious point that if animals do have experiences, they are probably not quite like ours. Interestingly enough, they went on to suggest that we could not know that other people's experiences are quite like ours either, a profound point to which we shall return later. Often, however, scientists obscure the deep moral issues by burying them under an avalanche of empirical data. A primatologist friend of mine attended a scientific conference on animal pain, expecting to hear discussions of morally relevant issues. Instead, he encountered three days of reductionist minutiae involving neurophysiology and chemistry, during which pain as a phenomenological event or morally relevant phenomenon was never even mentioned.

Common sense and animal consciousness

To the ordinary dog-owner, it is perfectly legitimate—indeed necessary—to attribute mental states to animals, states ranging from simple feelings and emotions to intricate emotions and even plans. No one who has spent any time around dogs can even begin to describe their behaviour without employing mentalistic predicates, predicates which refer in part to subjective states: 'The dog is hungry and is asking to be fed' is an obvious response to the animal's sitting at attention in front of his dish and barking; 'He wants to play' or 'is trying to get you to play' in response to his tugging at your cuffs or bringing you his ball. 'The dog is in pain.' 'He misses my son who has gone off to college.' 'The dog dislikes Reverend Unctuous.' 'The dog is dreaming.' And so on. And these locutions are not only used; they have predictive and explanatory power, for without them, we cannot make sense of dog behaviour.

It is true that the dog is an unusual example, since we have developed in close proximity with dogs, have evolved together, and have selected the dog because of its ability to fit in with humans. But none of this weakens the basic point, that it is perfectly permissible, indeed necessary, to speak of mental states in non-human animals, and that it *works*.

It is true that people often abuse subjective mentalistic locutions when dealing with dogs—'The dog knows that his birthday is coming and that he will get a lamb chop', for example—but this is irrelevant as regards the cogency of using such locutions in ordinary, transparent cases. We sometimes abuse mentalistic locutions in dealing with people as well—'I'm sure that Professor Midlife thinks about me all the time, even though he has never spoken to me and acted like he didn't recognize me at the grocery,' asserted by a female college student, for instance. But that doesn't render all such attributions untenable. Indeed, it is by virtue of the fact that mentalistic attributions to people and to animals work most of the time that we are able to identify far-fetched or implausible cases that don't work!

But as we said earlier, it would certainly be a strange universe if we could talk of mental states only in other people and dogs. The dog is part of the human family; we are around dogs far more than any other animals and encounter them everywhere we go. But it should not be forgotten that many people who enjoy a similar familiarity with other animals are able to make predictive and explanatory mentalistic judgements about these creatures as well. Whereas most of us are not terribly familiar with horses—so unfamiliar in fact, that we cannot tell Morgans from Arabs, males from females (this is true of many city folk), and colts from ponies—'horse people' can read their moods and emotions, know when they are hurting or tense, and use this knowledge to manipulate them and communicate with them. Anyone who has struggled to learn to ride a horse can attest to the ability of certain people to read these animals with no difficulty. They can catch the animal with no bother, whereas for you, catching them is a two-hour work-out. They can get the horse to cross a stream, whereas for you it merely dances, try as you might to replicate the kicks, pelvic thrusts, and sounds of encouragement of the experts. Still other people know cattle, which to most of us are as indistinguishable from one another as bananas, and about as scrutable.

Equally important, in addition to acquiring knowledge of general psychological traits of other species, many people working with animals get to know individual differences among animals just as we do among people, a point often forgotten. There are great variations in pain behaviour, for example, even among dogs of the same breed. Unfortunately, however, the common sense of science not only discourages any concern with animal mind, but is interested only in the universal, not the particular. Thus researchers are encouraged to see all mice, all beagles, all horses as examples of a type, and to look for similarities, not differences.

In searching the literature in the early 1980s for material on what is required to make domestic farm animals comfortable, I looked in vain for psychological profiles of these animals. I hardly expected to find a treatise on 'the mind of the cow', but I did think that I would find much material on what is fashionably called 'the behavioural repertoire' of such animals. However, I did not find much of that either; it turns out that ethologists and behavioural psychologists have taken relatively little interest in these creatures. I then set out to discover what those who work with cattle feel about their charges, and talked at length with dairy herdsmen, cowboys, ranchers, and farmers. They confirmed what I had suspected, that to those who are in close daily interaction with these animals, their minds are not inscrutable. I discovered, for example, that it is common knowledge among dairymen that the personality and attitude of the herdsman towards dairy cows is the single most important variable in determining milk yield and reproductive success—this has indeed been confirmed by a rather unusual piece of production-oriented research,[1] a nice example of science reappropriating common sense. I was informed that cows have widely differing personalities, and that the key to successful handling lies in knowing those personalities.

In the course of a lecture to a group of Colorado ranchers, I asked them if they thought that castration of cattle without anaesthesia really hurt, since I had sometimes heard from scientists that cattle didn't really feel pain or that if they did, it was trivial and fleeting. 'How'd you like yours cut off with a rusty pocket knife?', they shouted, pithily evidencing ordinary common sense. A friend who is a sophisticated senior scientist and veterinarian who has kept cattle for many years shared many of his insights into bovine mentation with me. As I thanked him for the information, I remarked that he ought to write all this up, since there was so little literature on cattle psychology. 'I have enough for a book,' he replied, 'and I have no doubts about its validity. But I would never say this stuff to any of my colleagues, or put it in writing, even though anyone can determine its truth for himself. They would dismiss me as a kook even if they have had similar experiences themselves.' This situation is extraordinary. It is not as if he had seen a flying saucer piloted by little green men or a unicorn in the north forty. What he had were common-sense data which could readily be checked;

but, given the common sense of science, such data cut no ice, and are not even permitted to count as facts.

Animal consciousness and Darwinian science

It has not always been this way in science, and prima facie, it is startling that it is this way today. After all, as we remarked earlier, *all* biological science today is worked out on an evolutionary stage—one simply cannot be a reputable contemporary biologist and not accept evolution. Evolution is, as it were, part of the official common sense of biology. Evolution entails continuity, and molecular biology has elegantly underscored that continuity at the cellular, subcellular, and biochemical levels. In so far as humans represent a step in the evolution of life, they share with other creatures enzymes, proteins, functions, and structures. It would be evolutionarily odd if consciousness had emerged solely in humans, especially in light of the presence in other creatures of brains, nervous systems, sense-organs, learning, pain behaviour, problem-solving, and so on. Continuity and small variation constitute the rule in living things. If someone wishes to violate the principle of continuity and assert quantum jumps between animals, while remaining a proponent of evolution, the burden of proof is on him. Recent notions of punctuated equilibrium refer to increased *rates* of evolutionary change, not to *de novo* appearance of traits. In any case, it begs the question to apply these notions to consciousness, since punctuated equilibrium was postulated to explain gaps in the fossil record, something clearly irrelevant to consciousness. By the same token, brain evolution does seem to proceed incrementally.[2] So too does the evolution of the physiological mechanisms underlying the human experience of pain, as we shall see. So it is a puzzle how denial of animal consciousness can be part of the same ideology which asserts evolutionary progression, and a challenge for the student of science to explain how this state of affairs came to be. As we shall see, this denial entailed the discrediting of *all* consciousness, animal *and* human, and the relegation of the notion of consciousness to the ash heap of outmoded, 'non-scientific' ideas. Most scientists could not of course subscribe to this view—no amount of compartmentalization can remove one *that* far from ordinary common sense, and no science can be purged of the need for communication among people, and consequently of the need for mentalistic attributions referring to subjective states in humans. But science can survive without attributing subjective mentalistic traits to animals and, indeed, as indicated earlier, becomes easier without them, for one is then relieved of the need to worry about animal pain, fear, grief, suffering, boredom, or happiness, and can proceed with impunity to write off the moral twinges that stem from ordinary, common-sense beliefs as relics of one's benighted pre-professional past.

For the half-century stretching roughly from 1870 to 1920, no such mental gymnastics were required. On the contrary, Darwinian science gave new vitality to ordinary common-sense notions that attributed mental states to animals, but which had been assaulted by Catholics and Cartesians. During these fifty years, the guiding assumption in psychology was one of continuity; so the study of mind became comparative, as epitomized by Darwin's marvellously blunt title for his 1872 work *The Expression of the Emotions in Man and Animals*, a title which brazenly hoists a middle finger to the Cartesian tradition, since Darwin saw emotion as inextricably bound up with subjective feelings. Furthermore, in *The Descent of Man* of the previous year, Darwin had specifically affirmed that 'there is no fundamental difference between man and the higher animals in their mental faculties', and that 'the lower animals, like man manifestly feel pleasure and pain, happiness, and misery'.[3] In the same work, Darwin attributed the entire range of subjective experiences to animals, taking it for granted that one can gather data relevant to our knowledge of such experiences. Evolutionary theory demands that psychology, like anatomy, be comparative, for life is incremental, and mind did not arise *de novo* in man, fully formed like Athena from the head of Zeus.

The work of George Romanes

Perhaps the most detailed exposition of the notion of continuity of mental life is found in the work of George Romanes, a scientist of independent means who later became secretary of the Linnaean Society. Enjoying the active support and encouragement of Darwin, Romanes may be viewed as speaking for Darwin himself, as two centuries earlier Samuel Clarke spoke for the other titanic figure of British science, Isaac Newton, in an extended debate with Leibniz on the nature of space and time. Just as Newton had supplied Clarke with the principles and details of his response to Leibniz, so too, Darwin provided Romanes with the outcome of years of enquiry into the question of animal mentation. In the preface to his 1882 volume *Animal Intelligence*, Romanes acknowledges his debt to Darwin, who, in his words,

> not only assisted me in the most generous manner with his immense stores of information, as well as with his valuable judgment on sundry points of difficulty, but has also been kind enough to place at my disposal all the notes and clippings on animal intelligence which he has been collecting for the last forty years, together with the original manuscript of his wonderful chapter on 'Instinct.' This chapter, on being recast for the 'Origin of Species,' underwent so merciless an amount of compression that the original draft constitutes a rich store of hitherto unpublished material.[4]

While Romanes's work focuses mainly on cognitive ability throughout the phylogenetic scale, he also addresses emotions and other aspects of mental life, all of which, for a Darwinian, ought to evidence some continuity across animal species.

Although mainstream, establishment science in its time, Romanes's work is now virtually forgotten, and has not been reissued for more than seventy-five years. If reference is made to his work today, it is invariably uncomplimentary, dripping with scorn for his 'anthropomorphism', his *naïveté*, his uncritical acceptance of anecdote, his failure to provide experimental justification for his claims, his mentalistic tendencies, his 'sentimental Darwinism', and his failure to anticipate Lloyd Morgan's Canon, according to which one should not invoke a higher faculty to explain a piece of behaviour if a lower faculty will do. Let us recall, as Thomas Kuhn has reminded us, that the history of science, like all history, is written and rewritten from the perspective of those who have triumphed.

While there may be much to criticize in Romanes's ambitious work, there is also much to admire. Romanes was far from being a gullible, naïve, unsophisticated, cracker-barrel naturalist who deserves to be patronized by today's experimentalists and their historians. He was quite conscious of the methodological and philosophical assumptions which underlay his work, and was quite willing to defend them, far more so than the majority of scientists writing on these issues today.

Romanes begins, ironically, by taking to task traditional works on animal intelligence, works of 'popular writers' whose motivation is presumed to be titillation of the imagination in order to gain a readership, who have, 'for the most part, merely strung together, with discrimination more or less inadequate, innumerable anecdotes of the display of animal intelligence. This has served to discredit the subject-matter, and exclude it from the hierarchy of the sciences.' In fact, he goes on to assert, 'the phenomena of mind in animals, having constituted so much and so long the theme of unscientific authors, are now considered unworthy of serious treatment by scientific methods.'[5]

No question, then, that Romanes sees himself as a serious scientist, not merely a 'stringer of anecdotes'. Further, there is no question that he was so viewed by his peers: not only was his work admired and cited by Darwin, but even Lloyd Morgan and J.B. Watson, often singled out as Romanes's most trenchant critics, respected and referred to his work. It is high time, Romanes tells us, that we realized that 'the phenomena which constitute the subject-matter of comparative psychology, *even if we regard them merely as facts in nature*, have at least as great a claim to accurate classification as the phenomena of structure which constitute the subject-matter of comparative anatomy.'[6] The phenomena of animal mentation are the subject-matter of facts, facts which have been common coin throughout human history, facts which are in principle no more problematic than any other kind of facts. We can know them as we know the facts of human mentation. There are no epistemological chasms to be bridged or metaphysical barriers to be scaled. The data pertaining to animal thought have been accumulating since the dawn of humanity; the methodological problem is separating the wheat from the chaff, the same sort of problem which exists in all areas of human knowledge. Fresh impetus has been given to this enterprise, Romanes says, by the advent of evolutionary theory, which gives greater credibility to these facts of common experience, explains them theoretically, and points to the 'high probability' of 'genetic continuity' between human and animal intelligence.[7]

The anecdotal method

But if what ensues is simply the chronicling of anecdotes relating to animal mentation along the phylogenetic scale, how does Romanes differ from the anecdote-mongers whom he deplores? In the first place, he would doubtless say, in seriousness of purpose, and second, in attempting to place his anecdotes within the context of evolutionary theory. But third, and more to the point, he differs in the care with which he provides criteria for selection of the facts which he addresses. As he puts it, 'considering it desirable to cast as wide a net as possible, I have fished the seas of popular literature as well as the rivers of scientific writings'.[8] Initially, he had intended to countenance as facts only what had been reported by observers known to be competent; but soon he realized that this was too rigorous, since the probability of 'the more intelligent individuals among animals happening to fall under the observation of the more intelligent individuals among men' was extremely low. So instead, he looked to other, less restrictive principles:

First, never to accept an alleged fact without the authority of some name. Second, in the case of the name being unknown, and the alleged fact of sufficient importance to be entertained, carefully to consider whether, from the circumstances of the case as recorded, there was any considerable opportunity for malobservation; this principle generally demanded that the alleged fact, or action on the part of the animal should be of a particularly marked and unmistakable kind, looking to the end which the action is said to have accomplished. Third, to tabulate all important observations recorded by unknown observers, with the view of ascertaining whether they have ever been corroborated by similar or analogous observations made by other and independent observers. This principle I have found to be of great use in guiding my selection of instances, for where statements of fact which present nothing intrinsically improbable are found to be unconsciously confirmed by different observers, they have as good a right to be deemed trustworthy as statements which stand on the single authority of a known observer, and I have found the former to be at least as abundant as the latter. Moreover, by getting into the habit of always seeking for corroborative cases, I have frequently been able to substantiate the assertions of known observers by those of other observers as well or better known.[9]

What is one to say of this method of Romanes, which to a working laboratory scientist would appear to be not much of a method at all? Where are the controlled experiments? The hypotheses and tests thereof under controlled conditions? Where is the distinction between the observations of trained scientists and those of laymen? Where does Romanes get off making the dubious assumption that animals have mental traits at all, rather than being merely complex mechanisms? Hearsay! Anthropomorphism! Anecdote! Not to be taken seriously!

Some of these objections are dealt with rather presciently by Romanes himself. Responses to others can be extrapolated from the logic of his position. However, none are terribly difficult to deal with once we have deduced the consequences of his assumptions. One of his assumptions is that, given evolutionary theory, it is hard to see how one can avoid postulating continuity of mental traits between humans and animals. Second, common sense across all ages and all cultures has seen and explained animal behaviour in terms of mentalistic attributions. Romanes even has an argument to this effect: namely, that if we do not allow appropriate animal behaviour to count as evidence of feeling and mentation, what right do we have to allow appropriate human behaviour to serve as such evidence? The only consciousness to which we have direct access is our own, the minds of other people being as inaccessible as those of animals. If we can argue by analogy in the one case, we can surely do so, *mutatis mutandis*, in the other. But if we choose to be sceptical about animal minds because we have no direct access to them, we must extend our scepticism to the minds of other people as well. And not only to other people, but also to the external world, for, in the final analysis, all we have are our own perceptions, which do not certify the existence of an external world existing inter-subjectively and outside perception. But in that case, physical science is no more coherent conceptually than mental science; so the science of bodies is no more defensible than the science of minds. In so far as ordinary common sense disregards this sort of scepticism as idle chatter, it must do so for mind as well as for body, animal or human. Romanes puts it this way:

> The only evidence we can have of objective mind is that which is furnished by objective activities; and as the subjective mind can never become assimilated with the objective so as to learn by direct feeling the mental processes which there accompany the objective activities, it is clearly impossible to satisfy any one who may choose to doubt the validity of inference, that in any case other than his own mental processes ever do accompany objective activities. Thus it is that philosophy can supply no demonstrative refutation of idealism, even of the most extravagant form. Common sense, however, universally feels that analogy is here a safer guide to truth than the sceptical demand for impossible evidence; so that if the objective existence of other organisms and their activities is granted—without which postulate comparative psychology, like all the other sciences, would be an unsubstantial dream—common sense will always and without question conclude that the activities of organisms other than our own, when analogous to those activities of our own which we know to be accompanied by certain mental states, are in them accompanied by analogous mental states.[10]

So much then, for metaphysical objections. But what of the other questions which today's scientist might raise? Even if we grant that there are facts of animal mentation, surely they are best studied by controlled experiment, not by sifting through anecdotes.

Is Romanes's anecdote-sifting scientifically invalid? Does it require that we weaken and suspend ordinary canons of proof? Quite the contrary. What Romanes recommends that we do with the vast hodgepodge of data relevant to animal mentation is to apply to it the same sort of reasoning which we employ when we reconstruct historical events, write biographies, assess people's motives, their guilt and innocence in trials, defend ourselves against accusations, make judgements about people of whom we hear conflicting stories, and so on. In all these cases, what we do is to measure data against standard rules or canons of evidence and plausibility. Does the data violate known laws or established evidence, as in the case of the parrot reported by Locke to be able to hold conversations? Does the source of the data have a vested interest in telling a certain sort of tale? Does it have an axe to grind? Does the data accord with other data from totally independent sources across time and space? In short, weighing data on animal mind is no different from weighing any other data,—for example, data pertaining to the personality of some historical figure. Thus, we discredit Frau Hitler's assertion that Adolf was a nice boy who wouldn't hurt a fly. If you start out with the a priori assumption that no data are relevant *vis-à-vis* the existence and nature of animal thought, then such a method is absurd. But if you begin with the common-sense (and evolutionary) view that the existence of animal pain, suffering, guilt, planning, fear, remorse, loyalty, and other mental states is self-evident, and that what needs to be done is to sort out the scope and the limits of such states, then this method is not only plausible, but, given a sea of data, inevitable, in order to separate the wheat from the chaff.

But surely, says today's scientist, such random observations must lack the credibility of data garnered by trained observers under laboratory conditions! This at least, we must surely concede. Not at all—no more so than we must concede that laboratory experiments provide a better guide to human motivation and human nature than ordinary experience. No controlled experiment will provide me with better evidence that one of my friends is a lecher or that people will cut corners to make money than does my ordinary experience. Laboratory experiments on animal thought (or behaviour) tend to focus on abnormal animals under highly abnormal conditions. Though the laboratory rat and the cat are among the most highly studied subjects in twentieth-century psychological research, much of the data pertains to their behaviour under the most unlikely conditions imaginable, as when they are being shocked, frightened, or presented with inescapable painful situations designed to create 'learned helplessness', or are crowded, blinded, confined, and so on. It is hardly surprising that little has come from all this which even begins to describe—let alone explain—'normal' cat behaviour. In a key sense, all laboratory examination is by definition extraordinary and likely to be misleading as regards ordinary life. To study normal behaviour is to study animals in their normal environments, without human interference or manipulation. And such study must necessarily be 'uncontrolled' and anecdotal. (Even some of the pioneering work of Lorenz and Tinbergen is often derided by ethologists and psychologists steeped in the common sense of science for not being 'controlled', and for being excessively 'anecdotal', mere 'natural history'.)

Anecdote versus laboratory experimentation

But controlled conditions are needed for replicability, say our scientists. True. But is not similar data from diverse, unconnected sources scattered over space and time a form of replication? Why does replication qualify only if it takes place in laboratories, not in nature or in ordinary experience? (If science truly insisted on data obtained only under laboratory conditions, Newton could not have explained the tides, and astronomy would not be a science!) In fact, most data produced in laboratories, while certainly replicable in principle, are rarely replicated in fact; after all, there is no funding for replicating other people's experiments, unless we have good reason to suspect their results. But observations of the sort reported by Romanes are also replicable in principle in a number of ways. One can go out and observe animals and see if they behave as reported. Or one can set up experiments to evoke the same kind of behaviour. The fact that an observation was reported outside a laboratory is no reason to reject it out of hand. To even the most hard-headed scientist, much of the data reported by Romanes could, at worst, be seen as a heuristic device for suggesting viable experiments. In fact, many of Romanes's findings, such as his reports on the problem-solving ability of squids and the ability of monkeys to use tools and seek revenge, have been verified by subsequent experimentation, and used even by his critics, like Watson.

But the scientists' distrust of naïve, natural-historical observations still lingers, and is difficult to shake. Let us, then, ask scientists the following: Is it more implausible to trust disinterested reports of people who do not stand to gain by the information they disseminate, who have no vested interest in coming up with a conclusion and no professional theoretical bias, who report what they observe and whose reports are echoed across the ages; or to trust those who stand to gain from the results they report, and who approach those results with strong theoretical prejudices? Obviously, it is less plausible to trust the latter. Yet the latter are your average academic scientists publishing research reports! Given the socio-economic situation in universities today, one can only progress in one's career as a scientist—indeed, one can only *have* a career as a scientist—if one publishes papers and draws research funding. Success as a teacher or even, in a human or veterinary medical school context, as a clinician does not suffice. Universities depend on research funding, and pass on responsibility for garnering such funding to individual faculty members. The effects of this state of affairs upon education are well known. Since teaching is seen as a necessary evil, senior scientists and clinicians do as little as possible, and undergraduate courses grow larger and larger, and are taught by junior faculty and graduate students, in the United States at least. Senior researchers take graduate students, but the latter, rather than receiving the full measure of the researcher's experience, usually end up narrowly trained, serving as cheap labour for whatever research happens to be funded. Scientists are pressed into doing research in recognized, safe, narrow, non-risky, non-boat-rocking, mainstream, accepted areas. Innovative or revolutionary research, which may fail or which may be seen as deviant by inherently conservative funding agencies and their review boards, is discouraged or selectively ruled out.

It is becoming increasingly recognized that such pressures must inevitably damage the quality of science. Not only do they encourage conservatism and narrow specialization; by creating an atmosphere of 'publish or perish', they prescribe a creed of results *über alles*. Thus we are witnessing an increasing number of scandals involving 'fudged data', falsified data, pirated data, sloppy data, and so on. Anyone randomly perusing the pages of *Science* over the past decade can regularly find lurid tales of senior researchers in all fields at top institutions caught in various acts of piracy, theft, falsification, and so on. And obviously, the reported cases are only the tip of the iceberg. From my own daily dealings with researchers, I know that a certain amount of 'data cooking' is simply accepted as an unfortunate fact of life. A recent study presented at the American Psychological Association annual meetings in 1987 reported that one out of three scientists at a major university suspect a colleague of falsifying data.[11]

In reality, pressures for success and survival are only the baser motives for cheating on results. There are much more noble and fundamental reasons built into the fabric of science. Commitment to a theory as manifestly true and consequent disregard or explaining away of recalcitrant data are probably necessary for scientists to forge ahead.

Though Newton's calculations were notoriously inaccurate in describing the behaviour of the moon, the resultant discrepancies were seen as a challenge for future research, not a disconfirmation of Newton. Recent work has shown that the likelihood of Mendel actually getting the results he reported in growing pea plants was one in 14,000, or 0.00007. Clearly Mendel was convinced of the truth of his position, and simply made the data conform. Similarly, when Einstein was asked by journalists what he would have said if the astronomical data which Eddington gathered had not supported the general theory of relativity, he replied, in essence, 'So much the worse for the data—the theory is correct!'

Likewise, it has been shown that the so-called Rosenthal effect, according to which investigators find what they expect to find, is operative in research. Thus, if students are told before doing an experiment that one group of rats is smarter than another, they will report that that group learns faster than a second group, even when the groups are randomly selected.[12]

The point I am making is this: scientists are humans, just like everyone else, even in their scientific moments, contrary to what positivistic ideology and the common sense of science would have us believe. As such, they are subject to economic pressure, pressure to produce, pressure to publish, pressure to advance, pressure to be first, bias in favour of their own theories, their pet theories, the accepted theories under which they labour and through which they have defined their life's project and from which grows their sense of self-worth.[13] Like other people, they are mainly honest and straightforward; but also like other people, sometimes they are not. Thus it is difficult to see why the published, generally unchecked results of a scientist whose success and even survival depend on publishing those results ought to be more credible than the disinterested observations of neutral observers with no vested interest in results, whose observations are corroborated by those of other observers with whom they have had no truck. This seems especially true when we are talking about something like animal behaviour, which seems to demand that it be observed, at least initially, under natural circumstances, and where a trained observer is not required in the sense that someone is needed to read radiographs, do electron microscopy or X-ray crystallography or mass spectrometry, or to detect lameness in a horse.

A recent book, *Betrayers of the Truth*, by two science journalists, Nicholas Wade and William Broad, who chronicle the activities of science and scientists for *Science* magazine and the *New York Times* respectively makes much the same point. As journalists, they have not been imbued with the common sense of science, though they are very aware of its existence and its influence; indeed, they lay bare various aspects of this ideology early in their discussion. In their view, as in mine, scientists (and others) forget that science is a competitive, human enterprise, subject to the same sorts of corruptions, foibles, and seductions as any other human activity. The authors deftly document the fact that fraud and falsification are not new phenomena in science, but that they have increased as a result of current pressures for success to which scientists are subject. One passage early in the book is noteworthy:

> Where the conventional ideology goes most seriously astray is in its focusing on the process of science instead of on the motives and needs of scientists. Scientists are not different from other people. In donning the white coat at the laboratory door, they do not step aside from the passions, ambitions, and failings that animate those in other walks of life. Modern science is a career. Its stepping-stones are published articles in the scientific literature. To be successful, a researcher must get as many articles published as possible, secure government grants, build up a laboratory and the resources to hire graduate students, increase the production of published papers, strive to be awarded a tenured post at a university, write articles that may come to the notice of committees that award scientific prizes, gain election to the National Academy of Sciences, and hope one day to win an invitation to Stockholm.
>
> Not only do careerist pressures exist in contemporary science, but the system rewards the appearance of success as well as genuine achievement. Universities may award tenure simply on the quantity of a researcher's publications, without considering their quality. A laboratory chief who has skillful younger scientists working for him will be rewarded for their efforts as if they were his own. Such misallocations of credit may not be common, but they are common enough to encourage a certain evident cynicism.[14]

In January 1987 a major article in *Nature* pointed to numerous irregularities pertaining to checking and publishing of scientific data by senior, respected researchers, echoing the major point made by Wade and Broad.[15]

Romanes's principles

There is, in sum, nothing odd about crediting the sort of information which Romanes gathered. Not only did he adhere to the principles outlined earlier, but he admitted that his work was just a beginning of a theory of mind across the phylogenetic scale. This is not to say he was totally devoid of theoretical commitment; as mentioned earlier, he was committed both to the notion of evolutionary continuity and to the analogical ascription of mental properties to animals built into the ordinary common sense of virtually all cultures. Yet even here, he made critical refinements. Early in his work, Romanes provided objective criteria for the ascription of mentation. In reponse to the question as to why we are willing to attribute mind to higher animals but not to inanimate objects, Romanes said that we do so because, in the case of inanimate objects,

the objects are too remote in kind from my own organism to admit of my drawing any reasonable analogy between them and it; and, secondly, because the activities which they present are of invariably the same kind under the same circumstances; they afford no evidence of feeling or purpose. In other words, two conditions require to be satisfied before we even begin to imagine that observable activities are indicative of mind: first, the activities must be displayed by a living organism; and secondly, they must be of a kind to suggest the presence of two elements which we recognise as the distinctive characteristics of mind as such—consciousness and choice.[16]

But these criteria are only preliminary. He reminds us that many actions which seem to involve intentional and conscious choice turn out in fact to be automatic:

Physiological experiments and pathological lesions prove that in our own and in other organisms the mechanism of the nervous system is sufficient, without the intervention of consciousness, to produce muscular movements of a highly co-ordinate and apparently intentional character. Thus, for instance, if a man has his back broken in such a way as to sever the nervous connection between his brain and lower extremities, on pinching or tickling his feet they are drawn suddenly away from the irritation, although the man is quite unconscious of the adaptive movement of his muscles; the lower nerve-centres of the spinal cord are competent to bring about this movement of adaptive response without requiring to be directed by the brain. This non-mental operation of the lower nerve-centres in the production of apparently intentional movements is called Reflex Action, and the cases of its occurrence, even within the limits of our own organism, are literally numberless. Therefore, in view of such non-mental nervous adjustment, leading to movements which are only in appearance intentional, it clearly becomes a matter of great difficulty to say in the case of the lower animals whether any action which appears to indicate intelligent choice is not really action of the reflex kind.[17]

Given the existence of such actions, one needs additional criteria for the attribution of mentation, and Romanes attempted to supply these.

Very early in his discussion, he argued that the only distinction between adaptive movements due to reflex action and adaptive movements due to mental perception, consists in the former depending on inherited mechanisms within the nervous system being so constructed as to effect *particular* adaptive movements in response to *particular* stimulations, whereas the latter are independent of any such inherited adjustment of special mechanisms to the exigencies of special circumstances:

> Reflex actions under the influence of their appropriate stimuli may be compared to the actions of a machine under the manipulations of an operator; when certain springs of action are touched by certain stimuli, the whole machine is thrown into appropriate movement; there is no room for choice, there is no room for uncertainty; but as surely as any of these inherited mechanisms are affected by the stimulus with reference to which it has been constructed to act, so surely will it act in precisely the same way as it always has acted. But the case with conscious mental adjustment is quite different. For, without at present going into the question concerning the relation of body and mind, or waiting to ask whether cases of mental adjustment are not really quite as *mechanical* in the sense of being the necessary result or correlative of a chain of physical sequences due to a physical stimulation, it is enough to point to the variable and incalculable character of mental adjustments as distinguished from the constant and foreseeable character of reflex adjustments. All, in fact, that in an objective sense we can mean by a mental adjustment is an adjustment of a kind that has not been definitely fixed by heredity as the only adjustment possible in the given circumstances of stimulation. For were there no alternative of adjustment, the case, in an animal at least, would be indistinguishable from one of reflex action.[18]

In essence, the working criterion for mentation is the ability to adapt to new situations and to learn from experience. Romanes admitted that this criterion merely fixes an 'upper limit' of non-mental action; it does not fix a 'lower limit' of mental action. In other words, if an organism displays the ability designated by these criteria, it must be said to enjoy 'mind'. But if it does not display these abilities, it could still possess some nascent subjectivity.[19]

As a philosophical account, this appears at first look to leave much to be desired. For one thing, it narrows the focus of discussion to 'intelligent' behaviour, and seems irrelevant to other mental states such as the ability to feel pain or experience emotion. Could not an organism be devoid of adaptability in its behaviour, yet still enjoy other mental states, as Romanes seems sometimes to admit? Perhaps, as

Lloyd Morgan argues, Romanes does put too much stress on intellectual ability when discussing consciousness.

Or is Romanes possibly suggesting that there is a conceptual link between any mental state whatsoever and intelligence or plasticity? Such a link would seem implausible. There is no logical absurdity involved in imagining an organism which can feel pain or anger, but whose responses to these feelings are fixed and invariable. But although this is conceptually possible, it would surely seem wildly implausible to an evolutionist like Romanes. For to a thinker in the evolutionary tradition, the value of emotions or pains for an organism lies in their being able to elicit variable, appropriate behaviour, depending on the circumstances. Pain or fear, for example, presumably stimulates the organism to consider and choose among the available strategies for diminishing or eliminating it. If the organism's behaviour were totally fixed, the subjective experience would be essentially irrelevant to its behaviour, and would thus serve no purpose. It is for this reason presumably that Romanes placed such a heavy emphasis on intelligence as the fundamental criterion for mind.

Alternatively, one may attribute to Romanes great philosophical sophistication and suggest that he had in mind some very subtle proto-cognitivist theory of the logical relationship between any mental state whatever—be it pain, emotion, or thought—and intelligence. Such a theory might suggest, as Kant seems to have argued, that any mode of awareness whatever must involve some propositional judgement. In other words, even to experience something as simple as pain, one would need at least to be able to entertain the proposition 'I feel pain' or 'Something is hurting me'. This in turn would require a concept of 'I' (versus 'non-I') and the attribution of some state or predicate to the 'I' (along, perhaps, with some notion of 'now'). But this would require an ability to discriminate or highlight certain parts of one's sensory input, and to relegate the rest to the background. Such an ability, however, would surely involve flexibility of the sort postulated by Romanes, for on two different occasions, given exactly the same stimuli, the organism might highlight the pain or not do so, depending on the context in which the stimuli were embedded. In short, one could construct an argument that any mentation at all, even pain or anger, is propositional or linked to cognition, and thus requires at least sufficient flexibility for the organism to be able to pick out certain features from the 'buzzing blooming confusion' rather than others, and thus that even rudimentary sensation requires 'intelligence' in Romanes's sense.

A related point is also interesting: namely, the omission from his criteria of any reference to subjectivity or subjective awareness and consciousness. Romanes certainly believed in these, as I indicated earlier, yet he fails to mention them here. Is he thereby playing into the hands of future behaviourists who argue that talk of private experience and subjective mental states is essentially irrelevant? And what of the mechanical devices and computers which have surfaced since Romanes wrote, which are capable of learning from experience? Would Romanes have to say that they have 'minds'?

Such objections are not as damaging as they appear, I think. First, it is not that subjectivity has dropped out for Romanes; it is rather that he is discussing objective methods for determining the presence of subjectivity. Indications of intentionality, plasticity, flexibility, and choice are objective signs of subjective awareness. As to whether machines which met these objective criteria would then have to be said to possess subjectivity, this would doubtless be dismissed by Romanes as idle scepticism-mongering. For though machines might display such behaviour, there is no strong physical analogy between them and us and no evolutionary continuity. The intentionality and plasticity of a machine is parasitic on and derivative from our intentionality and plasticity, and thus is irrelevant to the question of animals.

In fact, Romanes addresses this question almost directly. What we are doing in studying animal consciousness, he tells us, is beginning with the facts of our own consciousness, and extending them on the basis of analogy. The greater the analogy, the more plausible our ascriptions. Thus, when we attribute anger or affection to a monkey, the analogy is much stronger than when we predicate such locutions of an ant or a bee. By the time we get to a machine, the analogy is so weak as to be meaningless, and our confidence in such ascriptions to machines must be nil.

In any case, we have seen that Romanes is hardly the simplistic purveyor of anecdote that he is often depicted as being. Rather, he is a sophisticated, conceptually astute thinker, who is quite aware of subtle conceptual questions.

Romanes's method applied

In his discussions of particular examples of animal mentation, Romanes is, in fact, quite rigorous in applying the methodological principles we have outlined. In discussing protozoa, for example, despite evidence of adaptability and flexibility in their behaviour, he feels compelled to withhold mentalistic attributions, because of their place on the phylogenetic scale, their lack of a nervous system, and their general primitiveness and disanalogy to humans. Ironically, as we will see shortly, at least one zoologist, H.S. Jennings, an experimentalist who did classic work on protozoa and is generally seen as one of the precursors of behaviourism, in attempting to explain protozoan behaviour, felt compelled to draw the opposite conclusion, despite his commitment to experimentally based results.

The bulk of Romanes's book is quite reasonable; the majority of the anecdotes related do no violence to common sense or ordinary experience. A number of examples will give the flavour of his work. In discussing the general intelligence of cats, for example, he cites the case of a man who left his cat at home while he went away on a two-month trip. During his absence, his house was occupied by two men who teased and frightened the cat. Before his return, the cat had kittens, which she concealed behind bookshelves. Immediately upon his return, the cat brought the kittens to his room and deposited them on the same spot where she had given birth to and raised previous litters.

In discussing the intelligence (or problem-solving ability) of rats, Romanes cites several anecdotes describing how rats were capable of extracting edible oil from narrow-necked bottles by inserting their tails into the bottle and licking either their own or each other's tails. In his account of elephant mentation, he cites the examples of a once-tame elephant who could recall and respond to commands fifteen years after escaping from domestication, and that of an elephant 'bearing a grudge' who searched out his tormentor six weeks after the latter had given him food laced with hot pepper (as an experiment). He describes the sympathy which elephants display towards injured elephants. In discussing the intelligence of elephants, he cites two instances of elephants retrieving objects out of reach of their trunks by blowing them towards a nearby wall so that they would bounce back. To give the flavour of Romanes's writing, I cite one example *in toto*:

Of the elephant's sense and judgment the following instance is given as a well-known fact in a letter of Dr. Daniel Wilson, Bishop of Calcutta, to his son in England, printed in a life of the bishop, published a few years ago. An elephant belonging to an Engineer officer in his diocese had a disease in his eyes, and had for three days been completely blind. His owner asked Dr. Webb, a physician intimate with the bishop, if he could do anything for the relief of the animal. Dr. Webb replied that he was willing to try, on one of the eyes, the effect of nitrate of silver, which was a remedy commonly used for similar diseases in the human eye. The animal was accordingly made to lie down, and when the nitrate of silver was applied, uttered a terrific roar at the acute pain which it occasioned. But the effect of the application was wonderful, for the eye was in a great degree restored, and the elephant could partially see. The doctor was in consequence ready to operate similarly on the other eye on the following day, and the animal, when he was brought out and heard the doctor's voice, lay down of himself, placed his head quietly on one side, curled up his trunk, drew in his breath like a human being about to endure a painful operation, gave a sigh of relief when it was over, and then, by motions of his trunk and other gestures, gave evident signs of wishing to express his gratitude. Here we plainly see in the elephant memory, understanding, and reasoning from one thing to another. The animal remembered the benefit that he had felt from the application to one eye, and when he was brought to the same place on the following day and heard the operator's voice, he concluded that a like service was to be done to his other eye.[20]

Such cases are easily confirmed by the accumulated wisdom of individual and historical experience and common sense.

It is difficult to find a great deal to fault in Romanes's work. His information is quite unlike that obtained through subsequent behaviouristic psychological experimentation, which studies animals under highly abnormal conditions, subject to extremely artificial and anthropocentric concerns and strange notions of what counts as 'intelligence' (for example, being able to run the proverbial maze); which isolates behaviour from the normal life-experience of the animal; which employs highly artificial contexts and stimuli; which gives us no clear picture of how the animal's mental processes operate during its life under natural circumstances; and which, in the final analysis, has given us little insight into the 'mind' or even the normal 'behavioural repertoire' of the animal. Romanes's information, on the other hand, as one would expect from a thoroughgoing evolutionist, is often much closer to a prolegomenon to understanding the animal under normal circumstances, as it solves problems and manifests emotion in the normal course of its life, not in response to electro-shock, induced helplessness, underwater submersion, foot-pad injury, and other refinements of behavioural experimentation.

Furthermore, it is vital to stress that there is nothing in Romanes's method which requires, or even suggests, that one must stop with anecdotal information. Virtually every instance he cites is subject to experimental replication and verification. Indeed, Romanes designs and performs his own experiments when he feels that the anecdotal material is suspect, surprising, or open to scepticism.

For example, he obviously did not fully credit the anecdotes mentioned earlier concerning the ability of rats to extract food from narrow vessels with their tails. So he designed an experiment, the results of which he subsequently published in *Nature*:

It is, I believe, pretty generally supposed that rats and mice use their tails for feeding purposes when the food to be eaten is contained in vessels too narrow to admit the entire body of the animal. I am not aware, however, that the truth of this supposition has ever been actually tested by any trustworthy person, and so think the following simple experiments are worth publishing. Having obtained a couple of tall-shaped preserve bottles with rather short and narrow necks, I filled them to within three inches of the top with red currant jelly which had only half stiffened. I covered the bottles with bladder in the ordinary way, and then stood them in a place infested by rats. Next morning the bladder covering each of the bottles had a small hole gnawed through it, and the level of the jelly was reduced in both bottles to the same extent. Now, as this content corresponded to about the length of a rat's tail if inserted at the hole in the bladder, and as this hole was not much more than just large enough to admit the root of this organ, I do not see that any further evidence is required to prove the manner in which the rats obtained the jelly, viz., by repeatedly introducing their tails into the viscid matter, and as repeatedly licking them clean. However, to put the question beyond doubt, I refilled the bottles to the extent of half an inch above the jelly level left by the rats, and having placed a circle of moist paper upon each of the jelly surfaces, covered the bottles with bladder as before. I now left the bottles in a place where there were no rats or mice, until a good crop of mould had grown upon one of the moistened pieces of paper. The bottle containing this crop of mould I then transferred to the place where the rats were numerous. Next morning the bladder had again been eaten through at one edge, and upon the mould there were numerous and distinct tracings of the rats' tail, resembling marks made with the top of a pen holder. These tracings were evidently caused by the animals sweeping their tails about in a fruitless endeavour to find a hole in the circle of paper which covered the jelly.[21]

In 1887 Romanes published the results of his experiments on the sense of smell in dogs.[22] In these he assembled twelve men, and had each man walk in the track of the one before him. The object of the experiment was to test the power of Romanes's setter to track his scent. Romanes took the lead, with his gamekeeper bringing up the rear, the gamekeeper being the only other person whom the dog would follow. Romanes had the party walk 200 yards, whereupon he and five of the men turned right. The other six turned left. The two parties walked in opposite directions for a while, and then concealed themselves. The dog, after overshooting the divergence point, found her way back, and without hesitation turned right. This occurred despite the fact that Romanes's footprints were overlaid in the common track by those of eleven other people, and in the right track by those of five others, and despite the fact that the gamekeeper's scent, which she would follow in Romanes's absence, was uppermost in both the common and the left-hand track. As Washburn later pointed out, this indicated that the dog could analyse a composite of odours and attend to a single component, much 'as a trained musician analyses a chord'.[23]

Neither Darwin nor Romanes was altogether content with anecdotal information. Throughout the works of both men, much emphasis is laid on verification of any data subject to the slightest question. (It is not generally felt to be necessary, however, to verify what is universally known across time and space.) Darwin himself contrived some ingenious experiments to test the intelligence of earthworms, a notion which he clearly felt was far beyond the purview of anecdotal information, and which was sufficiently implausible as to require controlled experimentation. These experiments, now virtually forgotten, occupy some thirty-five pages of Darwin's *The Formation of Vegetable Mould Through the Action of Worms with Observations on their Habits*, published in 1886. The question Darwin asked was whether the behaviour of worms in plugging up their burrows could be explained by instinct alone or by 'inherited impulse' or chance, or whether something like intelligence was required. In a series of tests, Darwin supplied his worms with a variety of leaves, some indigenous to the country where the worms were found, others from plants growing thousands of miles away, as well as parts of leaves and triangles of paper, and observed how they proceeded to plug their burrows, whether using the narrow or the wide end of the object first. After quantitative evaluation of the results of these tests, Darwin concluded that worms possess rudimentary intelligence, in that they shared plasticity in their behaviour, some rudimentary 'notion' of shape, and the ability to learn from experience. Darwin is no romantic anecdote-monger; he clearly distinguishes the intelligence of the worms from the 'senseless or purposeless' manner in which even higher animals often behave, as when a beaver cuts up logs and drags them about when there is no water to dam, or a squirrel pats nuts on a wooden floor as if he had buried them in the ground.[24]

In short, the conclusion seems inescapable that the manner in which Darwin and Romanes were attempting to study animal mind is hardly the sentimental, mindless, anthropomorphic, romantic, gullible wishful thinking which it has been depicted as by the twentieth-century common sense of science. Their approach was based on a sound, replicable, sceptical, and critical methodology which is consonant with, and indeed follows in part from, the philosophical assumptions of evolutionary theory that dominate biological activity and theorizing today. If this is the case, we are then faced with a major puzzle: why, by 1930, had the Darwin–Romanes approach to mind virtually vanished from mainstream scientific activity, whereas Darwinian biology has flourished? It is to this question that we now turn.

3 Aspects of change in science and philosophy

How does science change?—the orthodox view

Anyone who has studied science at the undergraduate or graduate levels must have been struck by the ahistorical nature of scientific activity and education. While most scientists are willing to hazard guesses concerning the future of their respective disciplines, few have much to say about their pasts. The names attached to scientific laws and theories—for example, Ohm, Hardy-Weinberg, Boyle, and Charles—serve for identification purposes only, like the titles of abstract paintings, and do not, for most scientists, evoke genuine historical figures and events. The history of science is seen as the province of historians, not scientists, or at best, of failed scientists or of Nobel prize-winners in their retirement years. And this in turn marks a fundamental feature of what we have called the common sense of science: namely, that what is past is past because it is wrong and that theories and concepts are superseded because they were inadequate or incorrect. On this view, theories fall only if they are refuted, so, if particular theories are no longer entertained, they *must* have been refuted. It is of no interest to the working scientist to know how or why they fell; since they are now gone, they are *ipso facto* wrong. The history of science, to put it crudely but accurately, is seen as a history of errors which have been overcome.

Thus, if a view held sway in science at some point in history, but is no longer taken seriously, it is because some flaw was detected in it. Flaws may be detected in two ways, according to prevailing ideology. First, and most commonly, the discovery of new data or the generation of data in controlled situations may lead inexorably to a rejection of the view in question. Paradigmatic examples are the notion of phlogiston, the postulate of the inheritance of acquired characteristics, and the notion of spontaneous generation. Second, and more rarely, conceptual or logical mistakes may be discovered in their assumptions and bases. In this manner, Einstein overturned the Newtonian notions of absolute time, space, and motion, and Russell overturned the Fregean account of number. For the common sense of science, these two possibilities exhaust the whole story of how theories fall.

What, then, does the common sense of science say about the fact that, by 1925 or 1930, the Darwin–Romanes view of animal mind, which followed from the assumptions of evolutionary theory and seemed fruitful and explanatory, had essentially disappeared from mainstream scientific activity and discussion? Clearly, it directs us to look for either facts which refuted the view or logical flaws which were unearthed in its substructure. Without some such refutation, there was no reason for what was, after all, a mainstream view to vanish into total obscurity. Alternatively, if the Darwin–Romanes view was superseded for reasons other than factual disconfirmation or fundamental logical incoherence, this is well worth knowing, both because it casts doubt on the orthodox wisdom of the common sense of science and, more to the point, because it may well indicate that the Darwin–Romanes position is not as invalid as current ideology suggests, and that the contemporary unwillingness to treat animal consciousness as a legitimate object of scientific study and discourse is ill-founded.

When I first became interested in this question, I fully expected to find ingenious experiments or trenchant logical analyses which laid the Darwin–Romanes position to rest. To my surprise, I found neither. What I did find, in fact, was quite the opposite; that the evolutionary approach to animal consciousness had not been refuted empirically or logically, but rather, that it had been swept out of fashion by a confluence of factors, more rhetorical than logical, more sociological than empirical, more philosophical and valuational than 'scientific'. It appears that this approach to mind was caught in the brushfire of positivism, behaviourism, and empiricism which swept through Western thought, cauterizing it, simplifying it, reducing it, and purging it of metaphysical, valuational, and non-empirical taints. This fire so seared the scientific landscape, in fact, that its effects are still highly visible well over half a century later, for it is this movement which is a significant constituent of what we have called the common sense of science.

Shortly, we shall selectively examine the work of some of the major figures whose work is traditionally seen as providing the intellectual basis for the supplanting of the Darwin–Romanes point of view by positivistic behaviourism. Surprisingly, we shall see that not only do these figures not provide adequate justification for abandoning the Darwin–Romanes stance, but that in a number of cases their work points in the opposite direction, and that their own assessment of their work is more congenial than inimical to Darwin and Romanes. In most cases, their disagreements with Darwin and Romanes were intra-familial disagreements, disagreements within the paradigm, not attempts to establish a new one, to use Kuhn's terminology.[1] In other cases, they saw their own work as actually buttressing Darwin and Romanes. Furthermore, we shall see that some of the sources cited by historians as factors in the dethronement of the evolutionary view of animal consciousness conflict with one another.

An alternative approach to scientific change

Before proceeding, however, let us sketch briefly an alternative to the orthodox, common-sense view of science. We may begin by noting some features of philosophical change; for, as we shall see, scientific change is often rooted in changes in philosophical approaches and assumptions, which in turn are motivated by ethical and other valuational assumptions.

Few scientists would protest the claim that dominant approaches to art, music, or literature change as a matter of fashion, for all sorts of non-rational reasons.[2] And, most likely, few would cavil at attributing similar mechanisms of change to the waxing and waning of philosophical approaches and schools. After all, any student of the history of philosophy can attest to the fact that philosophical schools and positions are not abandoned even when contradictions are detected in their very bases; in fact most, if not all, of the great philosophical works contain gaping holes and major contradictions—Plato, Aristotle, Descartes, Hume, and Kant come readily to mind.

Let us take an example directly relevant to our discussion. Logical positivism, the philosophical position which provided the basis for the common sense of science, flourished well into the mid-twentieth century, despite the fact that it contained many logical incoherences and even contradictions. Among other things, it had no way of adequately accounting for inter-subjectivity in science, since, in its major version, it was locked into epistemological reliance on an individual's sensations as the building blocks of knowledge. Furthermore, its fundamental canon, the principle of verification, which enjoined the verifiability of all meaningful propositions and which was designed to eliminate so-called nonsense from discourse, was itself nonsense in its own terms, since it could not be verified empirically. Even more surprising, the sorts of objections to which logical positivism was eminently vulnerable had already been raised in the eighteenth and nineteenth centuries against its precursor, classical empiricism, and had never been adequately countered.

Eventually, positivism went out of style among philosophers, if not among scientists, and was replaced in Britain and America by a new orthodoxy of ordinary-language philosophy, which in turn went out of style for extra-logical reasons. It wasn't that anyone conclusively refuted ordinary-language philosophy, or discovered a fatal flaw in its assumptions. Rather, it more or less went the way of skirt-lengths. People became bored with it and wanted to try other things. In particular, major figures in philosophy and students became bored with it. Good philosophical minds became restless and went on to other things, and students voted with their feet. Vocationalism became dominant in higher education, primarily for economic reasons, and in its wake came a renewed concern with 'practical', normative questions. If you wished to sustain a philosophy department of any size in an age of financial retrenchment and neo-philistinism, you had to adjust to the new trends. And so was born what is barbarously termed 'applied philosophy': business ethics, environmental ethics, philosophy of technology, and the like, areas which would attract student and public interest and support.

Yet the same scientists who are quite willing to allow that the vagaries of fashion are operative in the humanities would be far less willing to ascribe these sorts of mechanisms to the development of science. To be sure, if challenged, they would probably admit that some such account is relevant to the funding of scientific research. The infusion of money into mathematics, physics, and engineering after Sputnik, the war on cancer, the allocation of vast resources to battle AIDS, all evidence the influence of extra-scientific factors on science. Most scientists would also acknowledge that there are inevitable, unstated funding biases against research which would shake the work and position of major figures in a given field, since it is the major figures who serve on the panels which allocate the resources. (Remember, after all, that scientists are human, and science is a career.) I once heard a physicist give a public lecture in which he argued that, if the young Einstein were alive and functioning in the United States today, it would be extremely unlikely that he would be funded, since his work questioned the assumptions shared by all major figures in the field and was highly theoretical, and since he had no reputation or academic affiliation and needed no money for equipment or graduate students. Research which 'falls between the cracks' of current disciplinary categories is also unlikely to get funded. A physiologist I know who does highly original research on toxicology of psychoactive drugs using planaria (flatworms) as a model cannot get funded, because toxicology review panels have no acquaintance with planaria, and neuro-psychology review panels have no interest in toxicology. His proposals are therefore ping-ponged from one group to the other.

As an experiment, I telephoned a number of colleagues who work in very diverse fields of research, ranging from microbiology to plant physiology. Each of them was instantly able to cite examples from his or her own field of fashionable areas of research, which had assumed prominence, dominated research funding, then faded into obscurity *for no clear rational reason*. The reasons my colleagues gave for such waxing and waning were extremely varied, but it was remarkable that the critical ones were what the common sense of science would surely term 'extra-scientific'. 'Sutherland won a Nobel prize for cyclic AMP work,' one colleague told me, 'and all of a sudden everyone was doing cyclic AMP—looking for it everywhere!' Research into various diseases, alternative energy sources, gasohol, environmental toxins such as lead and mercury, interferon, AIDS, colic in horses were all cited as examples of how research funding follows fashion. One microbiologist complained that 'areas of interest go out of style so fast—every five to ten years—that I must keep revising my research to keep it fashionable, and thus fundable.'

It is not hard to pick out many of the kinds of things which would determine fashions in research and research funding. We have mentioned Nobel prizes, and it is important to remember that Nobel prize-winners are not selected by God. All sorts of political and other valuational factors enter into the selection process. A book or film may seize the popular imagination—on environmental pollution, for example—and this in turn may give rise to demands for research in the area. The abuses of Nazism coupled with growing concern about racial equality have forestalled any funding of research into racial differences or the genetic determination of intelligence. As a spin-off from this state of affairs, it then became timely to question the entire notion of intelligence and its measurement. When major figures in a field of science adopt a certain approach to a question, that strategy becomes central to the field.[3] And so on.

But though scientists may be willing to allow that fashion can be operative in science with regard to funding, which is after all a political matter, they would be likely to distinguish this from science itself, and affirm that the latter is impervious to vagaries of fashion. The basic nature of science, it would doubtless be claimed, always remains the same. On my view, however, this is not the case. For, as we shall see, science is inevitably bound up with philosophical and valuational commitments and assumptions. Indeed, even what counts as a legitimate explanation or, for that matter, as a fact in science, is ultimately a matter of philosophical and valuational commitment, and like all such commitments can change in response to all sorts of non-logical and socio-cultural forces and pressures.

On the other hand, even if this is the case, one might reasonably expect scientists or philosophers to provide a *logos*, an explanation, a rational account, a case, a justification when their philosophical and valuational commitments change, to give reasons for such change, not merely to detail causes. Thus, if one is probing a question which is relevant to the basic nature of science, such as our question of why the scientific community suddenly decided that it is not legitimate to study animal consciousness and that the Darwin–Romanes approach must be jettisoned, we would surely expect such a *logos*. Yet, as we shall see, one cannot find an articulated set of well-developed reasons which served as a rational basis for the change in fundamental philosophical and valuational assumptions underlying the abandonment of the Darwin–Romanes approach to consciousness in animals. What I shall demonstrate is that here, too, something like a change of fashion figured heavily in what took place.

The philosophical and valuational bases for scientific change

All human activities rest upon certain assumptions. If they are intellectual activities like science, rather than non-cognitive activities like baseball, these assumptions will be simply taken as true. One does not prove the assumptions within a given field—in fact, it is the nature of the assumptions which determines what will count as a proof in the field in question. Consider the paradigmatic case of Euclidean geometry. One cannot prove the assumptions of geometry within geometry, for proof is predicated upon taking the assumptions for granted. To challenge the assumptions, one must leave geometry *per se* and adopt not a geometrical approach, but a philosophical one.

Similarly, in any area, as Aristotle points out, we must leave something unproved in any enquiry, or else we become embroiled in an infinite regress. In my view, a large part of what has been meant by philosophy throughout the ages has been the examination of notions and assumptions which are taken for granted in various areas and disciplines, including philosophy itself. The sort of examination of assumptions which philosophy typically engages in is to a considerable extent in terms of logic, for as Aristotle points out in the *Metaphysics*, logic is really the one thing that all parties to any intellectual activity at all must adhere to as a ground rule if they wish to be heard. (This is not, of course, to suggest that logic itself is immune to examination; it is not. The epistemological status of an examination of logic, however, is far from clear, since one uses logic to examine logic.)

So anyone taking a stand on fundamental assumptions in science, philosophy, or any other discipline must be adopting a philosophical stance. And such a stance involves examining the logic of a position and its assumptions. But what does conducting such a logical enquiry entail? In part, of course, what is meant is looking to see whether any formal or informal fallacies have been committed. Do the assumptions generate a contradiction? Has the position assumed what it set out to prove? Has it equivocated on a term, made a mistake in the scope of a quantifier or a modal operator, or affirmed the consequent? Few philosophical critiques restrict themselves to such formal notions, however. Most also look at the assumptions they are examining in terms of adequacy, coherence, simplicity, elegance, comprehensiveness, beauty, moral content, effect on human life, and so on, terms which are far vaguer than those of formal logic and far more difficult to agree upon, and which, indeed, are highlighted or eclipsed depending on what school of philosophy happens to be in vogue.

Thus, a philosophical critique of something will not merely examine fundamental assumptions and presuppositions; it will do so with respect to features which are valued by the individual conducting the examination. Aristotle's fundamental critique of Plato was, in the final analysis, based on a distaste for Plato's other-worldliness. In Aristotle's view, an adequate theory simply had to deal with the world of experience. But for Plato, to whom experience was deception, and only the rational was real, adequacy to the world of experience was not a fundamental criterion in the construction of a theory.

I am thus arguing that if one looks at the *reasons* which scientists might give for preferring one approach over another, for considering one area a legitimate object of study and denying that status to another area, for accepting one sort of explanation and rejecting another (for example, mechanistic versus teleological explanation in biology), rather than looking only at the fact that they do these things, they will be couched in terms of philosophical positions (not necessarily, or even probably, recognized as such) on the issue in question. And while some of the philosophical claims they make (or would or could make) are couched in purely straightforward, logical terms, others will devolve upon valuational commitments to such diverse notions as simplicity, elegance, adequacy to experience, mathematizability, minimization of randomness and chance (Einstein) or, alternatively, lack of concern about randomness and chance (Heisenberg), implications for social justice (witness the sociobiology and IQ controversies), political implications (for example, Soviet rejection of Darwinism), implications for control of nature, and so on. Thus, to understand the acceptance or rejection of theories, schools, views, and approaches in the sciences, one must not restrict oneself to looking at what empirical results confirmed what hypotheses. Indeed, what counts as confirmation or disconfirmation or even what counts as empirical data will itself depend on one's philosophical and valuational stance.

Before exemplifying these claims, it is worth reminding ourselves of the inescapability of *some* philosophical position. Obviously one must make assumptions in any activity in which one engages; in particular, one must make cognitive assumptions in order to generate intellectual activity. And, while a person is often not conscious of all of his or her assumptions, and indeed, in ordinary or professional life is typically unaware of even having made assumptions, in so far as one attempts to justify one's assumptions if they are called to one's attention, one must be said to have a philosophical position. Such a position, furthermore, will inevitably rest upon valuational commitments. What I have called the common sense of science is, in fact, the basic philosophical position which scientists imbibe with their science, and which they bring forth in response to certain basic questions about their activity. Contemporary scientists, of course, also share other philosophical and valuational commitments which we have not hitherto discussed. For example, as Paul Feyerabend has pointed out, they share the belief that a way of knowing which allows us to control and manipulate what is known is a *better* and more valuable way of knowing than an approach which does not increase our control. (As we shall see, this belief played a major role in the rise of behaviourism.) Indeed, Feyerabend argues that science is superior to other traditions, such as Tibetan mysticism or Navajo mythology, as a mode of knowing only if one assumes that a way of knowing is good in direct proportion to the extent to which it allows us to manipulate the world. If one's fundamental valuational commitment were, say, to living in harmony with the world as we find it, one might prize Navajo mythology over science.[4]

Before applying our analysis to the Darwin–Romanes view of animal consciousness and its demise, let us look at some better-known illustrations. As I mentioned earlier, Aristotle's natural science held sway for longer than any other explanation of the world. The key assumption of Aristotle's science is that 'what we see is what we get'—in other words, that the senses are a fundamentally reliable source of information about reality, and that the world of fish and flowers and trees and hot and cold and wet and dry and bitter and sweet and living and non-living matter is the *real* world. This assumption shaped Aristotelian and later medieval science, and determined what counted as a fact and what counted as an acceptable explanation. By the seventeenth century, however, this view had fallen from favour. The world of the senses was no longer seen as ultimately real, but at best, appearance, and at worst, illusion. And this change came about not because the Aristotelian view was empirically disproved, but because its philosophical and valuational presuppositions were challenged and supplanted.

Thanks to thinkers like Galileo, Descartes, and Newton, the fundamental philosophical and valuational assumptions about what is real and what is worthy of study changed. What is real came to be cast in a Platonic mode, as what meets the requirements of the intellect: namely, what is quantifiable and mathematical, what can be expressed in geometrical terms. In this way, 'modern science' came into being. No new facts forced this change. How could they? After all, 'empirical facts' all buttress Aristotelianism. What happened was that a change in metaphysical perspective intertwined with a change in value led to an alternative philosophical and valuational basis for science. Plato came to be valued over Aristotle, and Aristotelian categories came to be ridiculed. Reason came to be valued over experience, and sense experience came to be seen as appearance or illusion, as in Descartes's first meditation. (An additional, related philosophico-valuational assumption, of course, was that no right-minded thinker could value studying illusion or appearance when they could study reality.) Mathematics became valued over biology as the master science and root metaphor. Whereas Aristotle explained physics in terms of biological categories, Descartes explained biology in the categories of mathematical physics. Descartes, Galileo, Newton, and others do not prove that the Aristotelian perspective was invalid, nor did they disconfirm the assumptions upon which it rested. Rather, they picked away at these assumptions, discredited them, disvalued them, pressed an alternative, and located intellectual value in areas neglected by Aristotle, most notably mathematics.

To be sure, they could not have done this had myriad other conditions and causes not obtained in society which made their position socio-culturally viable. These factors have been abundantly chronicled by historians of science like Kuhn, Burtt, Feyerabend, and Koyré.[5] Intellectuals were drawn to Platonism for religious reasons; Kepler wanted to demonstrate the music of the spheres; voyages of exploration required precise astronomical reckoning; the advent of artillery demanded great precision in predicting projectile motion, and so on. The key point is that these are not new facts which disconfirm the Aristotelian, qualitative world-view, but rather, socio-cultural and valuational factors which disposed the culture towards a new, quantitative, philosophical perspective, and which explain why the philosophical position which valued and highlighted common sense and sense-experience, and the sorts of explanations in accord with them, went out of style and mathematical Platonism came in. I am not of course suggesting that scientific or philosophical change is just as arbitrary and whimsical as change in fashions in clothing. What I am saying is that many extra-logical, extra-rational cultural factors enter into and shape philosophical and scientific changes.

Such influences are patently obvious in biomedical science, as indeed the very term 'biomedicine' implies. Modern biomedicine has been shaped by cultural valuational decisions—in large measure, moral ones. We have the biomedicine we do for a variety of reasons. Because we have unreflectively argued that any life is better than death, we have developed incredible machinery (and knowledge) to prolong life, and have been willing to expend tremendous sums on heroic medicine even in terminal cases. Because we want to believe that evil can be regarded as a disease, as something overpowering us rather than caused by us, as something curable, we subscribe to the idea that alcoholism is a disease, wife-beating a disease, child-beating a disease.

On a less dramatic scale, treatments for illnesses follow the vagaries of fashion. Certain drugs and treatments go out of style, then come back in, for reasons that have little to do with science. The use of diet to control migraine, the use of sodium amobarbitol as a tranquillizer, and the almost standard removal of tonsils and adenoids are all examples of regimens which have been in and out of style. The saga of premenstrual syndrome (PMS) is another example, directly traceable to moral-valuational factors. Thirty years ago, PMS was viewed as a chronic, ubiquitous, incurable female genetic disease, whose presence allegedly helped to explain why women did not hold certain jobs that men did. ('How would you like a woman president faced with a nuclear crisis during *that* time of the month?') With the advent of women's liberation, PMS came to be regarded as a normal biological occurrence, one which had been turned into an illness by men in order to subjugate women. But now that women have gained access to jobs which would have been out of the question earlier this century, many feel sufficiently secure to admit that they suffer, and to demand that medicine provide relief.

Another interesting example of a dramatic philosophical and valuational basis for scientific change comes from recent scholarship by Paul Forman, who has demonstrated that it was changes in social values in Germany between 1918 and 1927 that made quantum theory possible. According to Forman, a major impetus for both the development and the acceptance of quantum theory was a desire on the part of German physicists to adjust their values and their science in the light of the crescendo of indeterminism, existentialism, neoromanticism, irrationalism, and free-will affirmation which pervaded the intellectual life of Weimar, and which was hostile to deterministic, rationalistic physics. Thus quantum physicists were enabled to shake the powerful ideology of rationalistic, deterministic, positivistic late nineteenth- and early twentieth-century science, with its insistence on causality, order, and predictability, as a result of the powerful social and cultural ambience in German society which militated in favour of a world in which freedom, randomness, and disorder were operative, and valued such chaos both epistemically and morally.[6]

I am arguing that one's notions of science and knowledge rest upon philosophical assumptions which are intertwined with valuational assumptions, both epistemic and moral, concerning what is worth knowing, what counts as knowledge, how it ought to be known, what ought and ought not be done to acquire that knowledge, and so on. As societal and cultural factors and values change, these assumptions change also, determining in turn what count as legitimate objects of study, methods of study, genuine facts, and even what counts as evidence for or against a hypothesis. For example, the fact that Newton's calculations were significantly wide of the mark in describing the behaviour of the moon was taken not as a disconfirmation of Newton; rather, the challenge was to explain the discrepancy.[7]

Thus science succumbs to the same 'foibles' as philosophy—inevitably so, because it is based on philosophy. Just as philosophical works and systems are revered and taken seriously despite the presence of fundamental contradictions and circularities within them, so too, scientific theories are adhered to despite the presence of holes and contradictions, and even in the face of empirical falsifications, as Kuhn, Lakatos, and other historians of science have shown.[8] By the same token, scientific positions, like philosophical positions, will change on the basis of value changes. These value changes may be moral or epistemic, or may result from changes in cultural values. We saw that the philosophy of science of Aristotle went out of style in part because of changes in epistemic values; similarly, the moral philosophy of Aristotle went out of style because of changes in moral values, notably the rise of egalitarianism. The value changes may occur within the philosophical community or within the culture at large. Thus, a revival of concern with the history of philosophy was primarily a function of a change within the philosophical community itself. On the other hand, the waning of Nietzsche scholarship after the rise of Nazism and the rise of 'applied philosophy' dealing with relevant social issues reflect general social pressures and value concerns. (One cannot, of course, clearly separate the two, since philosophers are also people, though compartmentalization, as in the case of science, often leads to this fact being ignored.) Similarly, the scientific revolution resulted from changes in epistemic values; the abandonment of race and IQ studies resulted from changes in moral values, as did the recent return of the scientific legitimacy of talking of animal pain. Once again, the value changes may be primarily internal to the scientific community, as in the scientific revolution's ignoring of qualitative facts, or external, as with social-moral pressures to change scientific procedures for the use of human subjects after the Nuremberg revelations and for the use of animals in the 1970s and 1980s.

In the wake of these reflections, it is worth noting as a corollary that ordinary common sense and its values can have a dramatic impact on science and its development. Unfortunately, this influence is rarely 'conscious of itself', to use a Hegelian idiom. For the most part, although science is shaped by social values and common sense, it resists this shaping, in part consciously and in part because it is so far removed from ordinary people, who are ignorant about it and stand in awe of it. Scientific illiteracy is endemic in our society, and even people very well educated in non-scientific areas feel no shame at, and indeed flaunt, their scientific ignorance, as C.P. Snow, showed long ago. Thus, as Feyerabend has demonstrated, the way is cleared for much of science to proceed for long periods of time impervious to public scrutiny and direction. The failure of people to understand science limits their ability to shape it rationally, to the ultimate detriment of science, and certainly to the ultimate detriment of the citizenry, who often respond with too little too late, and who do not even realize that much of their lives is shaped by scientific expertise to which they have never given their assent.

Relevance of this view to the science of animal consciousness: some preliminary remarks

We will now return to the demise of the Darwin–Romanes approach to animal consciousness. I shall argue that it was not refuted empirically, and that it was not abandoned because fundamental flaws were exposed in its logical structure. Rather, it was swept away in a new philosophical-valuational wave which did not allow the findings of Darwin and Romanes *vis-à-vis* subjective experience in animals to count as legitimate scientific data. Within the context of this new philosophical perspective, what could and should have been taken as emendations and criticisms *within* the context of the Darwin–Romanes approach came to be seen as fatal objections and definitive refutations. Furthermore, it is this new view, in its twentieth-century form, which gave rise to what we have been calling the common sense of science, thereby rendering it invisible to scientists.

Before examining the change, let us briefly summarize the status of the view that animals are conscious and that we can know their minds at the end of the nineteenth century. Such a position was, first of all, built into ordinary common sense. Then, as now, one would be hard pressed to find an ordinary person who would deny that animals feel pain or that we can know when an animal is happy, frightened, bored, grieving, and the like.

In addition, if one takes seriously the Kantian notion that all our experience is processed through and determined by our presuppositions and assumptions, coloured by our philosophical and valuational roots, and filtered through our conceptual lenses, one must say that what counts as a fact, as well as what counts as an explanation, must be rooted in these structuring mechanisms. As Kant and myriad others have taught us, and as even Hume showed when he attempted to deny it, we are not simply dartboards upon which experience falls and sticks. What we call experience (and *a fortiori*, knowledge) is a result of interaction between what our sense-organs receive and how we structure or interpret such data. How this works is the subject of endless debate and speculation. Theorists have variously claimed that experience is shaped by innate mechanisms, society, language, beliefs, expectations, theories, and so on. On the view which I have been outlining what counts as a fact in any world-view is at least in part determined by philosophical and valuational commitments of which one need not be aware, but which one inevitably has. In the world-view of ordinary common sense, animal consciousness is a fact, and the sort of fact which we experience directly and daily, just as we do human mentation. It takes a philosophy class or two to make an ordinary person doubt that he is directly aware of another person's pain, joy, anger, or grief in just the same way as he is aware of that person's body. Furthermore, the sort of doubt which a philosophy class can engender in a person about knowing another person's mental states is no different from the doubts which such classes engender about a tree really being there or another person's body being there. These are artificial doubts, which never touch common sense; for common sense, facts about mentation are as basic as facts about physical objects. (This I take to be one of the major points made by Ryle, Wittgenstein, Peirce, and Dewey.) Certainly one can't feel another person's joy, but one can't be a tree either, and that doesn't mean that one can't accept as fundamental data both that the tree is out there and that the experienced joy is out there. With regard to animal consciousness, we don't maintain that we know its nuances as well as we know the nuances of human mentation, but we certainly know something of the matter. It is as absurd to common sense to deny that a dog may be in pain as to deny that a person may be in pain.

In short, for ordinary common sense, there are facts of animal consciousness as well as facts of human consciousness. That we are aware of facts of consciousness in ourselves and others, human and animal, is built into the philosophical-valuational machinery of common sense. Correlatively, as we have seen, such a view is built into the presuppositions of Darwinism; hence Darwin and Romanes have little patience with contentious scepticism regarding it. If you doubt that we can know other consciousnesses, we have seen Romanes say, you must equally doubt that we can know physical objects. And so, despite multiple, deep conflicts between common sense and Darwinism, in this at least they are agreed, that one can know the minds of other people and of animals.

The rise of positivism

Nevertheless, the late nineteenth century saw an attack on this presupposition, or rather, the development of a new philosophical and valuational perspective which resulted in science no longer recognizing facts of consciousness (that is, subjective states). What developed was a reductive, physicalistic, positivistic ideology which infiltrated all aspects of thought, from art to science to philosophy, and which, in essence, ruled the study of consciousness out of existence.

By the end of the nineteenth century, numerous factors had converged to create a strong current of scepticism, positivism, and reductionism. In the first place, there were strong reactions against what were seen as the metaphysical excesses of idealism in philosophy. Talk about whether or not the absolute was happy and about the temporal unfolding of the Spirit was beginning to wear on many thinkers. Moreover, such philosophical excesses were seen as perversions not only in philosophy, but in science as well. In Germany, professors of physics at different institutions differed not only in their scientific approaches, but in their philosophical stances as well, which in turn reflected back on their scientific activities.[9] In the biological sciences, talk of entelechies and life forces had insinuated itself into scientific discussion.

Scientists had begun to react strongly against this tendency to freely mix empirical science and metaphysics. Helmholtz, in his *Popular Lectures* of 1884, asserted that 'a metaphysical conclusion is either a false conclusion or a concealed experimental conclusion'.[10] Following in Hume's footsteps, Mach argued that any concepts not reducible to experience ought to be eliminated from physics. For him, notions like atoms, particles, and causality ought not to be reified, and had, at best, only symbolic reality as shorthand for groups of experiential elements. Further, absolute space and time were medieval vestiges which should be eliminated from physics. Like Hume, Mach felt that the self was a fiction, a conceptual construct ultimately reducible to experiential components. Of course, he used introspection in his psychophysics, and relied explicitly on experience as the root of his epistemology. It was Comte who presaged the behaviouristic side of this reductionistic tendency when he denied the validity of introspection altogether.

Kolakowski, in an excellent book on the history of positivism, *The Alienation of Reason*, pointed out that a reductive tendency to 'get back to the most primitive, concrete datum, to a "natural" view of the world not mediated by metaphysical fictions' was characteristic of a good deal of European thought at the end of the nineteenth century, not merely of positivism. Kolakowski found parallel tendencies in Husserl and Bergson, and saw 'this whole period of philosophy as a relatively homogeneous development with points of contact in literary trends and world outlooks over the same years'.[11]

One can indeed find elements of this reductionistic, 'no frills' philosophy throughout European culture. By the end of the nineteenth century, art, architecture, design, music, and literature had become extremely extravagant. Egon Friedell wrote vividly of a typical bourgeois living-room:

> Theirs were not living-rooms, but pawnshops and curiosity shops . . . [There was] a craze for totally meaningless articles of decoration . . . a craze for satin-like surfaces: for silk, satin and shining leather; for gilt frames, gilt stucco, and gilt edges; for tortoise shell, ivory, and mother-of-pearl, as also for totally meaningless articles of decoration, such as Rococo mirrors in several pieces, multi-colored Venetian glass, fat-bellied Old German pots, a skin rug on the floor complete with terrifying jaws, and in the hall a life-sized wooden Negro.
>
> Everything was mixed, too, without rhyme or reason; in the boudoir a set of Buhl, in the drawing-room an Empire suite, next door a Cinquecento dining-room, and next to that a Gothic bedroom. Through it all a flavor of polychrome made itself felt. The more twists and scrolls and arabesques there were in the designs, the louder and cruder the color, the greater the success. In this connection, there was a conspicuous absence of any idea of usefulness or purpose; it was all purely for show. We note with astonishment that the best situated, most comfortable and airy room in the house—the 'best room'—was not intended to be lived in at all, but was only there to be exhibited to friends.
>
> Every material used tries to look like more than it is. Whitewashed tin masquerades as marble, paper maché as rosewood, plaster as gleaming alabaster, glass as costly onyx. . . . The butter knife is a Turkish dagger, the ash tray a Prussian helmet, the umbrella stand a knight in armor, and the thermometer a pistol.[12]

Anyone who has viewed a good deal of nineteenth-century academic painting has felt the same sense of excess. Much early twentieth-century culture can be seen as an attempt to eliminate or trim away that excess. The movement in visual art from Cézanne to Picasso and Braque and from them to Mondrian is a movement towards simplification, as are the changes in music initiated by Schoenberg and others. And in architecture and design one finds Adolf Loos's devastating equation of 'Ornament and Crime', as he called his essay on excesses in decoration. Loos's plea for a functional, simplified style in architecture and design found its culmination in the 'form follows function' theory and practice of the Bauhaus, and in museums like the Museum of Modern Art in New York, which has exhibited a ball-bearing as a paradigmatic aesthetic object.

In science and philosophy, this tendency reached its peak in twentieth-century logical positivism and behaviourism. Positivism was the self-conscious attempt to clearly demarcate science from non-science, to eliminate metaphysics from science and lay it to rest as 'nonsense'—witness Carnap's classic hilarious attack on Heidegger's assertion that 'Das Nicht nichtet' ('The nothing nothings').[13] Less well known is the desire of positivists to employ their methodology to do away with the bombast of political ideologies, for elimination of jingoistic nonsense was seen as a moral value, a precondition for democracy and tolerance. Behaviourism was the application of the same reductionistic spirit to psychology—psychology using Occam's razor to cut its own throat, one might say.

The claim I am making is this: that the elimination of animal consciousness from scientific concern can be seen as either a symptom or a result of a general reductionistic wave sweeping European culture at the end of the nineteenth and the beginning of the twentieth centuries, a movement which aimed at cleansing, purifying, and getting rid of unnecessary frills. And one of the things which got pared away in the twentieth century was consciousness, a development which gained unwitting support from various philosophies like Mach's which deny a self. (At the same time, and paradoxically, the denial of consciousness conflicted with the phenomenalism which lay at the root of much positivism and empiricism.) Consciousness came to be viewed as mystical, unscientific, unnecessary, obscure, and not amenable to study. Positivism wished all scientific description to be expressed in the language of physics. Introspective psychology was unacceptable; only behaviourist psychology met the requirements of positivism. Ironically, positivists overlooked the fact that the basic building blocks of their programme were experiences, sense-data, and sense-impressions, which can hardly be said to be verifiable physical entities. This, indeed, is one of positivism's great flaws; for how can one construct scientific objects out of sense-data if the former are public and intersubjective and the latter are inherently private? And how can a programme whose fundamental appeal is to the data of consciousness deny the reality (and amenability to being studied) of consciousness? (As far as science is concerned, the cash value is the same, whether you deny the reality of consciousness or merely deny that it can be studied.)

Positivism, behaviourism, and consciousness

This, then, is the general background to the overthrow of the Darwin–Romanes approach to animal consciousness. It fell victim to a reductive simplification which had as its *raison d'être* the elimination of unnecessary entities. But we have yet to examine how this reductivist movement of thought was brought to bear specifically on consciousness, and why consciousness was taken as one of those mystical, superfluous notions which needed to be purged. We shall see that the strong desire to pare away, start fresh, make psychology a science, and so on, served to eclipse good sense and common sense, and that many of the ideas and figures whom behaviourists later invoked as they looked back to justify the elimination of consciousness need not be taken in this way, and probably would not have been taken this way, were it not for the new, highly activist mode of thought.

During this period, 'soft' concepts were viewed as ripe for plucking, and radical excision was the order of the day. If one aspect of a concept is problematic, eliminate the entire concept. After all, is this not precisely what Einstein did with the venerable Newtonian notions of absolute space and time? True, there are disanalogies between psychology and physics here. Einstein had, after all, found logical flaws in Newtonian concepts, whereas no such flaws had been found in the notion of consciousness. On the other hand, consciousness has its own problems. In the first place, as mentioned earlier, we don't really have access to the consciousness of others; we can only experience their behaviour and infer their mental states. In the case of animals, we are even further removed, since we lack the possibility of linguistic verification of our hypotheses. We have access only to our own consciousness, and this too is fraught with difficulty. For what is it we have access to? Hume, Kant, and more recently Mach had shown that there is no knowable I, but only a flow of experiences. Ultimately we cannot even make sense of, let alone verify, statements about the mind.

On a less philosophical level, introspective techniques for studying consciousness, even when engaged in by 'trained introspectionists', yielded unsatisfactory results, which varied from investigator to investigator, and among which there existed no objective procedure for choosing. Behaviour had always been used, in both animals and humans, as the fundamental way of knowing about consciousness. (Language, of course, is a form of behaviour.) Given the *Zeitgeist*, a new notion inevitably suggested itself: that if we can't verify claims about consciousness, don't even know what we mean when we talk about consciousness (notoriously, no one has ever defined consciousness), and justify all our claims about consciousness solely by appeal to behaviour, why not stop talking about consciousness and talk only about behaviour? Consciousness is a frill, a useless decoration, an unnecessary entity, so pare it away! What, after all, do we lose? And what we gain is a real science, just like physics, in which every claim is verifiable. This, in brief, is the sort of argument generated by Watson in his polemical, sermonizing, rhetorical defence of behaviourism.[14] Give it a try; what can you lose?

Despite all we have said about the historical background, it is still hard to believe that behaviourism took hold. First and foremost, everyone knew very well that he or she had consciousness—for that matter, they knew that everybody else had it too. What am I to say of the images, thoughts, and ideas which I know I have? Am I to say that I don't really have them, but only think I do? Sometimes Watson and others came close to saying this; at other times they were more circumspect—maybe I have thoughts, maybe I don't, but anyway, it doesn't matter to the science of psychology. It doesn't matter because what we study is behaviour, we can't study thoughts. Doubtless many people had reservations: 'But isn't psychology there to study the mind? What about my thoughts? Haven't we thrown out the baby with the bath water? Isn't this a bit like solving one's marital problems by killing one's spouse?' And so on. 'Shut up and experiment' was the behaviourists' reply.

So psychologists experimented. After all, one could still study animals. Indeed, one was enjoined to study animals in controlled ways, as Thorndike, Watson, Yerkes, and others showed. Moreover, what had been given up in animal psychology? Only the theoretical nexus of Darwinism which said that when an animal screamed, it was hurting, it was feeling pain, as with us. Scientifically speaking, it was merely showing 'aversive behaviour'. And behaviourism was evenhanded, since neither animals nor humans had feelings in its theoretical universe, so Darwinian continuity was not, in theory, violated. The only trouble was that no human, however behaviouristically oriented he or she might be, could ever forget outside of the laboratory that people did have feelings. Every behaviourist, however dedicated, could not not think about rhinoceroses on command. Every behaviourist was forced in a million ways to recognize thoughts and feelings in others. Our moral and legal systems, not to mention our everyday interactions, are predicated on not seeing others as behavioural machines, but on regarding behaviour as indicative of a subjective mental life (as Thomas Reid argued long ago[15]).

But animal awareness did not force itself on the behaviourist in anything like the same way. So, in practice, Darwinian continuity was violated. Behaviourists were thus forced to see people as conscious, but were under no such constraints regarding animals, except perhaps, in the case of the family dog. (And, of course, an increasingly urbanized society made direct contact with animals less and less likely for more and more people.) There was nothing in the society to set valuational barriers against seeing animals as behavioural machines, for though common sense recognized animal mind, thought, and feeling, it didn't *care* much about them. Moral concern for animals was alien to most people. If only behaviour is real, and animals are models for the study of behaviour, then, from a scientific point of view, all that exists in the case of the animals is behaviour. We have pared away consciousness in theory, and are not forced to recognize it in practice, for our moral, and legal codes were, and still are, for the most part virtually silent regarding the treatment of animals. Certainly there were antivivisectionists who did assert the moral status of animals, but they did so on the basis of the very thing that behaviourism was created to do away with—namely, the presence of consciousness in animals. For that matter, even in the heyday of the Darwin–Romanes *Gestalt*, little concern was expressed by its leading proponents regarding the treatment of animals. Although everything in the Darwin–Romanes approach should have militated in favour of extending moral concern to animals because of their continuity with us, this did not happen. Romanes, in his books on animal mind, never mentions moral obligations to animals,[16] and Darwin, while opposing cruelty in research, never followed out the profound moral implications of phylogenetic continuity. Some Darwinians, like E.P. Evans,[17] did draw moral conclusions, but they were largely ignored in a society which depended economically on animal exploitation, and whose entire valuational and moral structure therefore neatly ignored the problem of animals. Or else they justified exploitation on the grounds of another aspect of Darwinianism, saying that if we are in a position to exploit animals, it is because we are superior evolutionarily, and therefore have every right to do so.

Thus, even before the rise of behaviourism, when science, like common sense, never thought to deny the existence of animal consciousness, it didn't care a great deal about the moral implications of the existence of such consciousness. Even pain in animals, the most obviously morally relevant mode of consciousness, was not of great concern to scientists. This is attested to by an 1896 veterinary textbook, Müller and Glass's *Diseases of the Dog*, in which the authors recommend general anaesthesia only 'in very serious operations accompanied by great pain'.[18] In the same vein, Liautard's 1892 *Manual of Operative Veterinary Surgery* asserts:

In veterinary surgery, the indication for anesthesia has not, to the same extent as in human, the avoidance of pain in the patient for its object, and though the duties of the veterinarian include that of avoiding the infliction of *unnecessary* pain as much as possible, the administration of anesthetic compounds aims principally to facilitate the performance of the operation for its own sake, by depriving the patient of the power of obstructing, and perhaps even frustrating its execution, to his own detriment, by the violence of his struggles, and the persistency of his resistance. To prevent these, with their disastrous consequences, is the prime motive in the induction of the anesthetic state.[19]

But in the first detailed discussion of anaesthesia in veterinary surgery, Merillat's *Principles of Veterinary Surgery* of 1906, the author indicts veterinarians for failing to worry about pain:

In veterinary surgery, anesthesia has no history. It is used in a kind of desultory fashion that reflects no great credit to the present generation of veterinarians. . . . Many veterinarians of rather wide experience have never in a whole lifetime administered a general anesthetic in performing their operations. It reflects greatly to the credit of the canine specialist, however, that he alone has adopted anesthesia to any considerable extent. . . . Anesthesia in veterinary surgery today is a means of restraint and not an expedient to relieve pain. So long as an operation can be performed by forcible restraint . . . the thought of anesthesia does not enter into the proposition.[20]

The exception noted by Merillat, the use of anaesthesia for dogs, bespeaks a sentimentalism regarding these creatures, rather than a thoughtful moral concern for animals in general, and this has persisted in ordinary common sense to the present. The notion of anaesthesia as primarily a chemical restraint, rather than a way of controlling pain, has also persisted, and has naturally received succour from the advent of the postivistic-behaviouristic component of scientific common sense.

Thus, morally speaking, behaviourism was socially tolerable. For even though both common sense and Darwinism saw animals as having consciousness and feelings, both failed to build this insight into their moral schema. So the behaviourist was in a position much like that of Cartesian or Newtonian scientists during the scientific revolution. *Qua* scientists, the latter had to adopt the ideology that the world of the senses was not ultimately real, or even that it was illusory, and that only what was quantifiable was real. This may not have presented them with a problem in their research activities: it is fairly non-problematic to look only at the quantifiable, mechanical aspects of falling bodies, rising tides, and so on. But when they went home, they had to change gears and concern themselves with secondary qualities just as vigorously as people who didn't know that such qualities were unreal. Common sense and living in the world made compartmentalization a practical necessity. On the other hand, whenever the new metaphysics didn't impinge on coping with the world, it probably did change the way they looked at the world.

And likewise with the behaviourists. They surely were not behaviourists at home, in their love lives, during bull sessions, and so on. They were behaviourists only in the laboratory, where they worked on animals. And even though there was a common-sense tendency to see animals as being conscious, nothing much hung on seeing them that way, so it wasn't hard to learn to see them as learning machines. This process is replicated today in the training of scientists who must use animals invasively; they quickly learn to see them as 'models' and scientific machines.[21] A similar tendency to reduce persons to bodies and disease to mechanical breakdown pervades medical education as well. One should never underestimate the power of high-level scientific training to eclipse ordinary common sense. On the other hand, as we shall see in a later chapter, it is ultimately, in large part, a change in social-moral valuation of animals—a change in ordinary common sense—which worked against the ideology of science, and forced scientists to again admit the facts of animal awareness.

4 The tortuous path from Romanes to Watson

Lloyd Morgan and his Canon

If one looks back at the steps which seem to have led from Romanes to Watson, or at least, which behaviouristically conditioned historians have claimed led rationally from one to the other, one finds that the progression was tortuous indeed, and the logic very forced. In an era in which the world-view was not so overwhelmingly reductive, the thinkers who provided the bridge from Darwinism to behaviourism would not have forced the logic in this way; at best, they would have fine-tuned the Darwin–Romanes approach. Indeed, it would be difficult to find a progression of ideas less rationally inevitable than that which spans the period between Darwin and Romanes and the overthrow of their approach. If we look at the figures who are traditionally adduced to explain what happened, we can see just how unlikely the path to behaviourism really was. Only by looking at these transitional figures from a very biased perspective can one pick up even a mood which is at all congenial to behaviourism. Indeed, the work of people like Lloyd Morgan, Loeb, and Jennings, all of whom are traditionally cited as providing the link to behaviourism (for example, in Boring's classic history of psychology[1]), provides more support for the perspective of Darwin and Romanes than it does for behaviourism, and could equally well have been used to *buttress* or refine the Darwin–Romanes view of animal consciousness.

Consider the work of Lloyd Morgan. In virtually any standard historical account of the development of psychological theory, we find the assertion that Morgan's Canon is one of the fundamental contributions to the overturn of the Darwin–Romanes point of view. As Daniel Robinson says in his Preface to a recent reissue of Morgan's *An Introduction to Comparative Psychology* (1894)[2], Morgan's Canon has been mastered by untold numbers of undergraduates over the past ninety years, many of whom have been taught, in essence, that

Morgan enunciated his principle as a logical truism, one which is undeniable, and one which the attribution of mentation to animals clearly violates. Thus, for example, Marx and Hillix in their *Systems and Theories in Psychology*, a standard textbook, assert that

> Romanes was demonstrating continuity by finding mind everywhere; Morgan also wished to demonstrate continuity, but suggested that it might be done as well if we could find mind nowhere. Morgan's appeal to simplicity and rejection of anthropomorphism would seem, from a modern perspective, to have made the development of a scientific behaviourism inevitable.[3]

The full irony of so reading Morgan can be appreciated only by those who actually read Morgan, who apparently are precious few, or this interpretation could not possibly have endured, or for that matter, have ever been promulgated. For Morgan's Canon is not a self-evident truth, is not a blow to the belief in animal consciousness, as Marx and Hillix suggest, and does not, in essence, touch the core of the Darwin–Romanes view of animal mind, a view with which Morgan was basically in agreement.

Let us look first at the Canon itself, in Morgan's own words: 'In no case may we interpret an action as the outcome of the exercise of a higher psychical faculty, if it can be interpreted as the outcome of the exercise of one which stands lower in the psychological scale.'[4] In an age of simplification and paring away of frills, such a principle fits right in. Indeed, it was seen to be a special case of Occam's razor: 'Do not multiply entities unnecessarily', intended by Occam to apply to universals, although it subsequently came to be applied to all sorts of explanations, and is still a mainstay of 'good scientific method'. There are, of course, many questions to be asked regarding any such principle of parsimony or simplicity, the simplest being, Why not? What logical absurdity is generated by multiplying entities? And secondly, what counts as 'unnecessarily'? But here we can avoid these troubled waters, and concentrate on Morgan's version.

The fundamental, crucial arguments against seeing Morgan's Canon as a refutation of the Darwin–Romanes view of the continuity of mentation and, consequently, as a denial of consciousness to animals are two: first, that it was not so intended by Morgan; and second, that even if it were so intended, far from denying animal mentation, it in fact presupposes it, in order even to make sense.

Regarding the first, Morgan, far from opposing Darwin and Romanes, specifically acknowledges his commitment to evolutionary theory and to a phylogenetic continuum of consciousness. Thus he writes:

We have . . . taken for granted the existence of consciousness, and the fact that there are subjective phenomena which we, as comparative psychologists, may study. We have also proceeded throughout on the assumption that subjective phenomena admit of a natural interpretation, as the result of a process or processes of development or evolution, in just the same sense as objective phenomena admit of such interpretation.[5]

This is hardly the language of a proto-Watson out to abolish consciousness and awareness. It sounds more like a colleague of Romanes, which Morgan in fact was. On the second page of his preface, Morgan refers to Romanes as a man 'to whom I owe much, and in many ways'. And in a touching footnote, he adds that 'The death of Professor Romanes, since this too brief acknowledgment of all that I owe to him was written and printed, has entailed a loss to science which is irreparable, and a loss to his personal friends which lies too deep for words'.[6]

Morgan flatly asserts his belief in animal consciousness and its evolution in his essay 'The limits of animal intelligence'. Even in the case of a baby chick pecking at food, he tells us, 'we have every reason to suppose that the performance of these actions is accompanied by feeling or consciousness'.[7] Later in the essay, he is even more explicit:

There can be very little question that the higher animals have impressions and ideas analogous to ours. When a dog sees before him a nice meaty bone, I have no doubt that he has a quite clear-cut and definite impression. And when he comes home hungry, after a long walk, and going down into the kitchen, looks up wistfully at the cook, I, for one, should not feel disposed to question that he has in his mind's eye a more or less definite idea of a bone.[8]

In his book on comparative psychology, he tells us that 'we are *logically* bound to regard psychological evolution as strictly coordinate with biological evolution'.[9]

Not only is Morgan not a behaviourist, he is a raving speculative metaphysician! He is a monist, a pan-psychist, a Spinozian, a double-aspect theorist, a believer that *everything* in nature has both a physical and a psychological dimension. This theory is even laid out as a prolegomenon to his book, prior to the first chapter, and is presupposed throughout the subsequent discussion. Indeed, it is only in the context of this metaphysical theory that we can even make sense of his Canon.

Throughout the book, Morgan accepts continuity of consciousness. Furthermore, he asserts the need for introspection as a precondition for comparative psychology, not only human psychology. 'In the systematic training of the comparative psychologist the subjective aspect is not less important than the objective aspect.'[10] The more an animal is like us physically and behaviourally, the more certainty there is in our induction to its mental states. So far, he sounds very much like Darwin and Romanes.

Then why the Canon? Simply because, in Morgan's view, other comparative psychologists have been too quick to attribute *reason* to lower animals. Contrary to the traditional division of mental intellectual faculties into instinct and reason, Morgan advocates a tripartite division into instinct, intelligence, and reason, which distinction he explains as follows:

> Animals and men come into the world endowed with an innate capacity for active response to certain stimuli. This is part of their organic inheritance. The response may be from the first an accurate and adequate response: in such cases we term it instinctive. But more frequently the responses have a variable amount of inaccuracy and inadequacy; in such cases the animal, as a matter of observed fact, has a power of selective control over the responses; and this power of selective control in the activities which are essential to daily life, is the first stage of intelligence. Now why do I say intelligence and not reason? Well, let us go back to Sydney Smith's description of the reasoned act: 'If I desire to do a certain thing.' he says, 'adopt certain means to effect it, and have a clear and precise notion that those means are subservient to that end, then I act from reason.' But have we any ground for supposing that a chick, a few hours old has 'a clear and precise notion that those means are subservient to that end?' Is it probable that the baby who is learning to put a crust into his mouth and not into his eye has any precise notions of the relation of means to ends? If not, then here we have a class of activities, and a very important class, those, namely, which are essential to daily life, which are perfected by means of a faculty which is not reason, and which I would term intelligence. To paraphrase Sydney Smith I would say, 'If I adopt certain means to effect a given end, but have no clear and precise notion of the relation of means to end, then I act not from reason but from intelligence.' And to modify a well-known statement of Mr. Romanes', I would say that, 'unlike reason, intelligence implies no conscious knowledge of the relation between the means employed and the ends attained, though it may be exercised in selective adaptation to circumstances novel alike to the experience of the individual and to that of the species.'[11]

More specifically, intelligence and reason are distinguished as follows:

Intelligence is the faculty by which, through experience and association, activities are adapted to, or, more strictly, moulded by, new circumstances; while reason is the faculty which has its inception in the true grasping of relationships as such. Intelligence is ever on the watch for fortunate variations of activity and happy hits of motor response; it feels that they are suitable, though it knows not how and why, and controls future activities in their direction. It proceeds by trial and error, and selects the successes from among the failures. Reason explains the suitability; it shows wherein lies the success or the error, and adapts conduct through a clear perception of the relationships involved. Individual experience, association, and imitation are the main factors of intelligence; explanation and intentional adaptation are the goal of reason.[12]

It is in reference to these notions that his Canon is developed. It is not meant to deny consciousness, awareness, mentation, thought, ideation, feeling, emotion, and so forth to animals. Rather, it is directed against over-zealous Darwinians who are too quick to attribute reason in the sense defined above to animals. Morgan does in fact believe that reason is unique to humans, but he holds this as a very tentative empirical hypothesis. It is worth quoting him one last time:

Incidentally I have expressed my opinion that, in the activities of the higher animals, marvellously intelligent as they often are, there is no evidence of that true perception of relationships which is essential to reason. But this is merely an opinion, and not a settled conviction. I shall not be the least ashamed of myself if I change this view before the close of the present year. And the distinction between intelligence and reason will remain precisely the same if animals are proved to be rational beings the day after to-morrow. For the distinction holds good between human intelligence and human reason, just as much as between animal intelligence and the possible reason of animals. It is no line of division which separates animals from men; but a distinction between faculties, one of which, at least (and perhaps both, though this I doubt), is common to animals and men.[13]

How ironic, then, that such a thinker should come to be seen as one of the fathers of behaviourism! We should remember, then, out of respect for his memory, that Morgan is not responsible for the use to which his ideas were put by people like Watson and other behaviourists, just as we cannot blame Marx for Marxists, Freud for Freudians, or Heraclitus for Cratylus. As Robinson puts it, we cannot blame Morgan that his 'Canon' became a 'Cannon' in such hands.[14]

There is one further point worth making as regards Morgan's Canon properly interpreted. We may raise the following question:

Why should we never interpret an action by a higher faculty if we can do so by means of a lower faculty? Presumably, in the name of simplicity and parsimony. But what is 'simpler' in such a context? If, for example, we can explain a person's or an animal's behaviour in solving a problem either by saying that the solution was arrived at by a process of reasoning *or* by citing a very complex chain of trial-and-error intelligence, how is it simpler to avoid the higher faculty? Is it simpler to have few principles and complex applications of those principles, or to postulate more principles, which can be applied more simply? Unless this question can be answered, it seems to me illegitimate to deduce Morgan's Canon from something like Occam's razor. Indeed, the sort of objection just sketched with regard to what is to count as simpler points up an area which needs elaboration in any principle of parsimony.

Loeb's mechanism

In any case, it is clear that Morgan's work was hardly inimical to the fundamental thrust of the Darwin–Romanes point of view, even if it was taken as such in the reductive context we have described. The work of Jacques Loeb, which has also been cited as a major historical factor countering Darwinism, is more deserving of such an accusation. Loeb's reductionism was mechanistic, more or less a continuation of Descartes's idea that animals are machines whose behaviour is totally explicable in physico-chemical terms. This mechanistic reductionism was also a significant force in nineteenth-century thought and indeed, has endured and flourished in the twentieth century, having received a shot in the arm from the development of molecular biology, and wrought much mischief in the study of animals and humans, especially in medicine. In 1845 Helmholtz, Brücke, Ludwig, and Du Bois-Reymond swore a blood oath to implement this truth:

> No other forces than the common physical chemical ones are active within the organism. In those cases which cannot at the time be explained by these forces, one has either to find the specific way or form of their action by means of the physical-mathematical method or to assume new forces equal in dignity to the chemical-physical forces inherent in matter, reducible to the force of attraction and repulsion.[15]

Loeb was in this tradition. Like the positivists, he saw an ethical imperative behind his philosophy of science—namely, the abolition of romantic nationalism, religious superstition, militarism, racism, and so on, which could be countered only by materialistic science.

This ethical veneration of mechanistic science as the only approach to truth led him to be highly suspicious of Darwinism. In the first place, evolutionary claims were not verifiable or duplicable in the laboratory. They were historical accounts, and Loeb seems to have shared the traditional mechanist's suspicion of historicism. (Loeb once remarked that finding a fossil in a lower stratum than other fossils only shows that the creature in question had the misfortune of falling into a deeper hole.)[16] Second, whereas Loeb advocated a philosophy which allowed for the description of human beings in mechanical terms, eliminating notions like free will in favour of determinism, Darwinism was muddying the waters by raising the notion that not only did humans have irreducible psychical properties, but that animals did too. Finally, Loeb saw Darwinism as generating its own mystical consequences; or, more accurately, he saw Darwin's followers as creating whole new obfuscatory, mystical, metaphysical schemes out of evolutionary theory. Haeckel, for example, like Morgan, saw all nature as suffused with mind, and attempted to substitute the notion 'ontogeny recapitulates phylogeny' for a mechanistic analysis of embryology.

Loeb's own work was thoroughly consistent with his philosophical ideals. His work on artificial parthenogenesis gave support to the notion that all life processes are mechanical, and that there is no need to invoke vitalism or religion to explain the miracle of fertilization; biochemistry suffices. More directly relevant to our concerns was his work on animal tropisms. In the 1880s Loeb worked with Julius Sachs, the man who attempted to introduce Helmholtzian mechanism into plant physiology. It was Sachs who studied the movements of plants—'tropisms'—mechanically elicited by such forces as light and gravity; and every schoolchild today is familiar with positive phototropism—that plants bend towards the light. For Loeb, this concept was the key to providing a mechanistic account of animal behaviour, at least as regards the lower animals. In such papers as 'The Significance of Tropisms for Psychology',[17] he tried to demonstrate, for example, that some animal movements in the presence of light for a whole range of animals, vertebrate and invertebrate, could be explained without invoking either psychic 'will' or physical instinct. Loeb objected to explanations in terms of instinct as much as to invoking will, on the grounds that instinct was covertly mystical. Thus, he writes: 'Instincts are defined in various ways, but no matter how the definition is phrased the meaning seems to be that they are inherited reflexes so purposeful and so complicated in character that nothing short of intelligence and experience could have produced them.'[18] To Loeb, instincts smacked of unsavoury teleology, vitalism, and the like. Indeed, commenting on the excesses of such notions, he informs us, that 'a prominent psychologist has maintained that reflexes are to be considered as the mechanical effects of acts of volition of past generations'.[19]

What is going on in tropistic reactions, according to Loeb, is purely physico-chemical. In the case of winged aphids, for example, the explanation of their moving towards the light is as follows:

> In the winged aphids the relations are as follows: Suppose that a single source of light is present and that the light strikes the animal from one side. As a consequence the activity of those muscles which turn the head or body of the animal toward the source of light will be increased. As a result the head, and with it the whole body of the animal, is turned toward the source of light. As soon as this happens, the two retinae become illuminated equally. There is therefore no longer any cause for the animal to turn in one direction or the other. It is thus automatically guided toward the source of light. In this instance the light is the 'will' of the animal which determines the direction of its movement, just as it is gravity in the case of a falling stone or the movement of a planet. The action of gravity upon the movement of the falling stone is direct, while the action of light upon the direction of movement of the aphids is indirect, inasmuch as the animal is caused only by means of an acceleration of photochemical reactions to move in a definite direction.[20]

Loeb was thus viewed as a thoroughgoing Cartesian mechanistic determinist who viewed animals as machines. And there is no question that this was his inclination. None the less, Loeb never denied the reality and existence of consciousness. In his book *Comparative Physiology of the Brain and Comparative Psychology*, he discusses the relationship between his programme and consciousness. He declares that 'we must find out the elementary physiological processes which underlie the complicated phenomena of consciousness.'[21] He debunks physiologists and psychologists who equate consciousness with purposefulness, and thus attribute consciousness too liberally along the phylogenetic scale. Machines too would have to be said to be conscious on this view, as would plants; and embryological development would have to be seen as conscious as well.

But Loeb emphatically does not deny the existence of consciousness in animals:

On the other hand, physiologists who have appreciated the untenable character of such metaphysical speculations have held that the only alternative is to drop the search for the mechanisms underlying consciousness and study exclusively the results of operations on the brain. This would be throwing out the wheat with the chaff. The mistake made by metaphysicians is not that they devote themselves to fundamental problems, but that they employ the wrong methods of investigation and substitute a play on words for explanation by means of facts. If brain-physiology gives up its fundamental problem, namely, the discovery of those elementary processes which make consciousness possible, it abandons its best possibilities. But to obtain results, the errors of the metaphysician must be avoided and explanations must rest upon facts, not words. The method should be the same for animal psychology that it is for brain-physiology. It should consist in the right understanding of the fundamental process which recurs in all psychic phenomena as the elemental component. *This process, according to my opinion, is the activity of the associative memory, or of association.* Consciousness is only a metaphysical term for phenomena which are determined by associative memory. By associative memory I mean that mechanism by which a stimulus brings about not only the effects which its nature and the specific structure of the irritable organ call for, but by which it brings about also the effects of other stimuli which formerly acted upon the organism almost or quite simultaneously with the stimulus in question. If an animal can be trained, if it can learn, it possesses associative memory. By means of this criterion it can be shown that Infusoria, Coelenterates, and worms do not possess a trace of associative memory. Among certain classes of insects (for instance, wasps), the existence of associative memory can be proved. It is a comparatively easy task to find out which representatives of the various classes of animals possess, and which do not possess, associative memory. Our criterion therefore might be of great assistance in the development of comparative psychology.[22]

Loeb's criterion for the presence of consciousness is a creature's ability to learn, what he calls 'associative memory', which he proceeds to equate (?) with a certain degree of physiological organization in the brain. This is hardly a denial of the existence of consciousness. Granted, it remains a problem for Loeb, as for any thoroughgoing physicalistic mechanist, to explain consciousness and its relation to the brain, and hence my question mark in the previous sentence. But that is not our problem. What is significant for our purposes is that, in the final analysis, Loeb has at most bequeathed a research programme, or made a prescriptive value judgement to the effect that one ought to study the brain if one wishes to understand consciousness. He has not denied the existence of consciousness, so he has hardly eroded the Darwin–Romanes approach. Let us recall that even Romanes, in all his alleged romantic sentimental anthropomorphism, denied consciousness to protozoa, and was agnostic with respect to consciousness in invertebrates. Thus there is really nothing in Loeb which is even incompatible with the position of Darwin and Romanes, let alone forces its abandonment. At the worst, Loeb has declared that explanations which 'evoke psychical properties are not ultimate—ultimate explanations must be physiological'. But such a position is promissory and optimistic; a letter of credit which has yet to be cashed.

H.S. Jennings

Let us recall that Loeb was a biologist, not a psychologist, and that the dominant theme in human psychology at that time was introspectionism. (We have seen that Morgan saw introspection as essential to animal psychology as well.) The next figure in the supposed chain leading towards behaviourism, or whose work at least historically provides some impetus for behaviourism, was also a biologist—indeed, a biologist reacting against Loeb. But, as we shall see, while Jennings made certain methodological commitments which are congenial to behaviourists, notably in his emphasis on controlled experimentation and his analytical separation of questions of consciousness from questions of behaviour, his conclusions strongly support the Darwin–Romanes view of the continuity of mental experience.

Jennings was one of the major figures in biology in the first half of the twentieth century, and his work encompassed a broad range of concerns. The work which is relevant to the issue before us, however, was done at the beginning of his career, and was published in 1906 as *Behaviour of the Lower Organisms*. In this book, Jennings attempted 'an objective description of the known facts of behaviour in lower organisms'.[23] His approach was designed to contrast with that of a number of earlier works which also dealt with protozoa, but which tended to give primacy to the psychic approach, notably Binet's *Psychic Life of Organisms* of 1889 (the same Binet now famous for his work on intelligence testing) and Lukas's *Psychologie der niedersten Tiere* of 1905.

As a zoologist, Jennings's basic concern was to separate out what was directly observable from what can be inferred theoretically from observations. In his view, consciousness is not directly observable; nor is it subject to direct experimental technique. As he tells us in the very first paragraph of his preface, 'Assertions regarding consciousness in animals, whether affirmative or negative, are not susceptible of verification.'[24] This does not mean that Jennings had no interest in consciousness, or that he excluded the study of consciousness from the domain of science. His precept was methodological: we must carefully separate 'this matter [of consciousness] from those which can be controlled by observation and experiment. For those primarily interested in the conscious aspects of behaviour, a presentation of the objective facts is a necessary preliminary to an intelligent discussion of the matter.'[25]

Jennings was thus interested in separating the gathering of data from the interpretation of that data. Given the prevailing disagreements about consciousness and what could be legitimately inferred from behavioural data, such a separation was necessary in order to achieve a common universe of discourse for researchers in the field. His rigid distinction between 'description' and fact-gathering and 'explanation', 'analysis', and 'theorizing' is reflected in the structure of his book. Parts I and II, which describe the 'objective' behaviour of protozoa and lower metazoa respectively, deal with such topics as the structure, movements, and method of reaction to stimuli of paramecia. These researches include responses to heat, cold, light, and electricity, and orienting reactions with respect to water currents, gravity, centrifugal force, and the like. Similar discussions are presented for amoebas, bacteria, coelenterata, and so on. These descriptions, Jennings tells us, are 'made as far as possible independent of any theoretical views held by the writer; his ideal was indeed to present an account that would include the facts required for a refutation of his own general views'.[26] This is done in contrast to those who pick and choose their facts to fit their prior theoretical commitments.

Although one must begin with objective description, Jennings sees that there is more to science than that, and in Part III he presents his analysis of the facts, his general conclusions, and his theoretical stance on his material. He stresses that this theoretical account must be seen as more tentative than a straightforward presentation of the facts. None the less, it is an equally indispensable part of scientific activity. Jennings's theoretical conclusions are of great interest. In the first place, on his view, the objective data do not support the sort of mechanistic explanation proposed by Loeb. There seems to be abundant evidence that the behaviour of lower organisms is not, under most conditions, a direct result of external mechanical forces of the sort postulated by Loeb. Behaviour does not require an external stimulus to be present, and may change without external cause. The major determinants of behaviour, even in protozoa, are internal or physical states. A well-fed paramecium behaves differently in the presence of a certain stimulus from one which has been deprived of food.

Secondly, and again contrary to Loeb, learning takes place even at the level of protozoa. Chapter 10 presents the results of detailed experiments with and observations of the unicellular organism *Stentor roeselii*, and these are invoked to justify the theoretical judgement that these creatures learn by both trial and error and experience.

> The reaction to any given stimulus is modified by the past experience of the animal, and the modifications are regulatory, not haphazard, in character. The phenomena are thus similar to those shown in the 'learning' of higher organisms, save that the modifications depend upon less complex relations and last a shorter time.[27]

All these movements are fundamentally regulatory and homeostatic; 'The behaviour of the animal under stimulation corresponds to its needs, and is determined by them.'[28]

The interesting features of Jennings's work discussed so far *vis-à-vis* the topic at hand may be summarized as follows: its emphasis on controlled experimentation of protozoan behaviour *as well as* observation under natural circumstances (unlike many later behaviourists, he was very sensitive to the artificiality of many experimental situations); the fact that it adduces evidence leading to a rejection of Loeb's tropistic explanations as inadequate even as regards the behaviour of unicellular organisms; and its conclusion that something like trial-and-error learning occurs even in lower organisms.

The most interesting theoretical conclusions, however, are presented in the penultimate chapter of his book, entitled 'Relation of Behavior in Lower Organisms to Psychic Behavior'. Here Jennings reveals himself as four-square in the theoretical camp of Darwin and Romanes, differing only in his insistence on separating 'fact' from 'theory'. His remarks are worth looking at in some detail, if only to show how a scientist of this period, thoroughly 'modern' in separating fact from speculation, balanced his methodological caution against his theoretical commitments.

Jennings points out that his descriptions of protozoan behaviour have been purely objective thus far—for example, that a paramecium draws away from heat and cold, even as we do. But in humans such behaviour is accompanied by consciousness, although we can experience consciousness only in ourselves. We can extend talk of consciousness to animals only by analogy, which may be suspect; therefore physiologists have tended to restrict themselves to 'objective' descriptions. Furthermore, physiologists are often committed to Helmholtzian physicalism, sworn to explain things only in physico-chemical terms. The disadvantage of seeing things only in these terms is that it suggests a radical discontinuity between humans and animals, one which is not concordant with evolutionary theory. So, as a first step, even on the level of objective description, as opposed to theorizing, we must ask whether the *descriptive* concepts we use *vis-à-vis* human behaviour have application to animals. From a behavioural point of view, says Jennings, concepts like perception, discrimination, choice, attention, fatigue, desire for food, pain, fear, memory, and habit can all have application to animal behaviour. One can talk sensibly of 'fear' in a sea-urchin, for example, without saying anything about its subjective state:

> The sea urchin reacts by defensive movements when a shadow falls upon it, though shade is favorable to its normal functions. Objectively, fear has at its basis the fact that a negative reaction may be produced by a stimulus which is not in itself injurious, provided it leads to an injurious stimulation; this basis we find throughout organisms.
>
> Sometimes higher animals and man are thrown into a 'state of fear,' such that they react negatively to all sorts of stimuli, that under ordinary circumstances would not cause such a reaction. A similar condition of affairs we have seen in Stentor and the flatworm. After repeated stimulation, they react negatively to all stimuli to which they react at all.
>
> The general fact of which the reactions through fear are only a special example is the following: Organisms react appropriately to *representative* stimuli. That is, they react, not merely to stimuli that are in themselves beneficial or injurious, but to stimuli which lead to beneficial or injurious conditions.[29]

Similarly with 'intelligence', the adaptive behaviour of organisms to changing circumstances and the ability to learn from experience proceeds along a continuum from the lowest animals to man: 'From the lowest organism up to man, behavior is essentially regulatory in character, and what we call intelligence in higher animals is a direct outgrowth of the same laws that give behavior its regulatory character in Protozoa.'[30]

Jennings thus adheres to a principle of continuity of behaviour which parallels morphological and physiological continuity. But so far he has talked only of 'objective phenomena'. What of consciousness? Jennings reiterates that we cannot directly verify claims about consciousness. On the other hand, one cannot avoid asking whether the behaviour of lower organisms is of the sort we would expect if they did have consciousness and acted from conscious states similar to those experienced by humans. Jennings's reply is interesting enough to quote in full.

Suppose that this animal *were* conscious to such an extent as its limitations seem to permit. Suppose that it could feel a certain degree of pain when injured; that it received certain sensations from alkali, others from acids, others from solid bodies, etc.,—would it not be natural for it to act as it does? That is, can we not, through our consciousness, *appreciate* its drawing away from things that hurt it, its trial of the environment when the conditions are bad, its attempting to move forward in various directions, till it finds one where the conditions are not bad, and the like? To the writer it seems that we can; that Paramecium in this behaviour makes such an impression that one involuntarily recognizes it as a little subject acting in ways analogous to our own. Still stronger, perhaps is this impression when observing an Amoeba obtaining food. The writer is thoroughly convinced, after long study of the behavior of this organism, that if Amoeba were a large animal, so as to come within the everyday experience of human beings, its behavior would at once call forth the attribution to it of states of pleasure and pain, of hunger, desire, and the like, on precisely the same basis as we attribute these things to the dog. This natural recognition is exactly what Münsterberg (1900) has emphasized as the test of a subject. In conducting objective investigations we train ourselves to suppress this impression, but thorough investigation tends to restore it stronger than at first.

Of a character somewhat similar to that last mentioned is another test that has been proposed as a basis for deciding as to the consciousness of animals. This is the satisfactoriness or usefulness of the concept of consciousness in the given case. We do not usually attribute consciousness to a stone, because this would not assist us in understanding or controlling the behavior of the stone. Practically indeed it would lead us much astray in dealing with such an object. On the other hand, we usually do attribute consciousness to the dog, because this is useful; it enables us practically to appreciate, foresee, and control its actions much more readily than we could otherwise do so. If Amoeba were so large as to come within our everyday ken, I believe it beyond question that we should find similar attribution to it of certain states of consciousness a practical assistance in foreseeing and controlling its behavior. Amoeba is a beast of prey, and gives the impression of being controlled by the same elemental impulses as higher beasts of prey. If it were as large as a whale, it is quite conceivable that occasions might arise when the attribution to it of the elemental states of consciousness might save the unsophisticated human being from the destruction that would result from the lack of such attribution. In such a case, then, the attribution of consciousness would be satisfactory and useful. In a small way this is still true for the investigator who wishes to appreciate and predict the behavior of Amoeba under his microscope.

But such impressions and suggestions of course do not demonstrate the existence of consciousness in lower organisms. Any belief on this matter can be held without conflict with the objective facts. All that experiment and observation can do is to show us whether the behavior of lower organisms is objectively similar to the behavior that in man is accompanied by consciousness. If this question is answered in the affirmative, as the facts seem to require, and if we further hold, as is commonly held, that man and the lower organisms are subdivisions of the same substance, then it may perhaps be said that objective investigation is as favorable to the view of the general distribution of consciousness throughout animals as it could well be. But the problem as to the actual existence of consciousness outside of the self is an indeterminate one; no increase of objective knowledge can ever solve it. Opinions on this subject must then be largely dominated by general philosophical considerations, drawn from other fields.[31]

Despite Jennings's cautiousness, or perhaps because of it, there is a great deal of philosophical perspicacity in his all too brief discussions. Later we shall have occasion to examine in greater detail some of the issues he raises almost in passing. Here I will merely point out a number of philosophically interesting points. First, Jennings, more than anyone else we have examined so far, makes a distinction between objective data-gathering and theorizing about that data, both scientifically and philosophically. He clearly adheres to a fundamental precept which was a mainstay of the spirit of logical positivism and which was absorbed into the common sense of science: namely, that one can collect data in a basically atheoretical manner and then generate theories based on those data. Second, Jennings seems none the less to have been aware, *unlike* positivists, behaviourists, and most scientists trained in the common sense of science, that science goes beyond the data, and seeks to explain things on theoretical and even philosophical levels, and that neither kind of explanation is directly deducible from the data, but is merely 'suggested' by it. As long as one realizes what level one is operating on, this process is not pernicious. (In this, Jennings is quite different from the behaviourists, who resisted theoretical accounts.) Third, on the specific issue of consciousness, Jennings does not regard notions relating to the awareness of others as fundamental data, but as entering at the theoretical level of explanation, where they represent the best explanation of certain phenomena, are useful in generating predictions, cohere with the evolutionary theory accepted in biology, and thus enjoy a pragmatic justification. Fourth, Jennings was extraordinarily prescient in anticipating the ideas of the later Wittgenstein, Ryle, and other thinkers traditionally called 'logical behaviourists'. Thus he saw psychological terms as deriving their meaning from objective behaviour manifested in appropriate contexts, and not from introspection. Of course this is equally true whether the terms are applied to animals or other humans. The key point is that men and animals are on equal terms as regards the applicability of mental terms.

Ironically, Jennings's work was attacked by supporters of Loeb as too subjectivistic and by introspectionists as too objective—that is, too willing to attribute consciousness to animals on the basis of mere behaviour. Even more ironically, one of his introspectionist critics was none other than John B. Watson writing in the *Psychological Bulletin* in 1907.[32] Shortly thereafter, Watson would father behaviourism, and, going much further than Jennings, would deny the legitimacy of talk about consciousness altogether.

But how are we to assess Jennings's work relative to the downfall of the Darwin–Romanes view of animal mentation? It appears that there is nothing in Jennings which is basically inimical to that perspective; in fact, his work is totally Darwinian in approach. We may see Jennings as refining the Darwinian perspective in his methodological insistence on introducing consciousness at the level of theorizing, rather than that of fact-gathering. One need only read the long passage quoted above and recall all Jennings's remarks on continuity to dispel any doubts concerning his acceptance of the meaningfulness of talk about consciousness, albeit not on the level of direct observation. His demonstration of the existence of learning at the level of protozoa was certainly evidence for phylogenetic continuity of mentation. At the same time, his refutation of Loeb and corollary emphasis on trial-and-error learning was also echoed in behaviourism's primary emphasis on learning. Finally, his insistence on the primacy of behaviour on the level of observation could not possibly have failed to influence Watson, who in fact studied with Jennings, as well as with Loeb.

In looking at Jennings and Loeb, we have identified some factors which historically were construed as contributing to the demise of the Darwin–Romanes view. We have seen that the work of these men not only did not necessitate this demise, but actually supported the continuity of consciousness. There was disagreement between Loeb and Jennings as to how far consciousness extended along the phylogenetic scale, of course, and doubtless Loeb would have been happy to eliminate consciousness altogether, had he seen a way of doing so; but he could not. So in the final analysis, the work of neither man represented any real undercutting of animal mentation.

Titchener's introspective psychology

Loeb and Jennings were biologists. We must now look at the psychological factors which contributed to the overthrow of the Darwin–Romanes *Gestalt* regarding animal consciousness. At the end of the nineteenth century and throughout the first decade of the twentieth, a good portion of mainstream American psychology was under the influence of principles established in Germany by Wilhelm Wundt and in the United States by his disciple E.B. Titchener. The dominant characteristic of their approach to psychology was introspection. According to Titchener, psychology, if it was to be a science, needed to stand independently of other disciplines, to have its own subject-matter and methodology. It was neither biology nor physiology. Whereas Loeb asserted that we study the body to study the mind, Titchener maintained that we study the mind to study the mind. The aim of psychology was the study of the (adult) human mind, which in turn meant the study of consciousness. And the way in which one goes about such study is through training in introspection and by using introspective methods to chart the laws, principles, and powers of the mind. This in turn, for Titchener, is simply the scientific method applied to mentation. Contrary to the current view of introspection as something engaged in by hippies, mystics, and drug-users, Titchener's view was that it was rigorous, thoroughly scientific, and experimental. Indeed, Titchener referred to his work as 'experimental psychology', and in so doing, he was not speaking metaphorically. 'As a general rule . . . and to the average student,' he tells us, 'an understanding of the introspective method either comes by way of the laboratory or does not come at all.'[33] On the frontispiece of his 1901 work *Experimental Psychology: A Manual of Laboratory Practice* is inscribed a quotation from Faraday: 'As an experimenta-list, I feel bound to let experiment guide me into any train of thought which it may justify. This work consists of prescriptions for a series of experiments, designed to teach psychology to students, and includes a student's manual and an instructor's manual. Since Titchener believed that 'the sensation *is* strictly subject matter for psychology',[34] his experiments were designed as training for introspective analysis of sensations. Thus he provided students with exercises for studying such topics as visual after-images, auditory interference, cutaneous sensations, visual space perception, tactual space perception, stereoscopic vision, optical illusions, and so on, many of which still appear in psychology texts—even behaviourists' texts, since a behaviouristic account of perception was never accomplished.

Given Titchener's primary concern with introspective study of human adult consciousness as a means of arriving at laws of mind, the study of animal mind must necessarily assume a secondary, derivative role in his psychology. In his *Text Book of Psychology*, first published in 1896, Titchener first addresses the problem of other minds as an objection to introspective psychology. If we can study only our minds, how can psychology be a science? His answer is simple and venerable: namely, that the existence of all our social institutions and especially of language bespeaks the existence of other minds, as well as their fundamental similarity to our own. This is buttressed by the results which have been achieved by introspective psychology, wherein there is fundamental agreement in introspective reports by different observers.

Social psychology, abnormal psychology, and animal psychology are derivative fields logically parasitic on the study of the normal adult human mind. They must be studied by methods derived from the method of introspection. In the case of animals, we must suppose, as did Darwin and Romanes, that consciousness operates along the whole phylogenetic scale.

If, however, we attribute minds to other human beings, we have no right to deny them to the higher animals. These animals are provided with a nervous system of the same pattern as ours, and their conduct or behaviour, under circumstances that would arouse certain feelings in us, often seems to express, quite definitely, similar feelings in them. Surely we must grant that the highest vertebrates, mammals and birds, have minds. But the lower vertebrates, fishes and reptiles and amphibia, possess a nervous system of the same order, although of simpler construction. And many of the invertebrates, insects and spiders and crustaceans, show a fairly high degree of nervous development. Indeed, it is difficult to limit mind to the animals that possess even a rudimentary nervous system; for the creatures that rank still lower in the scale of life manage to do, without a nervous system, practically everything that their superiors do by its assistance. The range of mind thus appears to be as wide as the range of animal life.[35]

The next step is to study animal behaviour in various contexts, and to apply the knowledge gained from introspection to the data obtained, in order to reconstruct the animal's consciousness:

Now the psychologist argues, by analogy, that what holds of himself holds also, in principle, of the animal. . . . He argues that the movements of animals are, to a large extent, gestures; that they express or record the animal's mental processes. He therefore tries, so far as possible, to put himself in the place of the animal, to find the conditions under which his own movements would be of the same general kind; and then, from the character of his human consciousness, he attempts—always bearing in mind the limit of development of the animal's nervous system—to reconstruct the animal consciousness. He calls experiment to his assistance, and places the animal in circumstances which permit of the repetition, isolation and variation of certain types of movement or behaviour. The animal is thus made, so to say, to observe, to introspect; it attends to certain stimuli, and registers its experience by gesture. Of course, this is not scientific observation: science . . . implies a definite attitude to the world of experience, and consists in a description of the world as viewed from a definite standpoint. None the less, it is observation, and as such furnishes raw material for science. The psychologist works the raw material into shape; he oberves the gesture, and transcribes the animal consciousness in the light of his own introspection.[36]

Note that there is nothing prima facie 'unscientific' in Titchener's claims, given what counted as 'scientific' at the time. If his method is valid, presumably it can be tested in terms of our ability to predict what animals will do under various circumstances. The claim that we can, on the basis of our own introspection, know the minds of animals, and *only* know them in that way, ultimately proved too much for many of his American contemporaries, as did his faith in our ability to know our own minds solely through introspection. As we shall see, the spirit of reductivism, positivism, and simplification could not ultimately countenance anything that could not *in principle* (whatever that means) be reduced to physical science, which meant abolishing consciousness altogether.

Thorndike

The first critics of introspectionism eschewed such a radical step. Edward Thorndike, for example, while unwilling to accept the notion that our knowledge of animal mentation must be based on prior knowledge of our own mentation, was clearly not attempting to show that animal consciousness was not real. Thorndike's seminal work, *Animal Intelligence*, published in 1911 and based on thirteen years of research, was in no way designed to attack the notion of phylogenetic continuity of consciousness. It was intended, rather, as a critique of introspectionist methodology, and as an exposition of an alternative methodology. Like Jennings and Titchener, Thorndike believed strongly in 'the experimental method'. Thorndike's work on animals was distinguished by being among the first to depend exclusively on systematic laboratory studies, which primarily involved his famous puzzle boxes.

The first chapter of *Animal Intelligence* is extremely interesting from our point of view, since Thorndike not only addresses the issue of animal consciousness, but does so with great philosophical subtlety, a subtlety which has gone virtually unnoticed in standard accounts of the history of psychology. According to Thorndike, statements made by psychologists about humans or animals may be either about consciousness or about behaviour, about 'inner life' or what the person or creature is to an 'outside observer'. Some psychological terms refer almost exclusively to inner states—for example, 'imagery'. Others refer almost exclusively to behaviour observed from outside—for example, 'imitation'. But most psychological terms are ambiguous, sometimes referring to inner states, sometimes to observed behaviour. Examples of such ambiguous locutions are 'perception', 'attention', 'memory', 'reasoning', and 'abstraction'.

According to Thorndike, most psychology during the last quarter of the nineteenth century looked only at inner states, and it looked at these by introspection. The aim of psychology was not to know what the subject did, but to help the subject know what he or she experienced. This, Thorndike argues, had a negative effect on the study of animal psychology, since introspectionists tended to be sceptical about the extent to which we could apply introspective analyses to animals (and children) and to denigrate mere behavioural studies.

Thorndike saw such a stance as wrong-headed. 'It is so useful, in understanding the animal, to see what it does in different circumstances and what helps and what hinders its learning, that one is led to an intrinsic interest in varieties of behavior as well as in the kinds of consciousness of which they give evidence.'[37]

Objections to this sort of approach are founded, according to

Thorndike, on a traditional assumption that there is 'a real discontinuity', 'an impassable gap', between 'physical and mental facts, between bodies and minds, between any and all of the animal's movements and its states of consciousness'.[38]

But in Thorndike's view, there is no real difference between how we study mental facts and how we study physical facts. Further, a subject has no privileged access to mental facts, and the statements he makes about his own mental states may be right or wrong. Someone may have privileged access to his body temperature through kinaesthetic sensations, but when he seeks 'efficient knowledge of his own body temperature, John does not use the sense approach peculiar to him, but that available for all observers. He identifies and measures his "feverishness" by studying himself as he would study any other animal, by thermometer and eye.'[39]

Even what are thought of as totally private states, like suffering from toothache or anxiety, have the same logical status. Outside observers make judgements about a person's toothache by reference to his 'gestures, facial expressions, cries, . . . verbal reports, and bodily structure and condition', and by asking questions like 'Where does it hurt?', 'How long has it hurt?' or 'How badly does it hurt?' The person himself, if knowledgeable about dentistry, checks the same data and asks himself the same questions. The only difference is that he can get additional data from the sense-organ in his teeth. In judging the intensity of the toothache, sufferer and external observer are in the same position: both need a scale of toothaches built up out of data of the sort listed above. Furthermore, 'well-trained outside observers might identify the intensity of John's toothache more accurately than he could.'[40]

Thus the gap between mind and body is bridged not by a metaphysical monism, but by an epistemological one. All judgements of psychological states, whether by the subject or by an observer, are on a par with each other, and on a par with judgements about physical facts. Privileged access provides additional data, not data of a different kind. In the case of anxiety, as of toothache, Thorndike tells us that the person who experiences it has an advantage in knowledge of that anxiety only if he reports his 'immediate experience' of that anxiety to himself not *qua* sufferer, but *qua* 'scientific student'. 'To really *be* or *have* the anxiety is not correctly to *know* it. An insane man must become sane in order to know his insane condition. Bigotry, stupidity, and false reasoning can be understood only by one who never was them or has ceased to be them.'[41]

The key insight pressed by Thorndike is that there is no gap between our knowledge of physical and mental facts. Scientific judgements about mind generally depend more on verbal reports made by the subject than do reports about the body, but not always. A doctor (studying a person's body) may depend more on verbal reports than a moralist (studying character). Besides, verbal reports are continuous with behaviour: 'They signify consciousness no more truly than do signs, gestures, facial expression and the general bodily motions of pursuit, retreat, avoidance, or seizure.'[42]

As against Descartes and the dominant philosophical tradition, Thorndike holds that there is no clear-cut separation between linguistic signs and bodily signs, a position which did not assume much philosophical significance until forty years later in the work of Wittgenstein, Ryle, and others. In a passage sounding remarkably like the British ordinary-language philosophers, Thorndike asserts the following:

> Nor is it true that physical facts are known to many observers and mental facts to but one, who *is* or *has* or *directly experiences* them. If it were true, sociology, economics, history, anthropology and the like would either be physical sciences or represent no knowledge at all. The kind of knowledge of which these sciences and the common judgments of our fellow men are made up is knowledge possessed by many observers in common, the individual of whom the facts is known, knowing the fact in part in just the same way that the others know it.
>
> The real difference between a man's scientific judgments about himself and the judgment of others about him is that he has *added sources of knowledge*. Much of what goes on in him influences him in ways other than those in which it influences other men. But this difference is not coterminous with that between judgments about his 'mind' and about his 'body.' As was pointed out in the case of body-temperature, a man knows certain facts about his own body in such additional ways.
>
> Furthermore, there is no more truth in the statement that a man's pain or anxiety or opinions are matters of direct consciousness, pure experience, than in the statement that his length, weight and temperature are, or that the sun, moon and stars are. If by the pain we must mean the pain as felt by some one, then by the sun we can mean only the sun as seen by some one. Pain and sun are equally subjects for a science of 'consciousness as such.' But if by the sun is meant the sun of common sense, physics and astronomy, the sun as known by any one, then by the pain we can mean the pain of medicine, economics and sociology, the pain as known by any one, and by the sufferer long after he *was* or *had* it.[43]

For our purposes, it is necessary to stress that there is nothing in Thorndike which either denies the reality of consciousness in humans or animals or asserts that consciousness cannot be studied. His emphasis on behaviour is directed largely against the introspectionists who would ignore it or minimize it. He is in the tradition of British associationist psychology of the sort espoused by Locke, Berkeley, and Hume, and most especially by Hartley and the Scottish Common Sense philosophers, notably Thomas Reid and Thomas Brown, for whom behaviour was a sign of mental states and for whom knowledge of minds was non-problematic.[44] In this context it is worth recalling that a dominant philosophical tradition taught in American universities during Thorndike's training was Scottish Common Sense philosophy, and he could not have failed to be influenced by this mode of thinking. It is also worth noting, especially as an antidote to those who see Thorndike as a proto-radical behaviourist, that his work *Animal Intelligence* was cast in terms of *associations* that animals form, not in terms of mere bodily movement: 'Behavior includes consciousness *and* action, states of mind *and* their connections. . . . Knowledge of the action-system of an animal and its connections is a prerequisite to knowledge of its stream of consciousness.'[45]

By 'connections', Thorndike means associations, principles connecting mental states, roughly what Hume called 'principles of association'. These are not themselves objects of consciousness. The stress on associations—that is, on the *theoretical*, non-observable principles of law-like connections which bind states of consciousness and explain their flow—is opposed to Titchener's 'structuralism', with its excessive emphasis on mere contents of consciousness. Thus one major thrust of his attack on introspectionism is not that it is mentalistic, but rather, that it neglects to define the workings of the mind. Introspectionism is excessively concerned with mental anatomy as it were; but proper understanding of mind also requires 'mental physiology'. It is in this spirit that Thorndike develops his famous laws of exercise and of effect, which are laws of mental causation.

There is one further point, to which Thorndike alludes only briefly in *Animal Intelligence*, but which is significant in foreshadowing a major valuational shift which was subsequently articulated by Watson and later carried on by Skinner. Whereas virtually all traditional psychology was an attempt to *understand*, Thorndike suggested that this is not enough. In fact, he maintained that mere understanding of mind has no moral justification, for 'There can be no moral warrant for studying man's nature unless the study will enable us to control his acts.'[46] It is this moral-valuational pronouncement as much as anything else which helps to explain the disappearance of the Darwin–Romanes approach to mind, and the ascendance of radical behaviourism.

Watson and behaviourism

We are now at the point of considering the emergence of behaviourism and the concomitant rejection of consciousness in the radical new ideas of J.B. Watson. But I must again stress that there is nothing in the work of the figures we have considered that necessitates, either empirically or logically, or compels psychologically the abandonment of the Darwin–Romanes view of the phylogenetic continuity of consciousness. On the contrary, the disagreements among the thinkers we have considered are at most family quarrels, involving refinements in the fundamental philosophical view which makes consciousness and thought evolutionarily continuous, amenable to study in humans and animals by both observation and experiment, and worthy of such study. Not one of them made, or even tried to make, a plausible case for abandoning talk about thought and consciousness in humans or animals. When the new view was promulgated by Watson and, shortly thereafter, widely adopted by the psychological community in the United States and later in Britain, it was neither to account for new empirical results which put the older view in peril nor to transcend major logical flaws unearthed in the older view. Nor was it the outcome of a historically inevitable line of thought. On the contrary, both empirically and logically, behaviourism as developed by Watson was less plausible than the old view, since it required that we ignore, even deny the existence of, obvious facts of consciousness, or at the very least, see them as irrelevant to a science of mind. Thus the rise of behaviourism certainly does not reflect the orthodox view that theories are transcended only for empirical or logical reasons.

In fact, even Kuhn's approach to the history of science does not begin to explain the replacement of the Darwin–Romanes view by behaviourism, for there were no *crises* in the Darwin–Romanes outlook in Kuhn's sense of fatal flaws. Quite the reverse. The Darwin–Romanes approach was in the process of being refined by the thinkers we surveyed, and was generating new, fruitful, explanatory data, experiments, and methods. What happened, as I mentioned earlier, was a change in value and philosophical commitment, which pushed the Darwin–Romanes view out of style, and brought behaviourism in. Watson was pushing a new philosophy, a new set of values, a new career for psychology, and was selling it, not proving it. But he was successful, for by the 1930s, virtually nothing was left of the old approach as a mainstream phenomenon.

One has only to examine Watson's own work to see that he was attempting to sell a new philosophical-valuational package. In 'Psychology as the behaviorist views it', Watson's 1913 manifesto, he urges psychology to 'throw off the yoke of consciousness'.[47] Only by so doing can psychology become a 'real science'. By concerning itself with consciousness, 'it has failed signally . . . to make its place in the world as an undisputed natural science' like physics and chemistry.[48] To be a 'real science', it must behave like a real science and study what is 'observable'. Thus he writes: 'Can image type be experimentally tested and verified? Are recondite thought processes dependent mechanically upon imagery at all? Are psychologists agreed upon what feeling is?'[49] Watson assumes, but does not demonstrate, that the answer to all these questions is negative. What is observable is behaviour. What we find in the world are 'stimulus and response, habit formations, habit integrations and the like'.[50] (In fact, of course, these are not directly observable; they are theoretical notions.) 'I believe we can write a psychology, define it as Pillsbury [i.e. as the science of behaviour], and never go back upon our definition: never use the terms consciousness, mental states, mind, content, introspectively verifiable, imagery, and the like.'[51]

Note that Watson does not prove that we *should* do this; he merely affirms that psychology will be more like physics and chemistry if we do and thus advocates it. (The physics Watson admired, of course, is nineteenth-century physics. Twentieth-century physics soon soared beyond mechanism.) To the objection that the abandonment of consciousness is a very heavy price to pay, a violation of what we all know to be part of the furniture of the universe (some would say the best-known of all the furniture), Watson said little, except that he didn't 'care' about consciousness.[52]

In most of his written work, Watson did not go so far as to say explicitly that there are no such things as consciousness, mental images, and the like; but it is clear that this is his bottom line. Throughout his life, he contended that thoughts, images, and the rest were 'implicit behavior', small muscular movements in the larynx or other organs which we would be able to detect if we had a more advanced technology. As I have put it elsewhere, Watson in essence, paradoxically held that 'We don't have thoughts, we only think we do'.[53]

Subjective mental states are at best dispensable psychic trash, at worst non-existent.

Watson's own deepest mental states are inaccessible to us, of course, not least since he is dead. But he was certainly interpreted as I have just outlined by his contemporaries and co-workers. One of his best friends and closest collaborators was Karl Lashley who, in 1923, published a famous article entitled 'The behavioristic interpretation of consciousness', which is of interest not so much for the insight it provides into Watson's thinking, but with regard to how Watson's ideas were promulgated. According to Lashley, there are three possible stances on consciousness for the behaviourist; first, that consciousness is real and capable of scientific treatment, but that the behaviourist is interested only in behaviour, a position he attributes to the Pavlovian Bechterev;[54] second, that facts of consciousness exist but that they cannot be studied scientifically and so ought to be ignored, the view found in Watson's 1913 paper, which Lashley called 'methodological behaviorism';[55] third, that

The supposedly unique facts of consciousness do not exist. An account of the behavior of the physiological organism leaves no residue of pure psychics. Mind is behavior and nothing else. This view is implied in much of Watson's writing, although it is not stated in so many words. For the most part he expresses a methodological behaviorism, but such statements as the following leave little doubt of his fundamental denial of the fact of consciousness, as described by the subjectivist. 'It is a serious misunderstanding of the behavioristic position to say, as Mr. Thompson does—"And of course a behaviorist does not deny that mental states exist. He merely prefers to ignore them." He "ignores" them in the same sense that chemistry ignores alchemy, astrology, horoscopy, and psychology telepathy and psychic manifestations. The behaviorist does not concern himself with them because as the stream of his science broadens and deepens such older concepts are sucked under, never to reappear.'

This is the extreme behavioristic view. It makes no concessions to dualistic psychology and affirms the continuity in data and method of the physical, biological, and psychological sciences. 'Consciousness is behavior.' 'Consciousness is the particular laryngeal gesture we have come to use to stand for the rest.' I shall speak of this doctrine as strict behaviorism, or for brevity simply as behaviorism, since methodological behaviorism is only a form of epiphenomenalism.[56]

If behaviourism is to be significantly different from other approaches which preceded it, including the Darwin–Romanes approach, it must deny the reality of consciousness in humans and animals, or at least its knowability. In this regard, Watson is his own version of a consistent Darwinian. In fact, in a bizarre dialectical turn in the 1913 essay, he accuses those who argue for a phylogenetic continuum of consciousness of anthropocentrism, because 'it makes consciousness as the human being knows it, the center of reference for all behavior',[57] just as Darwinian biology was anthropocentric in attempting first and foremost to describe the evolution of *Homo sapiens*.

Later behaviourists tended to vacillate between methodological behaviourism and the extreme behaviourism which denied consciousness in humans or animals altogether. But the net effect was the same either way: consciousness was ignored! Whether seen as not amenable to study or as non-existent, it ceased to be a legitimate object of scientific enquiry, and even to mention it was seen as a betrayal of the scientific method. Thus, when later behaviourists talk of positive and negative reinforcement, reward or punishment, they do so in terms of their effects, positive reinforcements being those which increase the probability of a particular behaviour, negative reinforcements those which decrease it. As to the common-sense idea that stimuli like electric shocks and rewards of food are negative and positive reinforcers because they *feel* bad and good respectively, this is dismissed as operationally meaningless and as scientifically irrelevant as invoking a soul. Watson's greatest living heir, B.F. Skinner, much like Watson, has been inconsistent in his position on consciousness; sometimes he has asserted that it is not real, sometimes that it cannot be studied, and sometimes that it provides data which cannot be ignored.[58] Summarizing the behaviourist revolution in 1960, D.O. Hebb remarked that it came to 'banish thought, imagery, volition, attention, and other such seditious notions'.[59]

Why did behaviourism triumph?

In any case, the obvious question which arises is why anyone (and later, almost everyone) took something as preposterous as the denial of all consciousness seriously. Certainly revolutions which have appeared equally strange have taken place in the history of thought—for example, the replacement of qualitative Aristotelian science by quantitative Galilean science, and the development of quantum theory and the overthrow of absolute space and time by Einstein. In both these other cases, however, there were at least some good logical and/or practical reasons why these revolutions succeeded. But what led people to fly in the face of both common sense and theoretical continuity with accepted Darwinian thought, and accept strict behaviourism?

Basically, people seem to have bought into behaviourism because it harmonized with new values and new philosophical approaches which were taking root. First, it was clearly in harmony with the reductivism and the tendency towards simplification which I discussed earlier. And, although Watson himself never thought that psychology could be reduced to physico-chemistry, but would always require irreducible behavioural notions for explanatory purposes, others, such as Lashley and the logical positivists, saw it as providing a philosophical hammer for smashing a major barrier to total mechanistic reductionism. Thus, given a prevalent, growing valuation for, and philosophical commitment to, a unity of science and to reductionistic physicalistic science, behaviourism became ideologically acceptable.

Second, even though behaviourism spat directly in the face of common sense, so too did recent revolutions in 'real science'—that is, physics and chemistry. What could be more counter-intuitive than relativity and quantum theory? If physics could ignore common sense, surely psychology could and indeed should, if by so doing, it could become more like physics. Moreover, behaviourist rhetoric kept hammering away at the point that only the silly commitment to consciousness stood in the way of psychology's becoming a real science. It is ironic, of course, that while psychology was earnestly attempting to become like physics, the latter was moving away from the positivistic, mechanistic dream towards acausality, possible entities and possible worlds, non-local causation, wave functions and all the other forms of 'quantum strangeness' we now take for granted.

A further irony may be found in operationalism, a version of positivism which enchanted behaviourists and which required that every legitimate scientific concept be defined by some operation. Thus, for example, the concept of length was exhaustively defined by how one goes about measuring it. Though operationalism was developed by a physicist, P.W. Bridgman, the impact of his book *The Logic of Modern Physics* on physics was relatively limited. In an article in the *British Journal of Psychology*, psychologist Cyril Burt points out that physicists regarded it as a creed borrowed from psychology! Burt cites A. O'Rahilly, an eminent physicist, who in his 1938 book *Electromagnetics* declared that 'In spite of laboratory vocabulary, the operational viewpoint is not a genuine product of physics; it's the echo, in the precincts of physical science, of that curious and typical American doctrine known as behaviorism'.[60] Gordon Allport, in his 1939 presidential address to the American Psychological Association, traced the influence of this notion in psychology and demonstrated the extent to which it had captured psychologists' imaginations.[61] It was common sense to these psychologists, of course, that notions like consciousness, awareness, and subjective experience could not be defined operationally, and thus operationalism fanned the behaviourist flame.

Third, and perhaps of greatest importance in explaining the success of behaviourism, given the prevailing American *Zeitgeist*, was the promise of *control* with which behaviourism seduced psychologists and the public, something which we have already encountered in Thorndike. Given that control, technology, practicality, and progress were major values at the time, Watson's rhetoric appealed mightily. Throughout his 1913 essay, Watson tantalizes with the promise of control: 'If psychology would follow the plan I suggest, the educator, the physician, the jurist, and the businessman could utilize our data in a practical way, as soon as we are able, experimentally, to obtain them.'[62]

He cites 'experimental pedagogy, the psychology of drugs, the psychology of advertising, legal psychology, the psychology of tests, psychopathology' as 'truly scientific [fields] . . . in search of broad generalizations which will lead to the control of human behavior'.[63] Not to be interested in such a psychology of control, says Watson, is not to be interested in human life.[64] Throughout his career, Watson hammered away at the same theme, that only a 'scientific' psychology of the sort he describes can create a better society and improved individuals. This theme was carried over into modern behaviourism by B.F. Skinner, and was criticized in Watson as early as 1914 by Titchener, who claimed that Watson was not after science, but technology.[65]

Fourth, Watson was able to capture the popular and scientific imagination both with his rhetoric and his actual achievements. At a time when many scientists shunned publication in the popular press, Watson thrived on it. Furthermore, Watson showed that behaviourism *worked*, first for the military, then in a long, lucrative career in advertising, which made him wealthy, and later in his work with children. (The fact that the techniques worked did not of course mean that the theory was true, but few people bothered to make such distinctions.) Watson used behaviouristic techniques to sell behaviourism, and they worked, so people began to believe in it. In a country like the United States in which people have always been fascinated by popular psychology (let the reader look in any American bookshop at the psychology section if he or she doubts this claim), Watson capitalized on this intoxication, and fed it, turning it to the advantage of the philosophy he espoused.

Finally, given the fact of compartmentalization, we can see that embracing behaviourism was not as radical as it might seem. One could be a behaviourist *vis-à-vis* people and animals in one's scientific moments (notwithstanding little allusions to conscious processes which, annoyingly, kept creeping into behaviouristic literature and theory), and go home, where one was still a human being, leaving one's behaviourism in the laboratory. After all, moral and practical considerations militated in favour of imputing consciousness to other humans and to selected animals, such as pets and Rin-Tin-Tin—no harm done. And indeed, no harm *was* done, except perhaps to the animals used in the laboratory, or to non-selected animals, where no widespread socially accepted moral theory forced one to think or worry about animal feelings, fears, pains, and other mental states. In the absence of moral pressure for concern about animals, scientists came to be neo-Cartesians, and, as far as animals were concerned, especially experimental animals in a research programme totally dominated by behaviouristic and positivistic ideology, consciousness did disappear, to be replaced by physical movements, which didn't involve 'hurt' in any subjective sense. All this was tremendously convenient for researchers doing things to animals which ordinary people would find horrible—given the behaviouristic *Gestalt*, animals had no feelings, and cries of pain were not cries of pain at all, but rather 'vocalizations.' As the amount of research on animals grew in all areas of science, behaviourism and positivism provided a convenient means for anaesthetizing one's conscience: ordinary compunctions against hurting, scaring, or tormenting animals were wiped away by an ideology which assured the researcher that mental states in animals were as mythical and non-explanatory as gremlins. So the behaviourist denial of animal consciousness spread to the common sense of science in general, where it made researchers' jobs much easier by allowing them to dismiss claims about animals' feeling pain, experiencing fear, suffering, and so on as non-scientific, unverifiable, *meaningless* rubbish. (We have already seen that even when science didn't deny consciousness in animals, it, like society in general, didn't care very much about the moral implications of animal thought and feeling.) Thus, moral categories became irrelevant.

Earlier in our discussion, I argued that, ideally, scientists should articulate reasons for changes in philosophical and valuational assumptions which govern scientific changes—in short, there should be philosophical justifications, even though they will not begin to exhaust the causes of such major scientific change, given the non-rational factors involved. Watson gave reasons; positivism gave reasons; behaviourists presented philosophical reasons for abandoning common sense and the Darwin–Romanes approach, although they were not very plausible reasons; nor were they soundly defended. Watson, for example, never put forward good arguments as to why consciousness was not amenable to scientific study, especially given that Titchener, Wundt, and Darwin had showed that it was. Nor did he (or anyone else) ever justify saying that there were no facts of consciousness, or why they should be ignored. The positivists did not explain how one could base science on phenomenalism, yet deny reality to consciousness. Nor did Watson even begin to discuss the moral implications of denying consciousness, in humans or in animals. (Skinner later discussed this *vis-à-vis* humans, but not animals.)

Lest it be thought that I have exaggerated the twentieth-century English-speaking psychologists' denial of consciousness or put too much emphasis on Watsonian idiosyncracies, I shall now demonstrate that this is not the case.[66] In his 1939 addresss, Allport described the widespread tendency of psychology to distance itself from common sense. But he went much further in his discussion, explicitly recounting the tendency on the part of mainstream psychology to dismiss subjective experience and its study altogether, under the aegis of scientific behaviourism:

So it comes about that after the initial take-off we, as psychological investigators, are permanently barred from the benefit and counsel of our ordinary perceptions, feelings, judgments, and intuitions. We are allowed to appeal to them neither for our method nor for our validations. So far as *method* is concerned, we are told that, because the subject is able to make his discriminations only after the alleged experience has departed, any inference of a subjectively unified experience on his part is both anachronistic and unnecessary. If the subject protests that it is evident to him that he had a rich and vivid experience that was not fully represented in his overt discriminations, he is firmly assured that what is vividly self-evident to him is no longer of interest to the scientific psychologist. It has been decided, to quote Boring, that 'in any useful meaning of the term existence, private experience does not exist'.[67]

Again, reading Allport, one gets a clear picture of why these psychologists were so philosophically naïve, why they wore blinders which blocked out what to the rest of us are obvious philosophical questions raised by their approaches and assumptions. Allport succinctly encapsulates the new common sense of science thus:

> The psychological system-builders of the Nineteenth and early Twentieth Centuries were filled with the lingering spirit of the Enlightenment which hated mystery and incompleteness. They wanted a synoptic view of man's mental life. If moral and metaphysical dogmatism were needed to round out their conception of the complete man, they became unblushing dogmatists. Yet even while their synoptic style flourished, the very experimental psychology which they helped to create was leading others into new paths. Their own students, in the very process of enhancing their experimental proficiency, came to admire not the work of their masters, but the self-discipline of mathematics and of the natural sciences. Willingly they exchanged what they deemed fruitless dialectics for what to them was unprejudiced empiricism. Nowadays, for one experimentalist to proclaim another 'superior to controversy about fundamentals' is considered high tribute.[68]

In actual fact, the philosophical, valuational assumptions pressed by behaviourism are not very good, even if they are extremely convenient for scientists using animals. High value was assigned to control, rather than understanding. Key moral questions which might have hamstrung or retarded dramatically painful psychological and physiological experiments on animals (and which ordinary common sense of the time, despite its lack of moral concern with animals, would have recoiled from), and which might perhaps have given rise to moral qualms about inflicting pain, fear, anxiety, terror, and so forth on animals were circumvented simply by denying the applicability of the notion of consciousness to animals. Furthermore, consciousness had never really found a place in the new science of matter in motion developed in the Renaissance. Despite Descartes's reliance on the *cogito* as his epistemological corner-stone, and despite positivistic phenomenalism, consciousness was an embarrassment to science. With behaviourism, it vanished from science as neatly as secondary qualities had from Newtonian physics. This was all very convenient, but was it rationally justifiable?

In the final analysis, then, behaviourism defeated the Darwin–Romanes viewpoint for reasons which are philosophically shaky, but which fitted the times. Behaviourism was both a cause and an effect of what we have called the common sense of science. It is an outcome of the fact that the common sense of science ignores moral and value questions, that it stresses observables as the only material of science, that it is ahistorical and has a simplistic positivistic bias, that it values science that leads to control, and exalts nineteenth-century physics and chemistry as the model to which all science should aspire, and emphasizes laboratory experiments. But behaviourism has also been instrumental in the development of the common sense of science and its abandonment not only of the Darwin–Romanes view but of ordinary common sense. For behaviourism has played a significant role in the evolution of the idea that we can't possibly study mental states, especially the mental states of animals, and that, *a fortiori*, we can't make moral judgements about what we do to animals, because such judgements must ultimately hinge on some imputation of consciousness, which behaviourism declares to be illegitimate. Most scientists in all fields involving animals in the twentieth century have been infected by behaviourism to some degree. Even ethology, which on certain basic issues is diametrically opposed to behaviourism, bought into the setting-aside of consciousness in favour of overt behaviour, and can be said to be behaviouristic in this regard. This infection is part of why for long enough we heard so little about moral questions occasioned by the use of animals. (The other part of the reason, of course, is that the common sense of science declared all moral questions to be meaningless.) Finally, behaviourism further augmented the compartmentalization between being a scientist and being a man, between ordinary common sense and scientific common sense. In the next phase of our analysis, we shall examine the moral and conceptual price exacted by science's adherence to the behaviouristic, positivistic dogma concerning animal mentation, specifically as it pertains to the study of pain and suffering in animals. I will show that this ideology has often successfully overridden even logic and coherence. We will then examine some of the arguments which have been adduced in defence of the ideological denial of pain and suffering. After that, we will explore the way in which scientific ideology, on this issue at least, is happily beginning to be eroded by burgeoning social moral concern for animals.

5 Animal pain: the ideology cashed out, 1

The common sense of science and the rejection of value questions

The common sense of science, rooted in positivism and behaviourism, has had enormous influence on twentieth-century science. In so far as scientists receive any exposure to philosophy in the course of their education, it is invariably to some version of the ideology we have been discussing, usually much distilled and simplified so that it fits readily into textbook prefaces and introductions designed to ease the student's transition from pristine ignorance to total immersion in unfamiliar, difficult material. And this ideology endured, thrived, and continues to endure for a variety of practical reasons. First, it provides—or appears to provide—a clear criterion of demarcation between science and other activities, notably speculative thought of any sort, be it philosophical or religious. (In this regard the ideology of science has always been critical of Freudian, Jungian, and other psychoanalytical theory, which it perceives as unverifiable, non-empirical, mystical nonsense.) Second, it provides a much-needed philosophical peg upon which to hang the notion of a 'scientific method', ubiquitous among scientists of all sorts. Third, it divests scientists of any professional responsibility for dealing with moral or other valuational questions, since such questions are seen as non-empirical, emotive, and non-cognitive. Science is value-free, says the common sense of science, and value questions, if they are meaningful at all, arise at the level of social consideration of the uses to which science and its practical findings are put, rather than in science itself. Thus science textbooks virtually never raise value questions inherent in the area in question. In eleven years of teaching and working with scientists on a daily basis, one of my greatest difficulties consistently lies in even getting them to acknowledge the existence of legitimate value questions or assumptions of any sort within the fabric of science. Such an admission, they often feel, would cheapen science as a way of knowing by introducing non-scientific notions and non-empirical questions. As a result, the implications, consequences, and effects of the valuational assumptions which are necessarily and unconsciously made in science go almost totally undetected, and are certainly never critically assessed. It is a herculean task even to get them recognized, since the blinders of scientific common sense render them invisible. In short, the ideology of science tends to infect its adherents with a blindness to value issues of any sort; and when the need for valuational assumptions inevitably arises, the decisions made often reflect the pressures and exigencies of the moment, and are rarely recognized as valuational. Hence it is always a rude shock to scientists when society questions their valuational commitments, since they are, in general, genuinely convinced that they don't have any.

In valuational matters, especially moral ones, ordinary common sense is generally far more sophisticated than scientific common sense, since morally problematic situations arise frequently in everyday life; they press in on one from all quarters, and there is no way to deny their reality (though one may refuse to deal with them). In science, on the other hand, they have been excluded by the accepted ground rules, and nascent scientists are trained to ignore them in much the same way that they are trained to ignore and suppress their religious predilections. The difference, of course, is that science does not inevitably involve theological presuppositions, whereas it cannot escape valuational assumptions.

The ethics of medical research on human subjects provides an interesting and salient case in point. Throughout the development of biomedical science in this century, the use of human subjects has been an invaluable tool in research. One need not be an expert in medical history to have some familiarity with the questionable forms this has sometimes taken: the use of prisoners to study disease and various therapies with and without consent or in exchange for favours, the use of indigents and derelicts in big-city hospitals for developing new forms of treatment, the use of ignorant foreign populations, the use of the insane, the retarded, students, military personnel, and other disenfranchised individuals, and so on.[1] The key point here is not the rightness or wrongness of these activities, but the way in which they were conducted, the fact that they were rarely seen as involving debatable moral assumptions, at least until non-scientists (and a few scientists) pressed such awareness on the researchers, but rather, as scientifically required and therefore valid. This is not to suggest that no scientists were morally aware during this time, but only to stress that such awareness was not built into the common sense of science. If it occurred, it did so, in spite of and contrary to the common sense of science.

Indeed, it is only the large-scale atrocities that emerged from the Nazi revelations after World War II which gave birth to full social awareness of the moral dimensions of research on human beings, and eventually stimulated the development of mechanisms designed to build moral concern for subjects into the scientific process, most notably by way of local review of projects by committees including lay persons chosen to serve as subject advocates.

In fact, the mandate for protecting human subjects was not promulgated in the United States until 1966, and even then, not without significant resistance from the research community, which tended to dismiss the Nuremberg revelations as extreme and irrelevant.[2] (The mechanisms for enforcing the mandate were not fully in place until the 1970s.) As was the case later with the thrust for regulation of animal research, the National Institutes of Health (NIH), the major source of funding for biomedical research in the United States, opposed any regulation on the use of human subjects as violating academic freedom. When NIH did begin to develop guidelines, it was not so much out of a recognition of the moral issues involved as out of fear of public criticism, which is much the same as what happened later in the case of animal research.[3]

Scientists trained in scientific ideology tended not to think in ethical terms at all. Jay Katz, a pioneer in the development of United States policy concerning human subjects, has remarked that prior to his immersion in this issue, he did not even think in moral terms about the subjects of a medical research project in which he was engaged—precisely the same thing I almost invariably hear from researchers when I begin to raise moral questions to them about animal research:

> During 1954 to 1958, two colleagues and I conducted experiments on hypnotic dreams, supported by grants from the National Institutes of Health. After we had worked out our research methodology, we became concerned that some of our volunteer-subjects, as a consequence of their participation, might experience sufficient emotional stress to require a brief period of hospitalization. While we believed this to be unlikely, we were sufficiently worried to ask the chairman of our department whether, if our fears materialized, he would make hospital beds available to our subjects free of charge. He readily

agreed. We were relieved, feeling that we had satisfactorily fulfilled our professional responsibilities to our research subjects. It did not even occur to us to wonder: Were we obligated to disclose to our subjects our concerns about the investigation's possible detrimental impact?[4]

Even so, many scientists viewed (and still view) the procedures designed to protect human subjects as yet another bureaucratic hoop to jump through. At an international meeting of medical researchers in 1984 I was appalled to hear a man from South Africa get up and urge his colleagues to consider moving their research to his country, where the regulations in this area are minimal, and a hospital administrator, he said, can sign informed consent forms for as many patients as a researcher needs. Amazingly, no one expressed indignation at his remarks.

Any reader who doubts the claim that I am making would do well to spend an afternoon talking to scientists about these issues. What one is likely to hear is instructive: namely, that ethical problems do occasionally arise, but that these are dealt with internally and are exceptional, and that a few bad apples shouldn't be allowed to spoil the work of the overwhelming number of decent people. In other words, ethics is associated with doing something wrong; thus, since most scientists don't do bad things, they are 'ethical'. Ethical questions arise only in a context of wrongdoing, as aberrations, not as part and parcel of scientific activity.

Alternatively, one has only to look at what until very recently were called 'codes' of medical and veterinary medical ethics to realize the extent of medical *naïveté* in this area. These documents were in fact codes of etiquette, designed to regulate intra-professional niceties such as advertising, fee-splitting, and the size of one's signs.[5] Nothing of substance was dealt with in these codes, yet they were appealed to as evidence of ethical concern and ethical behaviour in the profession in question. Starting in the 1960s and 1970s, this began to change, and human medicine, and later veterinary medicine, began tentatively to address the genuine moral questions inherent in their fields. Even then, however, much of the concern was reactive, inspired by widespread media coverage of and consequent public interest in medico-ethical questions. Many physicians, veterinarians, and researchers in these and other areas saw, and continue to see, this activity as essentially a matter of public relations, regarding these questions as having no real claim on scientists *qua* scientists.

In a startling study, Lindsey surveyed a group of psychological counselling professionals ranging from graduate students to licensed clinical psychologists with Ph.D.'s and years of experience.[6] He presented them with four counselling vignettes, two involving ethical issues much discussed in the literature, two involving ethical issues which had not been discussed. The first case concerned a student who revealed to a counsellor that a 15-year-old friend had related how her mother's boyfriend had repeatedly exposed himself to her. The second concerned a woman who had revealed that she intended to attack her husband's mistress. These cases have not only been much discussed, but there are legal requirements that counsellors report such cases. The third case concerned a counsellor who was told of a highly stressed medical resident who showed up late, missed drug sensitivities on patient charts, and so on. The fourth concerned a counsellor who was informed by a patient that she had genital herpes but was not informing her sexual partners. Astoundingly, 59 per cent of the counselling professionals surveyed by Lindsey did not perceive the cases as involving ethical problems. (Interestingly enough, Lindsey found that people who were taught ethics in a formal way did much better on identifying the issues.)

If it is difficult to convince researchers that ethical issues pertaining to the use of human subjects in research are significant or even real, one can imagine the degree of difficulty one encounters in attempting to raise questions about the use of animal subjects. Even to raise these issues is to run the risk of being branded a crank, a kook, a Luddite, anti-science, an antivivisectionist, a misanthrope, or worse. Though my writings are generally considered rational and moderate, as I have already mentioned, no less a journal than the *New England Journal of Medicine* saw fit to treat my book *Animal Rights and Human Morality* as if it were irrational and extremist.[7] An NIH official informed me that I was 'not acceptable' to the powers at NIH. (Ironically, although NIH consistently opposed the legislation regarding the use of animals in research that I was instrumental in drafting, it recently adopted the very same provisions as requirements for grantees.) After publishing an article of mine on the moral status of animals,[8] the journal *American Psychologist* promptly received a response which argued that my work 'promoted immorality', since it provided a moral ground for laboratory break-ins.[9] I cite these examples to show just how difficult bright and otherwise reasonable scientists find it even to entertain the possibility of thinking about animals in moral terms. A wonderful illustration of this point may be found in a recent textbook of abnormal psychology by two highly respected scientists, Martin Seligman and David Rosenhan. Among the illustrations is a photograph of a laboratory rat, captioned; 'For ethical reasons, animals are often used in laboratory experiments.'[10] What better, more eloquent evidence that, at least for these scientists and presumably for a large percentage of their readers, the scope of ethics does not include how we treat animals?

All this is made even more perplexing by the fact that most animal researchers, even the most hard-core, are quick to affirm that contemporary biology, medicine, and psychology are conceptually inseparable from the invasive use of animals, and that animals are entitled to 'humane treatment' and to not being abused. Yet in the same breath they will affirm that the use of animals does not reflect any moral assumptions, is not subject to moral analysis, is a matter of science, not ethics, and that animals are not to be discussed in a 'moral tone of voice'.

At a dinner I attended recently, I sat next to a prominent medical researcher who proceeded to castigate me for injecting ethics into science. I pointed out that ethical judgements are inseparable from science. 'After all,' I said, 'if only scientific considerations govern scientific practice, how come we don't do all of our research on children, who are surely better models for children and adults than animals are?' I had hoped to elicit his clear-cut agreement that it is in general wrong to experiment on children, whatever the scientific value of such research, and that therefore ethics does constrain science. But I was disappointed. Without any hesitation, he replied, 'Because they won't let us!'

For a long time, the common sense of science and its denial of the legitimacy of moral questions in science, coupled with ordinary common sense's apathy with regard to applying moral notions to animals, served to eclipse moral issues in animal use in science, and to protect scientists from having to deal with them. Here positivism and behaviourism could really run amok, for there was no counter-pressure. For, although ordinary common sense attributed thought and feeling to animals, it failed to draw moral implications from that attribution, since such moral awareness would call into question practices upon which ordinary life depended. Thus common sense could not serve as a counter-weight to the scientific perspective from which animals were merely behavioural or physiological machines. So the common sense of science could essentially ignore the morally relevant side of animal mentation—namely, the fact that animals experience pain and suffering, and that consequently what we do to them matters to them. And ignore them it did, not only at the expense of the animals but, as we shall now see, at the expense of the coherence of science itself.

Ideology and incoherence

What is truly remarkable in the area that we will be discussing is the extent to which ideology can eclipse logical coherence even in science, the way of knowing which prides itself on an overriding commitment to reason. (Though few scientists or philosophers can even begin to give an adequate account of reason, all would agree that logical coherence and consistency are at least part of what is entailed.) That such eclipsing may take place in political ideology and religious ideology is well known. One has only to read Orwell, Hitler, or Stalin to see it in the political sphere. And no one possessed of even minimal acquaintance with religion can fail to see it there—witness Tertullian's famous *credo quia absurdum*. But to find scientists systematically ignoring logical incoherences is remarkable. The only way this can be explained is in terms of the blinders we mentioned earlier. Since scientists are trained to be blind to value-notions, they are *a fortiori* blind to inconsistencies among such notions, even when those inconsistencies spill over into the practice of science.

Although I have been involved in animal issues for over a decade, I did not fully realize until recently the extent to which scientific ideology could eclipse reason. I was aware, of course, that scientists could be extremely emotional and irrational when it came to purely ethical questions having to do with animals, but I explained this—and still do to some extent—by their belief that ethical questions are merely emotive, by the threat to their life's work which criticism of current practices engendered in them, and by the fact that everyone on both sides tends to be irrational on animal welfare issues, since animals are not a vocal constituency able to press for reasonable resolution of the problems surrounding their situation.[11] Only now have I begun to comprehend fully the way in which ideological blindness on these issues not only colours ethical thought and results in unfortunate moral consequences, but taints scientific thought as well. I had myself been guilty of the error of separating science from value, since I, too, had been educated in positivism and scientific common sense.

The case of pain

I was first shocked out of this dogmatic slumber in March 1983, when I received a phone call from an eminent primatologist asking me if I would be willing to give the keynote address at a scientific seminar on animal pain.[12] My initial response was to express surprise that he was inviting a philosopher. Revealing a degree of conceptual sophistication rare among working scientists, he replied that the truly interesting and important issues concerning pain in animals are not purely empirical ones, but rather, moral, philosophical, and conceptual, and that the failure of science to engage or even acknowledge the existence of these issues discredited biomedical science and weakened its conceptual base. He went on to say what I already knew from interactions with scientists: namely, that animal pain was virtually ignored in scientific discussion, and that if it was discussed at all, it was in purely mechanistic terms, never experientially and never in terms of its moral implications. In short, the ideology of science had filtered out most of the interesting questions about pain, and had left a narrow, incoherent, and inconsistent body of beliefs and attitudes which had never been subjected to logical scrutiny.

The primary issue which needs to be brought out in this area is the fundamentally incoherent ambivalence displayed by the scientific community towards the reality and knowability of animal pain. As one might expect, this is first of all a consequence of the ideological denial of animal consciousness. But, moving to the other horn of the dilemma, it is also a consequence of an implicit belief that animals *do* experience mentation similar to humans. Thus, on the one hand, scientists are loath to speak of animals experiencing pain or pleasure (or any mental state), for such claims are thought to be unverifiable at worst (and hence unscientific) and anthropomorphic at best. On the other hand, scientists often find themselves incapable of doing and describing their own work if they do not presuppose animal pain (and other mental states), because they are using animals as models for human states (like pain) which contain essential experiential dimensions. This dilemma, which puts scientists in a logically incoherent position, is rarely resolved; most often, it is simply ignored, thereby generating the glaring inconsistency we alluded to earlier.

This was evident at an international seminar a few years ago on standards of care for domestic farm animals used in research, which I mentioned earlier. Participants found themselves unable to discuss the welfare of farm animals without talking about the conditions which make these animals happy; yet they were professionally reluctant to advance claims about animal happiness, for fear that they would be seen as unscientific. Most participants sailed over the difficulty by talking about what conditions make animals quote happy end-quote—a move which doubtless made them terminologically more comfortable, but which simply glossed over the issue we are discussing. (What, for example, as I asked at the seminar, is the difference between happy and 'happy'?) If the point of the quotation marks is simply to underscore the point that when an animal feels happy, what it feels is surely not the precisely same thing that I feel when I feel happy, this is perhaps true, but even if true, is trivial. For I am morally certain that when marathon-runners and quiche-eaters claim to find happiness in these activities, they are not experiencing anything that I understand by the term. Yet we do not put quotes around our attribution of happiness to other people. By the same token, I am morally certain that the pleasure a sweating horse feels when taking a cool drink on a 90-degree day is quite close to my own joy in the same experience. The quotes, in fact, are little more than an act of *pro forma* ritualistic obeisance to the ideology of science.

A moment's reflection will reveal that, on some level, science is absolutely committed to the belief that animals feel pain and experience other mental states which are, to a significant degree, analogous to what human beings experience. Otherwise it would be absurd to do research on them to learn about human pain, and to study dose responses to anaesthetics and analgesics. Not only does such research logically presuppose that animals feel pain, it even attempts to provide some quantitative measures of the pain and of its control by analgesics through such tests as the hotplate test, the writhing test, the tail-flick test, the skin-twitch test, the head-withdrawal test, pressure tests involving tails and digits, and electrical stimulation tests. Further, such tests also presuppose some meaningful analogies between animal and human pain, for they are employed in part to screen substances for potential analgesic effect in humans.

For those unfamiliar with these 'scientific' tests—and here the quotation marks *do* carry meaning—a parenthetical word of explanation is in order. All these tests are designed to provide measurable pain responses to controlled pain stimuli, and to measure the degree of pain alleviation provided by analgesics. In the writhing test, for example, a rodent is placed on a hotplate, and the degree of pain is measured by the extent of its writhing, the number of writhes per minute. The analgesic to be tested is then administered, and the extent to which it diminishes the frequency of writhing is noted.

A similar tension exists in psychology. On the one hand, after Watson, most American psychologists, even those who were not total behaviourists, rejected allegedly anthropomorphic and unverifiable ascription of human mental states like fear, anxiety, depression, and hopelessness to animals, as beyond the scope of legitimate scientific claims. On the other hand, the unstated presupposition of a great deal of psychological research on these notions was that these states are analogous in humans and animals, and that their study in animals provides valuable insights into their nature in humans. (A thoroughgoing behaviourist could, of course, and often did, deny that these mental states exist in humans as anything other than behaviour, thereby getting round the problem.) The currently popular work on hopelessness and learned helplessness, pioneered by the Dr Seligman mentioned earlier who refuses to acknowledge any moral status for animals, provides a very clear example of how psychologists who work with animals must presuppose analogous subjective states in humans and beasts. In fact, learned helplessness work has been severely criticized by some psychologists precisely for its anthropomorphism. At the same time, presupposing such analogies leads to the inevitable and much discussed 'psychologists' dilemma' (which is also the 'pain physiologists' dilemma'): namely, that if these noxious mental states in animals are sufficiently analogous to those in humans to provide adequate models for human experience, what right have we to induce them in animals? And if they are totally disanalogous, what is the value of studying them? This leads to a form of learned helplessness among psychologists; for, unable to resolve the conflict between their professed scepticism about animal consciousness and their implicit reliance on it, they avoid the issue altogether, and simply wish it out of existence.

Some years ago I came across an extraordinary case of the contradiction we have been discussing, one which has since become famous and which illustrates the point I have been making so well that I could not have orchestrated a better one if I had tried. I was on a panel on animal pain with a well-known pain physiologist, who spent the better part of an hour trying to show that since the electrochemical activity in the cerebral cortex of a dog differed dramatically from the electrochemical activity in the cerebral cortex of humans, and since the cerebral cortex is the information processing area of the brain, the dog didn't *really* feel pain in any sense that we do.

My time came for rebuttal. I said, 'Dr X, you are justly acclaimed for your work in pain. You do that work on dogs, do you not?' 'I do,' he replied. 'You extrapolate the results to people, do you not?' 'That is correct,' he said. 'In that case,' I said, 'I have nothing else to say!' In other words, either his paper was false, or his research work was; he could not have it both ways.

If the lack of intellectual coherence pervading this area of science which this anecdote illustrates were the only consequence of the scientific community's ambivalence regarding animal pain and awareness, that alone would justify a vigorous attempt to resolve the problem. But more is at stake than coherence. Moral issues are also involved, as we have seen. So to the ideology of science's inherent discomfort with moral questions is added a logical fallacy which still further eclipses moral deliberation in this area. It is of course a fundamental principle of logic that from a contradiction anything at all may be deduced, which is precisely what happens in this area. For whenever it is pragmatically expedient, one term of the contradiction is simply suppressed or ignored. When it is required to study pain mechanisms or analgesia in animals, it is taken for granted that animals feel pain. But when it is convenient or conscience-salving to ignore the painful consequences of one's research or teaching manipulations on animals, out comes the claim that one cannot really know what or even *that* animals experience. The study of pain is seen as the study of certain mechanisms or physiological states or responses having nothing to do with what an animal may feel, since we cannot study animal feeling scientifically. This in turn enables researchers to avoid having to deal with the infliction of pain as a moral problem.

Almost anyone who has been trained as a veterinarian in the United States will at some point have encountered the claim that animals don't really feel pain as we do, or if they do, it is purely momentary and transient. Not too many years ago, I am told by older veterinary colleagues, this was mainstream ideology. A leading veterinary pain expert recently told me that probably 75 per cent of veterinarians still view anaesthesia as chemical restraint, and I myself encountered this very dramatically at a veterinary college. I had just finished asserting that at least veterinarians couldn't doubt that animals feel pain; otherwise why would they study anaesthesia and analgesia. Up jumped the associate dean: 'Anaesthesia and analgesia have nothing to do with pain,' he said, 'they are methods of chemical restraint.' (An Australian scientist in the audience responded beautifully: 'He's daft,' he declared. 'What the hell do the animals need restraint for if they are not in pain?') If veterinarians can hold such views, it is not at all surprising to find human medical researchers who hold them, since the latter often see animals as being like test-tubes, and have little concern for them except as providing solutions to human problems. In February of 1985, I was told by a United States Department of Agriculture (USDA) inspector that he had recently inspected a medical school under the Animal Welfare Act and, in passing, had inquired about the pain dogs were experiencing in a particular experiment. To his amazement, the researcher responded by assuring him that 'Dogs don't feel pain—their nervous system is not sufficiently developed.' Even more dramatic than these cases, since it cannot be written off as an isolated instance, is the title of a 1980 bibliography of animal anaesthesia published by the federal government of the United States. Although its stated purpose is to discuss anaesthesia, its title is *Chemical Restraint of Reptiles, Amphibians, Fish, Birds, Small Mammals, and Selected Marine Mammals in North America.*

Moral consequences

The moral value or consequences of all this are readily documentable. Traditionally, little has been done in the course of animal research to control even 'unnecessary' pain—that is, pain not essential to the protocol in question. If animal pain as a subjective state that hurts is not real, why treat it? This philosophical position goes a long way towards explaining why so few protocols in the past embodied provisions for laboratory animal analgesia. Before the new guide-lines and laws addressing pain and suffering were promulgated in 1985 (for which, see Chapter 7), probably fewer than 1 per cent of NIH-funded painful protocols wrote in provisions for analgesia for laboratory animals; and if such provisions were included, it was virtually never for rodents. A recent article by Wright and Marcella points out that 'relatively few analgesics have been cleared for use in animals'.[13] Ironically, it has been estimated that fewer than 10 per cent of pain protocols would be compromised by use of analgesics, so there is no scientific basis for this cavalier disregard for pain.[14] It also explains why so few conferences have been held on laboratory animal pain and analgesia (the seminar on pain I attended was only the second ever held), and indeed, why the use of analgesia was so rare in ordinary veterinary practice and so little taught in veterinary schools. (An eminent veterinary surgeon informed me that in the mid-1960s a major veterinary teaching hospital did not stock narcotic analgesics or even bother to obtain a licence to do so.) It also helps explain why until very recently, few scientists have engaged the ethical questions arising out of inflicting pain on animals.

I encountered this dramatically not long ago while addressing a mixed audience of researchers at a major university. A medical researcher informed me that he had been in research for forty-five years, and that everyone *he* knew 'really cared about their animals'. In response, I asked why, if this were the case, so few federally funded projects wrote in provisions for laboratory animal analgesia, even when such use would not compromise the data. Up jumped the chief of veterinary surgery: 'Oh,' he said, 'that's because the use of analgesia isn't standard veterinary prac. . . . ' His hand flew to his mouth and he turned pale as he realized what he had said. 'Oh my God' he went on, 'I've been doing major surgeries like thoracotomies, for twenty-five years, and it never dawned on me that these animals surely experience post-surgical pain.'

In a beautiful comment on this state of affairs, Professor Lloyd Davis has made the following observation: 'One of the psychological curiosities of therapeutic decision-making is the withholding of analgesic drugs, because the clinician is not absolutely certain that the animal is experiencing pain. Yet the same individual will administer antibiotics without documenting the presence of a bacterial infection.'[15] One of my clinician colleagues informed me that throughout her veterinary training, when she was forced to do painful, repeat survival operations on dogs, including fracturing a femur and repairing it, she would ask for permission to use analgesia, since the animals were obviously in pain. Indeed, she would recount the signs of pain: whimpering, shivering, and so on. She was told, 'Those aren't signs of pain, they are merely after-effects of anaesthesia.' And in a long discussion I had with a noted pain physiologist, he remarked that I was correct in my view that most people who work on pain in animals are looking at pain *responses*, and dismissing questions of feeling.

In fact, as we said earlier, many researchers don't even see that the infliction of pain constitutes an ethical question. I recall one group of researchers telling me that the failure to use anaesthesia and analgesia in certain protocols has nothing to do with ethics; it is solely a scientific decision arising out of a desire not to introduce new variables which might skew the data! (As if that is not itself a questionable moral viewpoint.) It also explains why many scientists see the social demand for control of pain as an illegitimate intrusion by non-experts into their freedom of inquiry, rather than as a legitimate moral stance, and thus opposed even legislation which would require control only of pain 'not essential to the protocol'. It is ironic that rodents are the most infrequent recipients of analgesia in the course of research done on them, yet that virtually all analgesics have been tested on rodents, so dose-response curves are well known. It is also ironic that in an article published in the British journal *Laboratory Animals* in April 1984 entitled 'The Control of Pain in Laboratory Animals' (one of the few such articles ever written), the author feels compelled to introduce his topic as follows:

> Comparatively little progress appears to have been made in developing methods of reducing the degree of pain suffered by these animals which do undergo experimental procedures. . . . Several reasons for this lack of success can be identified. . . . The apparent lack of clearly identifiable responses to pain in animals has led to the suggestion that they do not experience pain in the same manner as human beings. Whilst this may be so, until such suggestions are supported by conclusive evidence, the argument must not be extended to suggest that attempts to provide pain relief in animals are therefore unnecessary. Until further progress is made in assessing the nature of pain in animals it should be assumed that if a procedure is likely to cause pain in man, it will produce a similar degree of pain in animals. Although such anthropomorphic views have been much criticized, no satisfactory alternatives have been so far proposed, therefore the relief of pain, particularly in the post-operative period, must be considered essential.[16]

In much the same vein, I know people in research and veterinary medicine who demonstrate great concern for the animals they use, and are scrupulous in their use of tranquillizers and analgesics. They stop short, however, of publicly asserting as a scientific claim that animals feel pain. At most, they claim, such a statement is 'subjective', a statement about their own impressions rather than about the world, rather like a religious perspective, and should not be part of scientific discourse.

To summarize, the philosophical problem built into the ambivalence about animal pain which we have been discussing may be restated as follows: Either animals feel pain or they don't. Science, in its activities, and in so far as its practitioners are people of ordinary common sense, has presupposed that they do, and current biological evolutionary theory suggests that this is surely the case. On the other hand, the positivistic-behaviouristic ideology of science has served to legitimate the claim that animals don't feel pain, or at least, that we can't know if they do, just as we can't know about any mental state in animals. (The latter two claims tend, of course, to slide into the former.) At the same time, of course, scientists are urged by their ideology to ignore ethical questions or to suppress them *qua* scientists. This in turn has been reinforced by society's long-time systematic lack of regard for moral issues pertaining to animals, a stance which better fits behaviourism than it does the ordinary common-sense view that animals can and do suffer. Furthermore, in the pursuit of scientific questions, it is tempting to subordinate all concerns to finding answers. In doing research, it is easy to forget about the moral status of people or animals, and to view them simply as tools for solving problems, just as in a competitive situation one tends to see one's opponents not as persons, but as game-pieces or as challenges to be overcome. Thus, even if one's natural inclination is to see one's research subjects as objects of moral concern, it is easy to forget or trump that stance under pressure of the chase. I recall discussing with a colleague a piece of research which I was interested in doing. I was proud of the design of my experiment, both because it was an original way of studying animal consciousness and because I believed it to be thoroughly non-invasive. My colleague pointed out that the research would be much easier if I kept the animals partially starved (at 80 per cent body weight, a common practice in psychological research), because they would be more inclined to perform more quickly. (Interestingly enough, research exists which shows that this practice actually skews learning experiments.)[17] Despite my commitment to animal rights, I could feel the pull of 'whatever is needed to make the experiment work elegantly', though I was able to resist it. But I could now better understand why animals have been allowed to suffer in research: not through cruelty, but rather, because consideration of suffering is forgotten in the thrill of the pursuit, by nature ultimately ruthless, complemented by an ideology which discounts the cogency of moral reflection in scientific activity and denies the meaningfulness of attributing feelings to animals, and is coupled with practical pressures. All this eliminates checks and balances from the system, and certainly circumvents any felt need for ethical theorizing about when infliction of pain is justifiable and pain control mandated. This, in turn, gets built into the educational system under which scientists are trained, further augmenting its hold, further entrenching the ideology, and making it even more implausible that convenience will be supplanted by ethical considerations.

Incidentally, the case of food deprivation and keeping the animals at 80 per cent of normal body weight provides an excellent example of the incoherence engendered by the ideology of science, as I discovered when I raised the issue at my own institution at an animal care committee meeting. The psychologists argued that such deprivation was necessary to motivate the animals to learn. Yet they also argued on the basis of the fact that humans quickly adjust to food deprivation and that hunger pangs disappear, that being deprived of food was not unpleasant for the animals. I pointed out that they couldn't have it both ways: either the deprivation was motivational because unpleasant, or else it was not unpleasant, in which case how did it motivate them? Other scientists, a zoologist, a nutritionist, and an agricultural scientist, quickly chimed in that hunger was a physical, mechanical state in animals, not an experiential one, and that my suggestion that food deprivation was unpleasant to the animals was 'anthropomorphic' and 'baseless', conveniently ignoring their own anthropomorphism in appealing to the human case! Still another scientist suggested that I was 'out of line' in raising the issue as a moral question, since 'the mechanisms of hunger are unrelated to the mechanisms of pain'—in other words, that only the infliction of pain is a legitimate question for a committee looking at issues of research animal welfare. Most astounding, perhaps, was the reaction of one committee member who banged his fist on the table, and asserted that he refused to discuss further such a 'non-issue'. 'What do you mean?' I queried. 'Recent evidence indicates that keeping animals underweight reduces the incidence of tumors,' he proclaimed; 'so the psychologists are doing the animals a favour!' With that parting remark, he stalked out of the meeting.

I should stress that in the preceding and ensuing discussions of pain and the ideology of science, I do not mean to suggest that ideological scepticism about the knowability of animal consciousness was always the only or even the primary factor leading scientists to ignore animal pain. At least as important are the ideology of science's disavowal of a need for science to concern itself with ethical questions and the fact that society in general, in its economic use of animals, didn't concern itself morally with animal pain, even in contexts where it was clear to common sense that such pain occurred—for example, in agricultural contexts, where cattle are castrated and dehorned without anaesthesia, ranchers do not usually deny that the animal hurts. So society in general did not force scientists to reckon morally with animal pain; and the ideology of science simply added another, more intellectual justification for disregarding animal pain.

The ideological scepticism about whether one can meaningfully talk about animals feeling pain was invariably just one ingredient in justifying activities already 'justified' by convenience and expediency, and by the claim that ethical questions were outside the purview of science. Thus the question of animals hurting in scientific experiments or in veterinary medical contexts was often simply not raised, unless someone (usually a student in science or medicine) challenged a given practice in those terms, at which point the ideological scepticism was trotted out as a retort. Most often, students simply got a clear message that they should not question a given practice, cued by the teacher's failure to discuss the practice in moral terms. Thus I am told by American veterinary graduates of the 1960s that when they were taught the widely used techniques of castrating horses using succinylcholine chloride, a paralytic agent used for convenience and ease of dosing, which restrains the animal by depolarizing the neuro-muscular junction but which has no anaesthetic properties, and which in humans and in animals actually increases pain and causes panic when used alone, no mention was made of ethical considerations—it was simply the way it is done.

Avoiding incoherence—the physicalization of pain and stress

In any event, the paradox remains—the contradiction built into scientists' presupposing animal pain as a basis for research on the one hand and ideologically denying its reality on the other. In practice, the scientific community has sometimes tried to smooth over this paradox. It has done so, in general, by treating pain as a mechanical physiological or neurophysiological event, rather than a mental state or mode of awareness. In this way, it has tried to avoid 'unscientific claims', and has also assuaged common sense and its concomitant moral reluctance to inflict uncontrolled real, felt pain on animals. On this essentially Cartesian view, pain responses are physical, mechanical states objectively amenable to study, rather than states of awareness. Such an approach admits the existence of physiological mechanisms of pain, while ducking the issue of how the animal feels. A good example is the aforementioned view of anaesthesia as chemical restraint. A more recent example may be found in the American Physiological Society symposium on animal pain held in 1983, published as *Animal Pain: Perception and Alleviation*, which, despite its title, deals in exquisite detail with the neurophysiology, neurochemistry, mechanisms, and (to a lesser extent) behaviour involved in animal pain, while almost never discussing the psychic, morally relevant component—namely, the fact that the animal *hurts*. (Physiologists refer to the working of the machinery or plumbing of pain as 'nociception'.) There is also a similar, extreme behaviourist approach to human and animal pain, which defines it purely in behavioural terms, arguing that appeal to subjective states adds nothing.[18] The author is at least consistent, applying his theory equally to animals and humans. Another conceptually problematic attempt to get around having to talk about states of awareness, especially unpleasant ones, by reducing them to mechanical processes, rather than admitting the existence of mental states, may be seen in the widespread, notoriously fuzzy use of the concept of 'stress' as a catch-all for fear, anxiety, and other sorts of misery. In the view of positivistically and behaviouristically or reductionistically oriented scientists, this concept allows one to admit that noxious situations have deleterious effects on animals, without invoking awareness or consciousness, and to study the mechanisms of these effects in the animal. On the other hand, the use of the term in the vast, rapidly proliferating literature is bewildering and ambiguous, and it is difficult to get a clear picture of exactly what is being discussed. While most of the work published in the area is primarily physiological and only secondarily behavioural, it seems evident that no sense can be made of the notion without implicit or tacit reference to the animal's state of mind or awareness—that is, to what the animal is *experiencing*.

The term 'stress' is, in fact, used in at least three distinct ways in the literature. Sometimes it is applied to an environmental situation: in this sense, scientists talk of cold stresses, heat stresses, noise stresses, and so on. Sometimes it is applied to the psychological state of the animal or person subjected to a noxious stimulus, as in talk of emotional stress or separation or isolation stress. And sometimes it refers to the effect of noxious situations on the physiology of the person or animal. In the latter use, stress is most often dealt with physiologically in terms of what Hans Selye, in his pioneering work begun some fifty years ago, called the General Adaptation Syndrome (GAS).[19] Selye's work distinguished between the physiological effects of short-term and long-term noxious stimuli. In the short-term case, what occurs is activation of the sympathico-adrenal axis—that is, the mechanisms mediated by the nerve hormones called catecholamines, namely, epinephrine and norepinephrine. This is what is commonly called the 'fight or flight reaction', which is evoked by perceived threats or dangers of short duration, as when someone jumps out at you and yells 'Boo', or when a cat is suddenly confronted by a dog. This aspect of stress was first described by Cannon.[20] With respect to long-term stress situations—for example, crowding or exposure to cold or to heat—Selye talked primarily in terms of the activation of the pituitary-adrenal axis—that is, the interrelationships between the pituitary hormone ACTH (Adreno-cortico-tropic-hormone) and the adrenal hormones called gluco-corticoids, steroids released by the adrenal cortex. Prolonged exposure to stresses resulting in long-term activation of the pituitary-adrenal axis can have incalculably damaging effects on the body and its health. Furthermore, a variety of other bodily systems are affected profoundly and directly by prolonged stressful conditions, the emphasis on the pituitary-adrenal axis being merely a historical accident. There is virtually no aspect of an organism which cannot be deleteriously affected, directly or indirectly, by prolonged stress. Cardiovascular health, blood pressure, shock response, susceptibility to infection, susceptibility to cancer, reproductive ability, gastro-intestinal activity, ulcers, post-surgical or other wound recovery, headache, migraine, colitis, irritable bowel syndrome, ability to tolerate toxic substances, skin disease, intellectual abilities, emotional states, nervous and mental disease, behavioural pathologies, kidney disease, arthritis, allergy, asthma, auto-immune disease, alcohol and drug abuse in humans (and in experimental animals) can all be affected, induced, and exacerbated by long-term stress.[21] And a great deal of research has been done on unearthing some of the physiological mechanisms by which this occurs.

Ironically, there is conclusive evidence that failure to control stress variables and their effects can wreak havoc with animal research results. Isaac has shown that virtually any stimulus can be a source of stress to the laboratory rat.[22] Known sources of stress which can profoundly skew all sorts of metabolic and physiological variables and thereby jeopardize research results, are heat, cold, noise, crowding, isolation, light, darkness, change in temperature or air quality, infection, restraint, trauma, fear, surprise, disease, and behaviour of the investigator or laboratory technician towards the animal. Relatively few researchers are even aware of these facts; even fewer control for them.

Indeed, researchers, especially medical researchers, are notoriously ignorant regarding many features of the animals they use, and receive little or no training in dealing with animals. Many know only that a particular animal is a good 'model' for a particular disease or syndrome which they are interested in studying. One can get an MD or a Ph.D. in an area of research involving animals, and never learn anything else about the animals. In one famous incident, a well-funded medical researcher asked the laboratory animal veterinarian at his institution, 'At what age do white mice grow up to be rats?' I know personally of a case in which guinea-pigs were dying of what the researcher thought was a wasting disease, but in fact turned out to be malnutrition stemming from mal-occlusion of the teeth, rendering the animal unable to chew! A laboratory animal veterinarian who was supplying beagles to a surgeon for research received an indignant phone call from the surgeon asking him how he dared to provide 'sick dogs'. 'Those dogs are perfectly healthy' was the reply. 'Nonsense,' snapped the surgeon, 'they are all febrile. They all have a temperature of over 98.6 degrees.' Environmental physiologist David Robertshaw estimates that only 40 per cent of researchers are aware that circadian rhythms can affect toxicity of substances by up to 80 per cent, depending on what time of day a substance is administered. Increasing amounts of data show what people in agriculture, most notably dairy farmers, have been aware of for centuries, that animals are incredibly sensitive to myriad variables. What an investigator may disregard as trivial can have major metabolic and physiological effects on an animal. To take one example: very few investigators think of, let alone control for, the effect of the behaviour and attitude of animal caretakers on the physiology of the animals. Yet Nerein and co-workers have shown that if two groups of rabbits are put on a 2 per cent cholesterol diet, and one group is petted and fondled while the other is simply fed and watered, the rabbits treated with tender loving care develop 60 per cent fewer atherosclerotic lesions.[23] Even more dramatically, Gärtner and colleagues studied the effects on 25 blood characteristics of simply moving rats in their cages or exposing them to one minute of ether. I quote their summary:

> The effects were observed of moving male, adult Han: Sprague rats in their cages or of exposure to ether for 1 min on the plasma concentration profiles of 25 blood characteristics linked with stress and shock reactions. 5 min after the stress, serum prolactin, corticosterone, thyroid-stimulating hormone, follicle-stimulating hormone, luteinizing hormone, tri-iodothyronine and thyroxin levels were elevated 150–500% compared with those in blood collected within 100 s of entering the animal room.

Heart rate (telemetrically recorded), packed cell volume, haemoglobin and plasma protein content were 10–20% elevated 2–10 min after cage movement or 2–20 min after ether confrontation over those of controls sampled within 50 s, indicating circulatory and microcirculatory shock reactions.

Serum glucose, pyruvate and lactate concentrations rose by 20–100% 2–5 min after cage movement and 1–15 min after ether exposure. Phosphate, calcium, urea, aspartate and alanine transferases, alkaline phosphatase and leucine arylamidase were not altered significantly by either stressor, while potassium and bound glycerol fell for 1 min and 5–20 min respectively.[24]

If something as apparently trivial as moving a cage can cause such profound effects, then it goes without saying that innumerable other intrusions on animals' natures can have similar consequences. In the United States the medical research community is talking more and more about using the pig to replace the dog as an experimental animal, yet there are few legally mandated standards of care for the use of any farm animal in research.[25] (As we shall see in Chapter 7, pigs and other farm animals are not animals according to the Animal Welfare Act!) One can only wonder about the value of the data which will be gathered under such circumstances. Similar results have been found regarding stress in food animals, demonstrating its adverse effect on weight gain, reproduction, disease resistance, and so on and the industry is beginning to appreciate the extent to which animals can be affected by stress.

At all events, it is easy to see that the physiological description of the animal body under stress does not *define* stress, and does not do what the common sense of science would like it to do—namely, provide a purely mechanical substitute for talk about states of mind. This is true for two basic reasons: first, physiological responses in the lab are activated by stimuli which, under normal conditions, we would never call stressful—for example, eating or exercising for fun, activities which in fact are often stress-reducing; second, the physiological descriptions are senseless without implicit reference to psychological states. In other words, when an animal responds to noxious situations like crowding or being frightened or isolated, it is surely because the animal is having an unpleasant sensation or conscious experience at least somewhat similar to what humans have under similar circumstances, which sensations are causally responsible for (or at least linked to) the activation of the mechanisms studied. That is, an animal would surely not show the physiological responses of stress under noxious or unpleasant conditions if it did not experience them as unpleasant. After all, fully unconscious animals do not react to these stresses.

It has also been shown that purely psychological stresses, like exposure to a new environment, can generate greater physiological stress responses than something clearly physical, like heat.[26] So it is clear that talk of stress involves covert reference to an animal's experience or consciousness or mental state to make it coherent and plausible, in which case it cannot serve as a way of avoiding mentalistic locutions. What much of the stress literature does, however, in its effort to avoid reference to animal thought, in accordance with scientific ideology, is to confuse the physiological signs and effects of stressful situations on the animal body—that is, the effect of those situations which are experienced as noxious or unpleasant—with stress (or the experiences) itself. This is analogous to defining eating beans as breaking wind. As we shall see later in Chapter 7 where we return to a detailed discussion of stress, it appears that at least some scientists have begun to see the limitations of treating stress as a purely mechanical physical response, and have begun to argue for an irreducible psychological and ideational or mentalistic component; but for most, stress is still a mechanistic escape-hatch allowing one to avoid having to confront the questions of consciousness, subjectivity, and how the animal *feels*.

I have argued that the common sense of science, notably with regard to animal thought, is rooted in what is ultimately a metaphysical position—that is, a basic view of reality—which is itself rooted in certain valuational commitments, including moral ones. It will be recalled that this metaphysical position is one which includes an epistemological commitment to physicalistic reductionism as well as a commitment to behaviourism. It appears that in so far as any attempt is made to tie these two strands together, the reasoning would go as follows: ideally, explanation of behavioural phenomena is to be neurophysiological or physico-chemical. But since science has not yet reached the stage where it can explain everything this way, looking at behaviour and regularities which obtain about behaviour is a legitimate place-holder. Ideally, behavioural facts are to be reduced to neurophysiological ones. Such reductions remain largely speculative, however, though some work has now been done which attempts to link physiological stress responses with behavioural stress responses. On the whole, though, behaviour is looked at on its own. Thus, another acceptable way of cashing out the concept of stress is to look at behavioural pathologies. Here, too, no reference is made to consciousness; what is examined is deviation in behaviour from a norm for most animals in a species. This approach is especially prominent in research on the effects of intensive confinement agriculture on farm animals. Stress-related or stress-induced behaviours include cannibalism and feather-pecking in chickens, tail-biting and cannibalism in pigs, homosexual behaviour in cattle, cribbing and weaving in horses, and stereotypical pacing in zoo animals such as caged tigers. Once again, these are surely *indicators* of animal misery, rather than definitions thereof.

Reprise: philosophy, value, and the neglect of consciousness

It is worth pausing to remind ourselves just how significant the philosophical and valuational components of the ideology of science are to the approaches we have been discussing—that is, the extent to which they shape what counts as legitimate data, what counts as real, what counts as fact, what counts as explanation, etc. We have already marvelled at the ability of scientific common sense to banish facts of consciousness. We have also discussed what allows this to be done so cavalierly in the case of animals—namely, the lack in both society and science of moral theorizing about or moral concern for the treatment of animals. But we should also recall that there is a strong, highly debatable metaphysical element involved in a commitment to patterns of explanation which affirm that the only things which are ultimately real, or the only things which are worth studying, are physico-chemical and mechanistic in nature. Ironically enough, this view, which we have discussed briefly as it appears in people like Descartes, Helmholtz, and Loeb, and which is as old as thought itself (witness the ancient atomists), is less evident today in physics than it is in the biological sciences. The quantum revolution has to all intents and purposes rendered it impossible for mainstream physicists to think in simple mechanistic terms. One look at Bell's theorem, according to which the behaviour of widely separated particles seems to be mutually and simultaneously interdependent even when there is no possibility of local causation, gives clear evidence of our claim. Where mechanism is most alive and most healthy is among molecular biologists. Individuals not trained in molecular biology, organismic biologists of all sorts, including non-reductionistic doctors, enjoy roughly the same status relative to molecular types in the scientific community as one-mile joggers do to marathoners in the running community. So the common sense of science regarding animal consciousness is perpetuated not only by psychological science, but by the mainstream of biological science as well. In this environment in which any explanation short of physicochemical is seen as mere natural history, consciousness does not have a chance. If organismic talk is suspect, mentation talk has surely been sentenced to death!

In fact, in their zeal to measure up to their ideal of being 'real scientists', biologists sometimes carry the ideological commitments we have been discussing to incredible lengths. A zoologist recently took me to task for saying that horses prefer the taste of rolled oats with molasses to that of ordinary rolled oats. 'You can't scientifically say that,' he told me. 'At most you can say that there is a mechanical process which leads them to be drawn towards the molasses, much as a thermostat is affected by temperature.'

Let us recapitulate our discussion of animal pain so far by reminding ourselves that what happened in the twentieth century was the same thing that happened with Descartes in the seventeenth namely, that philosophy was invoked in order to overcome common sense in favour of a reductionistic biology, and in the absence of strong moral concern for animals. Descartes said that animals were machines with no souls, minds, or feelings, thereby reconciling in one master stroke the prescription of Catholicism that animals do not have souls with the demands of the growing science of physiology, which was forced in its quest for knowledge to perform what common sense called atrocious and painful procedures on animals without any way of controlling the pain. No need to control the pain, said Cartesian physiologists, since it is not experienced pain, but rather, merely mechanical responses.

My colleague Ron Williams has made a fascinating point about the denial of felt pain in twentieth-century science. He shares my view that positivism and behaviourism are instances of a sweeping reductivistic mode of thought that pervaded all areas of culture at the turn of the century. But he goes further, and points out that there are analogies between what happened in philosophy of science and philosophy of art which further buttress and explain our main thesis. Just as much reductivistic aesthetic theory was formalistic, so too was positivistic-behaviouristic philosophy of science. Formalists like Bell and Fry in aesthetics scorned content and meaning in paintings and sculptures and argued that what was aesthetically essential was pure form. Similarly, positivists were very concerned with the logical form of the sciences—witness Carnap's axiomatization of physics, and the work of others, like Woodger in biology, in the same vein. From this formalistic perspective, coupled with a reductivistic behaviouristic tendency, argues Williams, it is easy to see how scientists could ignore subjective experience (mental content), just as aestheticians ignored pictorial content, looking instead only at the formal characteristics of pain, suffering and other mental states—that is, at the mechanisms which underlie them.

Ideology and human pain

Before turning to a more detailed discussion of some of the philosophical arguments used to buttress the denial of and lack of concern with pain in animals in contemporary science, it is worth mentioning that a similar state of affairs existed in the nineteenth century with regard to human pain. As in the case of animals, lack of social moral concern with certain classes of people led to a failure to control manageable pain in such people, which lack was justified by philosophical-ideological claims to the effect that such individuals didn't 'really' feel pain, or that, if they did, it didn't really bother them. The specific occasion for this state of affairs was the development of anaesthesia in the mid-nineteenth century. In a brilliant study, a historian has chronicled the response of the medical community, and its philosophical and ideological assumptions, to the ability to control surgical and other pain by anaesthesia. In his account, Pernick has persuasively demonstrated the extent to which the use (and non-use) of anaesthesia was guided by ethical and philosophical assumptions. He shows, for example, that some doctors did not see control of pain as a good but, rather, saw pain as good, because it was curative or useful in developing 'macho' traits in men, or 'natural', or a just punishment for transgressing natural or religious law.[27] In the case of obstetric pain, it was argued that such pain was a necessary prerequisite for a mother's love.[28] It was also argued that anaesthesia did not let the patient participate in his or her own treatment and gave the physician too much control over the patient.[29]

Rules for its use tended to conform to philosophically and morally questionable positions. It was asserted, for example, that wealthy, educated, civilized people needed anaesthesia more than lower-class or uncivilized ones; that women felt pain more than men (but needed less anaesthesia); that infants felt either no or only momentary pain; that immigrants, Irish, blacks, country people, uneducated people felt less pain than non-immigrants, whites, city-dwellers, and educated people.[30] Nor was this mere ideology; actual use of anaesthesia seems to have conformed to these ideological pronouncements.[31]

Philosophical and valuational claims were thus used to justify differential use of anaesthesia. For example, the ability to withstand pain was said to be part of true virility and essential to a soldier, and thus that 'real men' didn't need anaesthesia. An American Medical Association (AMA) report at that time asserted that an infant didn't need anaesthesia because 'the fact that it has neither the anticipation nor remembrance of suffering, however severe, seems to render [full anaesthesia] unnecessary'.[32] As we shall see, the same argument is still used today to justify withholding anaesthesia and analgesia from both infants and animals.

Level of civilization was seen as metaphysically relevant to the ability to suffer. Savages were said by Silas Mitchell, founder of American neurology, to experience things differently from civilized men: 'In our process of being civilized we have won . . . intensified capacity to suffer. The savage does not feel pain as we do.'[33] This view was even carried over to animals; thus domestic animals were said to feel pain more than wild animals.[34] By the same token, a major textbook on anaesthesia, Bigelow's *Surgical Anesthesia* of 1900, declared that 'an intelligent dog' felt more pain during surgery than a 'Bushman or Digger Indian'.[35]

In the context of research, the treatment of disenfranchised humans paralleled the treatment of animals. Thus blacks, who were morally insignificant in the nineteenth-century American South, viewed as 'primitive', and felt to have 'thicker skin', were the subjects of excruciating medical experiments.[36] Ironically, this fact tends to undercut the twentieth-century belief, widespread among animal welfare advocates, that if animals had language, scientists would be less likely to use them in hurtful ways; for, if ideology can overcome logic and our responses to natural signs of pain, it will be unmoved by conventional signs of pain as well. A telling quotation from a nineteenth-century medical journal asserts that 'Negresses . . . will bear cutting with nearly, if not quite, as much impunity as dogs and rabbits'.[37] Obviously, a person capable of ignoring the howls of dogs or rabbits when they are hurt—rabbits usually vocalize only when in extreme agony—is equally capable of dismissing the most articulate complaints of humans. Thus does ideology override common sense, not to mention decency, whether one is dealing with human or animal pain.

Infliction of pain on animals without anaesthesia or analgesia in the twentieth century is often justified by the claim that the procedure in question is 'minor'. Thus, I have heard ovario-hysterectomies, castrations, abdominal incisions, rumen and other operations in cattle, and liver and bone biopsies referred to in research and veterinary contexts as 'minor procedures' which don't involve significant, or any, pain; where anaesthesia is used primarily as a restraint, and analgesia is never used. By the same token, nineteenth-century medicine treated a whole host of procedures as minor; indeed, textbooks of surgery labelled virtually everything as minor, and thus not requiring anaesthesia, and one textbook, Smith's *Minor Surgery*, included amputations of limbs in that category as well.[38] This cavalier attitude towards pain—whether in humans or in animals—puts one in mind of the old Jewish joke defining major surgery as surgery *I* undergo, minor surgery as surgery *you* undergo.

The persistence of the attitudes described by Pernick has affected not only animals in the twentieth century; it has continued to affect disenfranchised humans. Most dramatic are recent revelations that neonates are regularly subjected to a variety of painful surgical procedures without anaesthesia, including circumcision, thoracotomies, lumbar punctures, supra-pubic bladder taps, and other procedures.[39] The journal *Birth* describes two major operations performed on awake infants using a paralytic, curariform drug for restraint.

> The first baby had a $1\frac{1}{2}$ hour operation to ligate a patent ductus arteriosus and install a chest tube. These procedures required incisions on the right and left of his neck and in his right and left chest and another from breastbone to spine with subsequent rib spreading. The second infant was shunted for hydrocephalus via scalp, neck, and abdominal incisions; craniotomy; and tube insertion from his brain into his chest and on through to his abdominal cavity.[40]

Furthermore, non-use of anaesthesia has been by and large accepted as state of the art among neonatal surgeons, and has only just become a controversial topic. It was only in September of 1987 that the American Academy of Pediatrics addressed the issue in a report which acknowledged some of the points I have been making.[41]

Behind these practices are the same sort of moral and metaphysical notions which we have discussed in connection with animals. Children in general, and infants in particular, traditionally have not been full objects of moral concern. Like animals, they have been seen as property. Killing an infant is not usually punished as severely as killing an adult, and even seriously abused children are often returned to the parents who abused them. Indeed, the first cases of child abuse in the United States were prosecuted under animal cruelty laws! A society in which abortion is widespread is predisposed to see neonates as less than fully human, as did ancient tradition. Furthermore, since neonates lack language, and thus concepts, one can invoke the claim cited earlier that pain is momentary and transitory.

6 Animal pain: the ideology cashed out, 2

Arguments for the insignificance of animal pain

Once the foundation for scepticism about animal consciousness had been laid in the early twentieth century, a superstructure of palliative arguments was erected upon it, essentially designed to teach young scientists and veterinarians that concerns for animal pain rooted in common sense were largely sentimental and were scientifically unwarranted anthropomorphism. Many older veterinarians were taught that animal pain is merely momentary, as evidenced by the fact that a cow will eat immediately after surgery, which allegedly indicates that the animal can't really be hurting (since *we* wouldn't be eating directly after surgery). Aside from the unjustified kind of anthropomorphism built into this claim, which we will discuss shortly, such arguments fail to mention that it is a selective evolutionary survival advantage for a cow to eat regardless of how it feels. A cow that didn't eat would be weakened considerably; a cow that didn't graze with the rest of the herd when hurt would be flagged as vulnerable to predators. By the same token, it was sometimes said that dogs didn't really feel pain because they would be up and moving about directly after abdominal surgery. Such an argument, it has been pointed out, neglects the basic anatomical fact that abdominal musculature in dogs does not support visceral weight in the same way that it does in humans (since the viscera are suspended from a mesentery 'sling') and that for this reason surgery on that area would not hurt as much as in a human. Lawrence Soma, a prominent veterinary anaesthesiologist, has pointed out that similar anatomical or other biological explanations can be given in all cases in which animals do not appear to hurt when we would, without having to resort to the notion that animals don't feel pain.[1]

The claim that animals lack concepts

In a similar vein, it was sometimes said by scientists and philosophers that animal pain, while perhaps present momentarily, was insignificant. The reasoning behind this claim was as follows: Since animals lack concepts enabling them to anticipate and remember, the kind of suffering engendered in us by worrying about and anticipating going to the dentist, which makes the pain of dental work so much worse, simply does not arise in animals.

It is worth pausing to examine this contention, which, in its own way, has done much to shore up the common sense of science's view of animals. Historically, it is rooted in Descartes's claim that only language, a 'universal instrument' as he called it, can evidence mind and go beyond immediate particularity. The equation of thought with the ability to universalize and generalize and go beyond the particulars given in sensation was made explicit by Kant, who made thought propositional, and rooted thinking in the organization of sensory data by concepts.[2] This tradition has assumed that since animals lack language, they must lack concepts, and are therefore trapped for ever in the momentary. Only a linguistic being has concepts, and only concepts enable a being to universalize, generalize, refer to what is absent, counter-factual, non-existent, past, future, and so forth. Since animals lack language, they must lack concepts; and since they lack concepts, they can live at best only in a world of isolated, fragmented, momentary particulars: William James's 'buzzing blooming confusion'. This claim is pithily captured in a poem by the twentieth-century poet Edwin Muir:

The animals

They do not live in the world,
Are not in time and space.
From birth to death hurled
No word do they have, not one
To plant a foot upon,
Were never in any place.

For with names the world was called
Out of the empty air,
With names was built and walled,
Line and circle and square,

Dust and emerald;
Snatched from deceiving death
By the articulate breath.

But these have never trod
Twice the familiar track,
Never never turned back
Into the memoried day.
All is new and near
In the unchanging Here
Of the fifth great day of God,
That shall remain the same,
Never shall pass away.

On the sixth day we came.[3]

I would venture to guess that if there is anything like a philosophical orthodoxy in the twentieth century, that is it. Philosophers like Davidson, Bennett, Frey, and innumerable others have constantly rebottled the same wine.[4] Extraordinarily, even Wittgenstein, the most anti-Cartesian of all philosophers, shares the Cartesian bias against animal mentation by virtue of the absence of language in animals. His works are peppered with cryptic, sceptical remarks about predicating mentalistic attributes to animals. In one famous passage, he tells us that if a lion could speak, we couldn't understand him; in another he suggests that it is conceptually impossible for an animal to smile. He also suggests that a dog cannot simulate pain or feel remorse, that an animal cannot hope or consciously imitate, and that a dog cannot mean something by wagging its tail, and a crocodile cannot think.[5]

Wittgenstein's reasoning is somewhat different from Descartes's, of course. For Descartes, language expresses thought and codifies it, but an individual human being in a solipsistic universe logically *could* have thought even in the absence of a public language. This is true for Kant as well; an individual human has an a priori conceptual apparatus that is logically independent of public language. Furthermore, as the *Critique of Pure Reason* implies, this apparatus is and must be the same for all beings who synthesize sensations to create experience and knowledge, and thus Kant never feels the need to address the problem of other minds.[6] For Wittgenstein, on the other hand, thought is constituted by the social system of conventional signs one is brought up in; without such a system there is neither thought nor concepts; there can be no 'private language', for there are no publicly checkable criteria and rules for correct and incorrect application of concepts in a private language, and if there is no way to be incorrect, there is also no way to be correct. Since animals lack a system of conventional signs, they lack the fundamental tools for a mental life. Second, language is a 'form of life' which both expresses and shapes the nature of one's *Umwelt*. The comment about the lion suggests that since animals have such a radically different form of life, we could not become privy to it even if they did have a rule-governed language.

All these arguments contain a great deal of implausibility, and the fact that they have endured virtually unchallenged attests to the power of ideology in philosophy, as in science. I say *virtually* unchallenged because, as we saw earlier, associationists like Hume, at once both the great challenger of common sense and its strongest supporter in practical matters, considered it patently obvious that animals think and feel more or less as we do. This sort of appeal to common sense was echoed by the Scottish Common Sense philosophers.

Any argument which equates thought and language and which denies any sort of significant thought in the absence of language, be the argument Cartesian or Wittgensteinean, must be hard pressed to explain how humans ever acquire language in the first place.[7] The acquisition of language entails that experiences and thoughts be processed at some stage without language. Even if one believes, with Chomsky, that the essential skeleton of language is innate, so that linguistic competence is native rather than acquired, it must still be triggered and fleshed out by non-linguistic experiences, which determine the particular version of universal language that the child learns. Further, as Thomas Reid pointed out, understanding of reference and meaning requires some non-linguistic comprehension of the linkage between sign and what is signified (such as ostension) prior to the acquisition of language; otherwise the entire process would never get off the ground.[8] In short, language requires a peg of non-linguistic experience on which to be hung.

Given the logic of the Kantian position equating having a mind with having concepts to organize the particulars given in sensory experience, denying this ability to animals on the grounds that they do not have language to betoken the concepts is self-defeating. For any careful reader of Kant will recall that he is not ever doing psychology. In ethics or epistemology, he believes himself to be doing conceptual analysis. In his ethics, he tells us that nothing he has to say depends upon or uniquely pertains to human nature, for ethics must be true of 'any rational being in general'.[9] Similarly, in his epistemology, Kant clearly asserts that everything he is saying about knowledge and experience is true for any being possessed of an 'ectypal intellect'—

that is, an intellect which depends on material received from outside itself for its experience. No one, including Kant, can deny that animals with sense-organs perceive objects; after all, dogs regularly pursue rabbits, fetch bones, avoid cars, and jump on little boys. But if he admits this, he is undone. For the arguments which Kant marshals against an atomistic, associationistic epistemology of the Humean sort regarding human experience would apply equally well against any empiricistic account of animal perception. If animals have perceptions of objects and causal relations, they must be doing something other than merely sensing. For as Kant himself points out, the senses supply only momentary, ever changing fragments. To experience, to perceive, one must tie these particulars together—'synthesize' them, in Kant's terminology. But this in turn means that there must be some internal mechanism for synthesis. The essence of Kant's argument against the atomistic empiricism of Hume was the insight that we cannot possibly be passive dartboards upon which atoms of unrelated sensation fall. While it is surely true, says Kant, that our experience is *composed* of sensory atoms, the end product is not fragmented, but is, rather, the experience of objects which endure and interact, notably causally. Clearly the sensory atoms end up being organized into wholes, showing that the experience is active rather than passive. The principles of synthesis are, for Kant, concepts—a priori concepts by which our sensory atoms are moulded into objects, standing in relationships with one another.

But by parity of reasoning, animals must be in precisely the same situation! As mentioned earlier, they obviously experience objects and causal interaction. By the same token, their access to the world is via sense-organs which are extremely similar to ours, and which, in and of themselves, can provide only fragmented atoms of experience. Therefore, as Kantians, we must conclude that they, too, possess a priori concepts fairly similar to our own. And since they learn from experience, even, as Hume points out, from single experiences—for example, to avoid hot objects after being hurt—they must surely possess a mechanism for generating empirical concepts. After all, an organism with no power of generalization and abstraction, which could experience only particulars, could neither learn nor survive. As Hume said, all this is no surprise to common sense, which knows quite well that animals possess at least some concepts—dogs, for example, clearly have concepts of food, water, danger, play, stranger, dog (or scent of other dog), and so forth. As to how these concepts, or abilities to pick out common features of the world, are symbolized, that is an open question; but it is plausible to suggest that animals have some mental tokens or images which serve in this capacity.

One can take Kant's logic a bit further. According to Kant, having innate concepts allows one to have a 'transcendental unity of apperception'—that is, a unity of self-awareness. What this means is that in order for a being to have unified experience of objects in relations, it must be the same consciousness which experiences the beginning of an event as the end, or the top of an object and its bottom. In other words, if it were not the same you that viewed the top of a tall building as the bottom and the middle, there could be no experience of 'the tall building'. But this same point must hold true for animals too; they must be able to realize that an event is happening to them in order to learn from it. We are surely licensed, I believe, to assert that animals have a sense of self as distinguished from world; what we do not know is what form it takes. Once again, common sense assumes that animals know the difference between what happens to them and what doesn't. The efforts of animals to protect themselves certainly supports our claim, and if one is willing to admit that animals feel pain, it follows that pain would not be of much use were it not referred to a self. In a now classic piece of research done earlier this century, the philosopher and physiologist Buytendijk showed persuasively that an octopus could distinguish between actively touching something and being passively touched, and concluded from this that even octopuses have a mental image, betokening a concept, of self and other.[10] (Incidentally, further research indicates that octopuses and squid can solve problems, learn, and be anaesthetized!)

The Wittgenstein version

But what of Wittgenstein's point that mental images cannot serve as markers for concepts, since there is no public check for correctness of application? Let us recall that to have a concept, so the argument goes, there must be rules for the use of the concept which can be checked publicly. That means that there must be ways in which one can conceivably *misapply* the concept, and be detected and corrected. This is possible only when the vehicles of the concept are public and accessible to others, who can see how one is using a concept, and who can correct deviations from proper use. Thus, consider a child learning the concept 'dog'. He may try to group a cow with dogs. But when he calls the cow a 'dog', someone corrects him. In the absence of a public way of expressing a concept and of being corrected, however, what is to stop a person from using it differently each time? What criteria does one have for deciding whether a concept does or doesn't apply to some new case? Such a state, says Wittgenstein, is comparable to a game in which one makes up the rules as one goes along. Without some fixed, externally verifiable rules, the activity is not really a game at all.

This is a strong argument. But does it really count against animals having concepts? Let us see. A person has words for his concepts. These words can be checked against what other people say. This is supposed to be different from the case of animals, who presumably only have some ideas in their heads, or perhaps certain perceptions to use as marks of their concepts. For example, an animal may have only some memory of the appearance of water or the visual appearance of the water itself—for example, its shimmering—to serve as a mark of his concept.[11]

Is there ultimately a difference between the two situations? On the surface, yes. But in a deeper sense, perhaps no. According to the private-language argument the animal must rely on memory and thus has no way of being shown to be wrong. But suppose, as we all know happens, a puppy sees me rattling a martini-shaker and approaches me, thinking that it is about to be fed. I say, 'No, that's not for you,' and don't give it any food. Its initial concept of 'dish rattling—food time' is thereby corrected. I see no relevant difference between this case and the case of the child who calls a cow a 'dog'. Nor does this process of public correction require a human being. Let us return to the shimmering perception which serves as the visual sign of water. An animal may see shimmering on asphalt and believe it to mean water (even as we do), but he is 'publicly' corrected when he reaches the road and finds no water there. In other words, the fact that the animal is an active *agent* can serve as a basis for correction.

If the private-language theorist is persistent, he may say, 'But how does the animal know the next time that he is using the sign or idea in anything like the way he did before? The animal has only memory, we at least have other people.' The answer is simple. If we can be sceptical about memory, we can also be sceptical about other people's memory, and ask how we ever really know that they are using a word or concept the way they did before? So public checks don't really help in the face of extreme scepticism.

This discussion has, of course, presupposed that memory without language is possible, and that animals can remember without language. Aside from the fact that behavioural evidence supports this claim, it is obvious that we humans must be able to remember without language; otherwise we could never learn language in the first place.

There is an even deeper philosophical response to the private-language argument. The possibility of publicly checking linguistic concepts itself depends on leaving certain linguistic concepts unchecked. For example, let us return to the case of the child who calls a cow a 'dog' and is corrected by a parent who says, 'No, cow.' The presupposition here is that the child's concept of 'No' is correct. How do we check that publicly without presupposing some other concept that we cannot check without presupposing some other concept, and on and on, *ad infinitum*? We do not, of course, worry about this; we take subsequent correct behaviour as evidence that the child has the concept straight. But if that is so, why do we not take an animal's correct behaviour in context as evidence that it, too, has concepts? Its concepts are, as Berkeley said, learned from the language of nature, and are certainly not as complex or abstract or variable or (sometimes as) precise as ours, but they still seem to be concepts—that is, some sort of intellectual capability that allows a creature to recognize repeatable features of the world and to synthesize experiences.

It is hugely ironic that Wittgenstein, the philosopher who stressed the sagacity of common sense and ordinary language and wished to protect both from taint by philosophical mischief-makers, should have had such a blind spot *vis-à-vis* animal thought. For if, as he said, 'ordinary language is all right as it is', he should surely have acknowledged that for ordinary language, as for common sense, animals unquestionably do have full mental lives. As we said earlier, ordinary people simply could not discuss animals without using terms like 'bored', 'is hungry', 'wants to play', 'doesn't like the mailman', 'is depressed since the kids left home', and so on. In the same vein, classic research by D.O. Hebb showed that zoo attendants simply could not do their job if they were barred from using mentalistic locutions about animals.[12]

Wittgenstein's second point, that since language separates humans from animals, and since language is a 'form of life' which both shapes and is shaped by one's *Umwelt*, we could not understand a lion if it spoke, seems implausible. I venture to suggest that our forms of life are not all that dissimilar: both the lion and I have interests in eating, sleeping, sex, avoiding encroachments on our environments, and so forth, about which we could doubtless make small talk. He might lose me if he went off on how one anticipates a gazelle's next turn, and I might lose him if I raised questions about animal rights (specifically gazelle's rights), but that sort of thing often happens when I'm talking to fundamentalist ministers or accountants, whose forms of life and language games are also incomprehensible to me. Lloyd Morgan once asserted that about the only human beings he could be sure of understanding were other upper-class educated Englishmen, and that he despaired of comprehending the minds of primitive men.[13] I doubt that most of us could understand the mind of an SS man, even if he spoke English; a lion would be *much* easier.

As to Wittgenstein's claims that an animal can't hope or simulate pain, these are truly perplexing. What else can one say of a dog when it sits at attention while you are eating but that it is hoping you will give it a scrap? As to simulating pain, any pet-owner and any veterinarian can relate cases in which animals simulated pain in order to get attention, avoid punishment, and so on, especially if they have been fussed over in the past when they had an injury.

But let us return to our main discussion of pain. We were examining the claim that since animals lack concepts which would allow them to anticipate and remember, and since a good part of pain is anticipating and remembering, animal pain is momentary and insignificant. We have just seen that there is little reason to deny concepts to animals; that standard philosophical gymnastics, however tortured, simply cannot touch the plain, commonsensical fact, embodied in all cultures of all ages, that animals do anticipate and remember, and that that is how they learn and fear. And if this entails having concepts, as surely it does, then animals have concepts, which should come as no surprise if they are to deal with the world. They certainly behave *as if* they had concepts, and the best explanation for this behaviour is that they do have them, especially since the aforementioned arguments totally fail to show that this is impossible. In general, the most powerful reason for believing in animal mental states is that they constitute the best way of explaining what animals do, how they behave, and how they survive, both philosophically *and* scientifically. The fact that we cannot experience these states directly is of little consequence, of as little consequence as the fact that we cannot directly experience the particles of microphysics or the past is to the explanatory value of postulating particles and a past.

Pain and the intellectual limitations of animals

In terms of countering the pernicious moral power of the claim that animals can't anticipate and remember pain and that therefore their pain is insignificant, the most relevant point has little to do with the presence or absence of concepts. It comes rather from the following insight: that if animals are indeed, as the above argument suggests, inexorably locked into what is happening in the here and now, we are all the more obliged to try to relieve their suffering, since they themselves cannot look forward to or anticipate its cessation, or even remember, however dimly, its absence. If they are in pain, their whole universe is pain; there is no horizon; they *are* their pain. So, if this argument is indeed correct, then animal pain is terrible to contemplate, for the dark universe of animals logically cannot tolerate any glimmer of hope within its borders.

In less dramatic and more philosophical terms, Spinoza pointed out that understanding the cause of an unpleasant sensation diminishes its severity, and that, by the same token, not understanding its cause can increase its severity.[14] Common sense readily supports this conjecture—indeed, this is something we have all experienced with lumps, bumps, headaches, and, most famously, suspected heart attacks which turn out to be gas pains.

Spinoza's conjecture is thus borne out by common experience and by more formal research. But this would be reason to believe that animals, especially laboratory animals, suffer *more* severely than humans, since they have no grasp of the cause of their pain, and thus, even if they *can* anticipate, have no ability to anticipate the cessation of pain experiences outside their normal experience. At least one major pain physiologist, Professor Kitchell, one of the co-editors of the aforementioned pain symposium volume, supports this conjecture. According to Kitchell, following a suggestion of Melzack, response to pain is divided into a sensory-discriminative dimension and a motivational-affective dimension. The former is concerned with locating and understanding the source of pain, its intensity, and the danger with which it is correlated; the latter with escaping from the painful stimulus. Kitchell speculates that since animals are more limited than humans in the first dimension, since they lack human intellectual abilities, it is plausible to think that the second dimension is correlatively stronger, as a compensatory mechanism. In short, since animals cannot deal intellectually with danger and injury as we do, their motivation to flee must be correlatively stronger than ours—in a word, they probably hurt more.[15]

At the risk of provoking an avalanche of indignation at my 'anthropomorphism', I would like to suggest the following thought experiment. Consider an animal, say a dog, which has spent its life as a pet, experiencing nothing worse than an occasional reprimand or slap on the rump. Let us suppose that the animal is turned in to the pound and ends up as a research subject in a learned helplessness experiment, as is permitted in the United States, though not in Britain, wherein he is subjected to *inescapable*, painful electrical shock to see if he develops the helplessness syndrome which is alleged to model human depression. If one is at all willing to admit any consciousness in animals, one must surely affirm that this animal feels pain and fear. Furthermore, and disanalogously to the case of a human being, the animal has no notion whatever of why and how the pain comes. Any human put into such a situation would at least be able to formulate hypotheses—for example, 'I am being tortured for political reasons,' or 'I am being used for research.' But the animal cannot even begin to plug his pain into any of his categories of understanding. His cognitive tools simply don't fit. As a result, there is no possibility of cognitive moderation of the pain and fear. Whereas a person can say, 'Perhaps I can reason with my captors,' or 'Perhaps this is of short duration,' the animal cannot begin to get a purchase on any aspect of the experience. So the pain which is experienced must surely be deepened and rendered more extreme by its total incomprehensibility. As we all know, the unknown is by its very nature terrifying. And given an animal's intellectual limitations, most of what is suffered at our hands in research contexts must be totally incomprehensible. This insight is probably the basis for people's objections to the use of what were formerly pet animals as research animals.

Anthropomorphism and animal pain

In any case, let us turn our attention to what I consider to be the most ironic and perverse argument of all in attempts to justify lack of concern with animal pain in the common sense of science. It is often claimed that worry about animal pain is misplaced anthropomorphism, for in circumstances in which humans would be screaming and writhing, many animals show very few signs of extreme pain. Aside from the point made earlier that stoic behaviour doubtless confers a selective advantage on animals, we can make a more subtle point. It is not the people who impute pain to animals who are anthropomorphic; they have good evolutionary, physiological, and behavioural reasons to do so. It is, rather, those who *deny* pain to animals on the grounds that their behaviour is unlike ours who are anthropomorphic; for who else besides someone guilty of the grossest anthropomorphism would expect expressions of animal feelings to be precisely like ours, would expect a cow in pain, for example, to run about beating its breast and bellowing 'Oy Vay'? (Even Gentiles don't do this.) Animals do show unique pain behaviour. It just doesn't happen to be human pain behaviour. But then why should it be? We would expect its behaviour to be appropriate to its *telos*, the unique, evolutionarily determined, genetically encoded, environmentally shaped set of needs and interests which characterize the animal in question—the 'pigness' of the pig, the 'dogness' of the dog, and so on.[16] People who deal with horses a great deal and who follow the dictates of experience and common sense are aware that in some cases mere tightening of the palpebral (eyelid) muscles eloquently bespeaks great agony, but obviously not to the person who is expecting the full range of human pain behaviour from the horse.

In one extraordinary case, a veterinary student working with the department of wildlife in a western state was shocked to learn that some members of the department were routinely doing Caesarean sections on moose which had been rendered immobile by injection of succinylcholine chloride, a curare-like drug which paralyses all muscles by blocking neuro-transmission across the neuro-muscular junction, but has no anaesthetic or analgesic properties. On the contrary, reports from humans on whom it has been used indicate that it heightens pain response, given the extraordinary panic which accompanies total paralysis (including respiratory paralysis), even when one understands *exactly* what is happening and why. In the case of an animal, one can only begin to imagine the utter black terror experienced. In any event, the student's objections fell upon deaf ears. 'Those animals ain't hurtin',' he was told. 'If they was, they'd be hollerin' '—no mean feat when totally paralysed.

This is an extreme case, but not all that extreme. Succinylcholine has been routinely used for castration of horses in the American West, and until recently, was the drug of choice for 'chemical restraint' in procedures like stereotaxic (cranial) experimental surgery, which require that the animal be conscious—whatever that means to a behaviourist—yet immobile. As mentioned earlier, many researchers and veterinarians do not distinguish between anaesthesia and chemical restraint. Amazingly enough, the USDA has used succinylcholine to 'euthanize' thousands of pigs afflicted with cholera, and has then frantically back-pedalled and funded research to prove that this was really 'humane'.

Why do such arguments thrive?

Many of the philosophical moves we have been discussing in this chapter do not seem very plausible. The arguments, as we have noted before, have been perpetuated as a result of a variety of extra-logical influences on science of the sort which scientists refuse to admit have any relevance to what they do. One has been the fact that for many years society didn't particularly care about how animals were treated, especially animals in science, whose use and abuse, it was widely believed, were absolutely essential to human welfare, and unavoidable in the quest for knowledge. For this reason, people often did not wish to know 'what horrible things go on in laboratories', just the way they do not wish to know what happens to dogs given to 'shelters' for 'adoption', or to cattle at slaughterhouses. Humane society workers report that though relatively few pound dogs are adopted, and though owners who relinquish their animals are told this, owners persist in believing that *their* dog will be adopted, and even specify the type of home they want it placed in! Another factor which helped perpetuate these arguments against taking animal pain seriously was the undeniable convenience to scientists of treating animals as biochemical mechanisms. Still another point harks back to our earlier discussion of compartmentalization. Researchers engaged in invasive work on animals are, for the most part, as decent as the rest of us, and thus needed a bail-out mechanism to help them survive having to do to animals what, if done in other contexts—say, at home, to the family dog—would be construed as monstrous sadism. To genuinely believe that the animals were really hurting might force them to look into an abyss from which there is no easy return.

Logical behaviourism and the ascription of pain to animals

At this point in our argument it seems reasonable to assert that philosophically the simple denial of thought, awareness, and pain to animals on the grounds that we can't experience them is not cogent. In the first place, if we are positivistic enough to claim that we cannot know anything which we do not experience directly, then we can make no claims about anyone's mental states but our own. Therefore, we cannot claim to know that other humans have minds and pains and other private experiences either. Of course, behaviourists have been perfectly willing to deny subjective experiences to humans, but common sense won't allow it. Much of our ordinary discourse depends on a tacit assumption of mental states in others, as the Scottish Common-Sense philosophers were given to pointing out. So, at best, behaviourists set aside the study of consciousness in the case of humans, while declaring it non-existent by fiat in the case of animals. In fact, as we have mentioned, much of their work was incoherent without the presupposition of mental states. For example, how does one ultimately make sense of positive and negative reinforcement without postulating pleasant and unpleasant subjective experiences?

The so-called logical behaviourists, most notably Gilbert Ryle in his *Concept of Mind*, but also the later Wittgenstein (according to some interpretations) and some of his followers, were more subtle on this point. While not denying the existence of subjective experience, they seemed to argue—Ryle is never clear on this point—that mentalistic attributions can be made and mental predicates applied to persons without any reference to subjective states. On this view, we learn words like 'angry', 'in pain', 'day dreaming', and so on by seeing what sorts of behaviour in what sorts of contexts these words are applied to. If subjective states do exist, they are logically irrelevant to the meaning and attribution of mental words. Thus we need not say that subjective states don't exist, only that they don't concern us.

This view, contrary to that of Wittgenstein, would not forbid the application of mental locutions to animals, as long as such attribution involved the proper behaviour on the part of the animal. It further has the virtue of (at least theoretically) taking seriously the ordinary-language application of mental terms to animals. (I say 'theoretically', because often the language examined by these philosophers was hardly ordinary; it was the language of the Oxford don in the street, not that of the 'plain man'.) In any case, I do not believe that logical behaviourism dispenses with subjective experience as neatly as it wishes to. While certainly making an invaluable point in showing, against extreme sense-data-type empiricism, that the meanings of words, especially mental words, cannot reside solely 'in the head', it does not totally free us from the view that what goes on in the head is somehow relevant to meaning. To wit, it is true that my learning to attribute pain to others is based on observing their behaviour, be it howling, teeth-gnashing, hopping around, or gritting their teeth. But while this may be necessary to explicating the meaning of my assertion that someone is in pain, it is surely not sufficient. Implicit, but indispensable, in my attribution, is some notion that the subjective state which the person is in is unpleasant, or, more accurately, is experienced as unpleasant. And the way in which this, in turn, is understood is by reference to one's own experience, but with a built-in 'fudge factor' allowing for individual differences. In other words, part of understanding or learning to assert 'He is in pain' is indeed observing that he is yelling and hopping around. But part of it also is affirming tacitly that he is feeling something which I cannot feel, but which is reasonably like, *mutatis mutandis*, what I experience in the face of something unpleasant. This is not to say that one is in a special privileged state of knowing totally incorrigibly about one's own pain; as we saw in discussing Thorndike, a person's own assessment of his pain can be corrected by others—doctors, say. I know my own subjective state better than you do, but you may understand my behavioural and physical symptoms better than I do, and since both are involved in predicating pain, neither aspect alone is incorrigible and conclusive.

Is the case of animals radically different from that of other people? Our first impulse is to say yes, quite emphatically. Other persons are, after all, more like us. But is that true, and if true, is it relevant? First, is it important that animals belong to different species from us? Perhaps not in those cases where they behave very like we would in similar contexts. Much of the behavioural evidence which licenses us to attribute experienced pain to other humans is present in animals. Animals cry out when injured, are tender at the point of injury, cringe before blows, avoid electrical shock and heat, and so on. Anger, too, is very similar in many animals, as are various other psychological states.

Furthermore, though this may sound like heresy, certain mental states of certain animals are easier for us to understand than certain mental states of other humans. To take a shocking example: I believe that I can better understand experientially what a male animal is feeling when he is copulating with a female than I can understand the pleasure of a human female when she is engaged in sexual intercourse. (I am abstracting here of course from emotion-laden sexual experiences and looking at only purely physical encounters.) After all, I have something like the animal's biological equipment; I have nothing much like the woman's. Language helps not at all—I cannot comprehend her metaphors. Similarly, and I am quite serious about this, I can better understand the pain and surprise of a dog who yelps indignantly when I step on his foot than the alleged pleasure that a marathon-runner feels when he 'breaks the pain barrier'—whatever that means—or the satisfaction a masochist feels when he is whipped or humiliated.

Variability of pain experience in humans and animals

Unquestionably, animal pain responses are not always the same as ours—compare a horse's wincing rather than crying out when hurt—but, for that matter, human pain responses vary across different cultures and subcultures and among individuals. While the pain-detection threshold—that is, the minimum stimulus which, when applied to a person or an animal, gives rise to the perception of pain—is relatively uniform across human beings,[17] the pain-tolerance threshold—that is, the maximum pain that a subject can endure before the stimulus is perceived as unbearable—varies widely from individual to individual, depending on a number of factors.[18] Indeed, it varies widely even in the same individual, depending on the context of the pain, the person's mood, anxiety level, and so on. Athletes are notoriously good at ignoring great pain. What else would one expect when locker-rooms are festooned with giant signs declaring 'No pain, no gain' and quoting from football great Johnny Unitas that 'If pain bothers you, you have left your game in the locker-room.'

Ordinary observation, as well as classical research, supports the view that socio-psychological and and socio-cultural factors loom large in shaping the experience of pain in humans. Compare, for example, the way in which middle-class New York Jewish children are encouraged to express pain and are reinforced when they do so when they fall down as toddlers, in contrast to children raised on ranches in the West who are taught to ignore the pain and get on with their business. Research has shown that northern Europeans are less susceptible to painful stimuli than southern Europeans—that is, have a higher pain-tolerance threshold—and this is explained by cultural, rather than biological, differences.[19] Stoicism, control, and keeping emotion in check are valued among northern Europeans; volubility, effusiveness, and emotional display are accepted and encouraged among southern Europeans. Other studies show similar results. When told that Jews are less able to tolerate pain than Christians, Jewish subjects were found to increase their pain tolerance significantly; but when told that Jews can take more pain than Christians, they showed satisfaction with the status quo and did not increase their tolerance.[20] Other studies indicate that Jews are much more concerned about future pain than Italians, whose concern tends to be with immediate relief and who forget about the pain after it has been relieved, while Jews continue to complain even after the pain is alleviated. This is explained by the notion that Jews tend to be culturally future-oriented, as well as pessimistic.[21] This dovetails with the well-known fact that fear and anxiety increase pain, and indeed may cause it.

In his now classic work,[22] H.K. Beecher showed that wounded front line soldiers required less analgesia than non-military surgical patients, even though the injuries they sustained were far more massive. He explained this by reference to the fact that the soldiers saw real benefit in the wound—that is, no longer having their lives at risk on the battlefield—whereas the surgical patients focused more on their pain. Facts like these have sometimes been used to claim that animal pain is qualitatively different from human pain, because it is unaffected by psychological and cultural factors. However, clinical veterinarians and common experience dispute this. A wounded animal seems to suffer less pain in the presence of its owner, less when treated at home or in familiar surroundings and when reassured, less when rapport is established with the clinician, less when stroked, and so on. 'Cultural' differences appear here as well. Dogs accustomed to rough treatment and minor injury (like athletes) show far less pain behaviour than highly pampered lap-dogs, even members of the same breed, poodles providing a notable example. In addition, various breeds of dogs or other animals selected for certain purposes appear to tolerate pain better than others selected for other purposes. Thus pit-bulls, bred for fighting, seem to tolerate pain better than lap-dogs. Furthermore, it is widely known that dogs learn to utilize pain behaviour to manipulate people; Fox and others have reported feigned injury by animals which is clearly designed to elicit better treatment.[23]

As mentioned earlier, I myself experienced this vividly in a Great Dane I had when living in New York. After a day of romping in the park, she developed a severe limp, which my wife and I proceeded to overtreat with massage, heating pad, treats, and a great deal of attention. Predictably, I now realize, she continued to limp until, ten days later, we took her to a major veterinary hospital, where she was diagnosed as having a degenerative cartilage disease which would inevitably cripple her permanently and which required immediate surgical intervention. I decided to take her to the park for what could well be her last run with her friends. We slowly made our way down the hill to Riverside Park, she limping bravely and pathetically at my side. We finally arrived at the park. The dog spotted her playmates and began to wag her tail. Choking down my emotion and steeling myself to the inevitable pathetic scene, I slipped off her leash and—to my utter amazement—watched her take off like the proverbial bat out of hell to join her peer group. That was the end of the limp forever, with one notable exception. Two years later we were in the Canadian North Woods, and, spotting a rabbit, the dog took off, heedless of my demands that she come and heel. Twenty minutes later, she bounded back to confront her enraged master. As I roared 'Bad dog!', she suddenly began to limp pathetically, desisting as soon as I convulsed with laughter. Melzack and Scott have provided another interesting example of 'social' influence on animal pain. They showed that puppies raised in total social isolation did not react to painful stimuli like burns and pinpricks as if they felt pain.[24]

All this raises an important, often neglected point about animal pain. As with humans, one should not forget about individual differences; what hurts one person significantly may seem irrelevant to another, even one closely related. (This is equally true, of course, of what gives pleasure to humans or animals.) Hereditary and environmental factors are both operative here; even 'macho' dogs like pit-bulls can be rendered very sensitive to pain if pampered. Thus, if one wishes to be able to recognize pain in animals, or humans for that matter, one can try to develop general criteria which apply to the type of creature in question. But one must also know the individual animal (or human). One needs to know some people quite well to become aware that extreme quiet, rather than voluble outcry, is a sign of pain in that person. (In the same way, excessive cheerfulness can be a sign of depression in certain people.) All of which leads to an inevitable result: namely, that to be morally responsive to pain in animals, one must ideally know animals in their individuality. The practical problems involved in doing this in research colonies or feedlots are obviously monumental. In fact, I believe that these practical barriers are a greater obstacle to recognizing pain in animals than the epistemological ones posited by the common sense of science. Farmers or caretakers (uninfected with scientific ideology) who deal with small group of animals have no problem identifying pain in their animals. But one can conjecture that the large numbers of animals dealt with today in science, animals identified by number not name, have contributed to the hold of the ideology of science. The animals become as indistinguishable as grains of sand, which in turn weakens both our sensitivity to signs of pain and our moral response. This is what happened in the concentration camps; shaved, starved prisoners, identically dressed, lacked individuality in the eyes of their captors and were thus more easily perceived as part of an endless flow of clones, about which one need not feel moral concern.

The scientific incoherence of denying pain in animals

Let us refocus our discussion. Our concern has been with demonstrating that there is no good reason, philosophical or scientific, to deny pain in animals. The indubitable fact of such denial has to do with the powerful ideology we have been discussing, which saturates scientists with a questionable philosophy while disavowing that it is doing any such thing, and makes a value out of denying a legitimate place in science for value questions. Thus scientists can officially repudiate the legitimacy of talking about animal pain, at the same time as they presuppose it in their research.

We have seen that this ideology is powerful enough to eclipse the value of consistency in science, and to submerge coherence as well. For, as Darwinians recognized, it is arbitrary and incoherent, given the theories and information current in science, to rule out mentation for animals, particularly such a basic, well-observed mental state as pain.

We have already mentioned the tendency of scientists to acknowledge pain in animals only in terms of the machinery, or plumbing, of pain. One can well believe that only by thinking of animal pain in terms of Cartesian, mechanical processes devoid of an experiential, morally relevant dimension, could scientists have done the experimental work which has created the sophisticated neurophysiology we have today. But given that science, the neurophysiological analogies that have been discovered between humans and animals, certainly at least the vertebrates, are powerful arguments against the Cartesianism which made it possible. In a dialectical irony which would surely have pleased Hegel, Cartesianism has been its own undoing, by demonstrating more and more identical neurophysiological mechanisms in humans and animals, mechanisms which make it highly implausible that animals are merely machines if we are not.[25]

Pain and pleasure centres, like those found in humans, have been reported in the brains of birds, mammals, and fish; and the neural mechanisms responsible for pain behaviour are remarkably similar in all vertebrates. Anaesthetics and analgesics control what appears to be pain in all vertebrates and some invertebrates; and, perhaps most dramatically, the biological feedback mechanisms for controlling pain seem to be remarkably similar in all vertebrates, involving serotonin, endorphins and enkephalins, and substance P. (Endorphins have been found even in earthworms.) The very existence of endogenous opiates in animals is powerful evidence that they feel pain. Animals would hardly have neurochemicals and pain-inhibiting systems identical to ours and would hardly show the same diminution of pain signs as we do if their experiential pain was not being controlled by these mechanisms in the same way that ours is. In certain shock experiments, large doses of naloxone have been given to traumatized animals, reversing the effect of endogenous opiates, and it has been shown that animals so treated die as a direct result of uncontrolled pain.[26] In 1987, it was shown that bradykinin antagonists control pain in both humans and animals.

Denial of pain consciousness in animals is incompatible not only with neurophysiology, but with what can be extrapolated from evolutionary theory as well. There is reason to believe that evolution preserves and perpetuates successful biological systems. Given that the mechanisms of pain in vertebrates are the same, it strains credibility to suggest that the experience of pain suddenly emerges at the level of humans. Granted, it is growing increasingly popular, following theorists like Gould and Lewontin, to assume the existence of quantum leaps in evolution, rather than assume that all evolution proceeds incrementally by minute changes. But surely such a hypothesis is most applicable where there is evidence of a morphological trait which seems to suddenly appear in the fossil record. With regard to mental traits, this hypothesis might conceivably apply to the appearance of language in humans, if Chomsky and others are correct in their argument that human language differs in kind, as well as degree, from communication systems in other species. But in other areas of mentation—most areas apart from the most sophisticated intellectual abilities—and surely with regard to basic mental survival equipment like that connected with pain, such a hypothesis is both *ad hoc* and implausible. Human pain machinery is virtually the same as that in animals, and we know from experience with humans that the ability to *feel* pain is essential to survival; that people with a congenital or acquired inability to feel pain or with afflictions such as Hansen's disease (leprosy), which affects the ability to feel pain, are unlikely to do well or even survive without extraordinary, heroic attention. The same is true of animals, of course—witness the recent case of Taub's deafferented monkeys (monkeys in which the sensory nerves serving the limbs have been severed) who mutilated themselves horribly in the absence of the ability to feel. *Feeling* pain and the motivational influence of feeling it are essential to the survival of the system, and to suggest that the system is purely mechanical in animals but not in man is therefore highly implausible. If pain had worked well as a purely mechanical system in animals without a subjective dimension, why would it suddenly appear in man *with* such a dimension? (Unless, of course, one invokes some such theological notion as original sin and pain as divine punishment—hardly a legitimate scientific move!) And obviously, similar argument would hold for discomfort associated with hunger, thirst, and other so-called drives, as well as with pleasures such as that of sexual congress.

So not only does much scientific activity presuppose animal pain, as we have seen *vis-à-vis* pain research and psychological research, it fits better with neurophysiology and evolutionary theory to believe that animals have mental experiences than to deny it. Outside positivistic-behaviouristic ideology, there seems little reason to deny pain (or fear, anxiety, boredom—in short, all rudimentary forms of mentation) to animals on either factual or conceptual grounds. (Indeed, research indicates that all vertebrates have receptor sites for benzodiazepine, which, in turn, suggests that the physiological basis of anxiety exists in all vertebrates.)[27] One may cavil at attributing higher forms of reason to animals, as Lloyd Morgan did, but that is ultimately a debatable, and in large part empirically decidable, question.

The alleged unobservability of mental states

The one lingering doubt which positivism leaves us with concerns the ultimate unobservability of mental states in animals. After all, we cannot experience them, even in principle. Perhaps there is something fundamentally wrong with admitting such unobservable entities to scientific discourse, for would we not be opening a Pandora's box containing such undesirable notions as souls, demons, angels, entelechies, life forces, absolute space, and the rest? Would we not be giving up the hard-won ground by which we demarcate science from other forms of knowing, and opening ourselves to a dissolution of the line between science and metaphysics, science and speculation? Surely that consideration must far outweigh any benefit of admitting animal mentation into legitimate scientific discourse.

Since this is the key (official) reason behind the common sense of science's refusal to talk about animal mentation, it is worth examining in some detail. The first assumption behind this view is obviously that science is, or can be, and surely ought to be, totally empirical—in other words, that science ought to make no assumptions, postulate no entities, and countenance no terms which cannot be cashed empirically. This is, indeed, a mainstay of classical, hard-line positivism.

Unfortunately, things are not that simple. As progress in both science and philosophy has shown, and indeed as much of our argument so far is designed to illustrate, science is incapable of operating under such restrictions. In the first place, science makes certain assumptions which cannot be verified empirically. To take a simple, but essential, example, it assumes that there exists a public, intersubjective world of physical objects to which all of us have more or less equal access. When I am doing biology or physiology or physics, I am talking about external objects which are the furniture of a public world, not about mental states of my own (phenomenalistic positivism notwithstanding). But such a claim is not an empirical one; empirically, the data I gather are equally compatible with solipsistic phenomenalism. Correlatively, science also assumes the existence of other subjects whose access to these objects is more or less the same, who can check and correct each other's experiences, who must in principle be there to be able to duplicate my experiments and observations, who hold beliefs, understand meanings, and manipulate symbols to convey information. None of these assumptions can be verified empirically; yet, without them, science is rendered pointless, as well as impossible. In the same vein, science must assume that there is a past, and that we can reconstruct it and make correct and incorrect judgements about it, despite our inability to experience it (if we could experience it, it would not be past). Russell pointed out that we have no scientific way of knowing that God did not create the whole universe *in toto* a few seconds ago, complete with fossils and us with all our memories.[28]

All the above are metaphysical or philosophical assumptions which cannot be verified empirically. In a real sense, the possibility of empirical verification rests upon these assumptions. In the same manner, as we discussed earlier, the scientific revolution which focused enquiry upon the quantitative, mathematically expressible aspects of phenomena as being what are truly real, and relegated the qualitative aspects of things to illusion, was a change in metaphysical perspective intertwined with a change in value. The assumption that 'science ought to study only what is describable in mathematical terms' is surely not empirically verifiable; it is, in fact, constitutive of what counts as legitimate empirical study. For Aristotle, with his commitment to common sense and to a metaphysical perspective which said that ordinary objects were ultimately real, mathematical science was the study of the most general features of all things, and was hence trivial. Further, Aristotle argued against the atomists and other reductionists of his period by forcefully asserting what is clearly a metaphysical and valuational stance—namely, that there can be no one master science of everything, but rather, that science must take account of the basic laws of each *kind*. This is not an empirical position: nothing, not even the extraordinary predictive and manipulative success of modern physico-chemistry, would have convinced him that concern with mechanical and efficient causes, however successful, renders superfluous the search for formal and final causes. His unwillingness to substitute 'how' questions for 'why' questions reflects a metaphysical stance, not an empirical one, a certain a priori picture of what counts as reality which determines and shapes how that reality should be studied.

So far we have been arguing that science cannot be wholly empirical because it must inevitably make metaphysical assumptions about what is real and value assumptions about what is worthy of study and what counts as a legitimate fact, a satisfactory explanation, or indeed, as what science *ought* to be doing. In Aristotle's view, mathematical facts were not to be sought; mathematical explanations were not adequate to physical phenomena; and unification of all the sciences was not a value. For Galileo, it was quite the reverse. For positivism and behaviourism, facts of consciousness don't count; they are excluded in advance by a certain metaphysical perspective and set of values.

One of the mainstays of the version of positivism informing the common sense of science is that theoretical terms must be built up out of observational terms, that facts are, as it were, given to us, and that we build theories out of them, building-block style.[29] Facts, in good empiricist fashion, are *there*, whereas theories are constructs of facts. This sort of foundationalism obviously presupposes the logical priority of facts—in other words, that one can catalogue facts in a totally non-theoretical or atheoretical fashion (what one may call the dartboard theory). It is no surprise that for many such positivists, facts were sense-data, direct observations, qualia, or whatever. In short, many positivists were phenomenalists. (I must stress the oddity of phenomenalists being behaviourists, for surely sense-data are private contents of consciousness.) It is also easy to see how this led positivism up a blind alley, for as a basis for science, which is after all generally viewed as the intersubjective study of real objects 'out there', phenomenalism, with the inherently solipsistic, logically private nature of its objects, will not suffice. That aside, a number of problems are of great concern here, points raised by Kant's and other philosophers' critiques of Hume's phenomenalistic proto-positivism.

First, if every legitimate scientific judgement and every legitimate theoretical term must be cashed out in factual or observational terms, how can that injunction itself be cashed out? Surely it is not itself an empirical judgement. After all, the positivists have not examined every possible scientific judgement and theoretical notion, and shown that they are observational at base. Furthermore, even if the set containing every possible scientific judgement and term were not infinite (which it surely is), what empirical test could be used to exclude any? Suppose I affirm, as Driesch did, that judgements about entelechies are legitimate scientific judgements. The positivist wishes to exclude them because they are not the sorts of things which we can observe directly. Perhaps I might admit that, but claim at the same time that such judgements do not lack the property of scientific legitimacy. How do we move from 'non-observable' to 'unscientific'? Certainly not by observation—only, in fact, by buying into the positivists' *values*, by accepting their 'ought'. For the statement about scientific legitimacy is a value judgement, a statement about what ought to count in science, a statement growing out of a particular metaphysics and epistemology, not out of simple data-gathering. The positivists tell us what ought to count as data—namely, what we can observe—but they can hardly meet a challenge to that exhortation by affirming that our challenge does not meet their condition!

It is for this sort of reason that, when I teach classical twentieth-century philosophy of science, I point out that such philosophers are not concerned that their rules do not fit the actual practice of most scientists, for, in response to such a point, they would surely reply that their concern is not with what science *is*, but with what it *ought* to be.

If phenomenalistic positivism is correct, science is not correct, for science believes in public objects, and, for positivism, statements concerning such objects cannot legitimately be made. Furthermore, if phenomenalistic positivism is correct, it is hard to see why judgements about mental states are illegitimate, since mental states (sense-data) are the epistemological basis of positivism. And since to even begin to have science, we must accept other people's mental states (perceptions) without being able to experience and observe them, why not those of animals?

Mental states as a perceptual category

So there is no good 'scientific' reason for acquiescing to positivism's demand that only observables be permitted in science; first, because that demand cannot itself be observationally proved, and second, because it would exclude all sorts of basic things like other people and intersubjective physical objects in which science has much stake. Furthermore, it has become increasingly clear since Kant that there is no good reason for believing that facts can be gathered or even observations made independently of a theoretical base. Kant showed that sensory information must be 'boxed' before it becomes an object of experience, that even the notions of 'object', and 'event' are brought to sensation, rather than emerging from it. Indeed, examples demonstrating the role of theory in the broadest sense in perception are endless. To see a fracture or a lesion on a radiograph requires an enormous amount of theoretical equipment and a great deal of training, for which the radiologist is very well paid; though he gets the same sensations on his retina as you or I do, he doesn't see the same thing as we do. This is equally true of the woodland tracker, who spots a trail or some sign of an animal's passage; the artist, who sees dozens of colours in a person's face, whereas most of us see only 'flesh colour'; the horse *aficionado* who sees a thoroughbred, while we see only a horse. And what we see in the standard ambivalent figures like the one which can be seen as a vase or as two faces or the duck-rabbit or the young-old woman depends on what we are thinking about, expecting, hypothesizing, and the like.

Indeed, returning to our main concern, what we perceive or observe in science or consider worthy of calling a fact must be in large measure determined by our metaphysical commitments and their associated values. Aristotle saw the world as an array of facts of function and teleology. Galileo saw a mathematical machine. In so far as common sense has a metaphysics and an epistemology, it surely maintains that we perceive the mental states of others. Contrary to the stories that many philosophers have told in the twentieth century, and which we discussed earlier, I do not believe that common sense uses mental terms like 'happy', 'afraid', 'bored', and so on only to refer to overt behaviour in appropriate contexts. Nor do I believe that common sense simply *infers* mental states by analogy from overt behaviour. I think, rather, that common sense *perceives* mental states in others in exactly the way that it perceives physical states or objects. (I don't infer the externality of objects of experience; I simply perceive them as external until I am given reason to doubt their externality on a given occasion—for example, if others don't see them.) In other words, part of the theoretical filter by means of which common sense organizes experience *is* the notion of mental state. One can indeed be wrong when it comes to judging mental states in others, but one can equally be wrong in judging physical states in others, just as one can be wrong, as Thorndike brilliantly argued, in judging these in oneself.

On this view, perceiving in terms of mentation is one of the categories by which we commonsensically process reality, and mentation is a fundamental plank upon which our common-sense metaphysics is built. And the reason why this notion is so geographically and historically pervasive is because it works so well. Nothing can disconfirm our attribution of mentation to other humans *in general*, because mentation is a (if not *the*) fundamental cognitive category by which we map other humans. (In certain cases, of course, special circumstances can dispel this presumption of mentation, just as it can dispel the presumption of the externality of the objects of perception, as when we learn that the person is sleep-walking or under the control of a hypnotist or some drug.) One need only ask oneself how one could communicate with and understand the behaviour and speech of other humans if tacit or explicit reference to mental states were not allowed.

If one suddenly discovered that the grieving individual with whom one is commiserating is in fact a robot of naugahyde and aluminum (cf. the sort in fact contrived by Walt Disney for the 1964 World's Fair), one would never attribute sadness to it, however pathetic its behaviour. And this is precisely because, rightly or wrongly, our attribution of mentation does not—at least at the moment—extend to machines, not because their behaviour can't be perfectly appropriate, but because we are predisposed to assume that they have no subjective experiences. Contrary to Ryle and other logical behaviourists, belief in and consequent attribution of subjective experience to other humans is not some insignificant limpet inconsequentially and by happenstance attached to our mental talk; it is a presuppositional and essential part of it. (Likewise, Descartes's talking of animals as machines was, for him, logically equivalent to denying that they had any functioning *res cogitans*, or mind.)

When we use words attributing passion, rage, sadness, joy, depression, and the like to other people, we surely do not do so in the absence of behaviour relevant to and expressive of these mental states. On the other hand, we are not just talking about the behaviour; we are unavoidably referring to what the behaviour is directly and essentially tied to—namely, a feeling which, while perhaps somewhat unlike mine, serves the same function in the other person's life as the feeling in question does in mine.

This is not to suggest that we can't be wrong about the presence of the feeling in another; hence our development of ample categories for feigned emotion, such as 'crocodile tears', notions which mean primarily that the requisite feeling is absent in the person, whatever their behaviour. Nor I am trying to deny the (usually) very close connection between appropriate behaviour and the feeling it reflects. In the eighteenth and early nineteenth centuries the Scottish Common Sense philosophers called attention to the fact that actors and actresses who begin by feigning an emotion may well end up feeling it. My point is rather that, given the metaphysics of common sense and ordinary language, we are prima facie committed to the assumption that human beings have subjective states which are portrayed by their behaviour. This presumption is neither infallible nor immune to being over-ridden, but this is equally true of the metaphysical presumption that what we see betokens an external object. In either case we may encounter evidence more powerful than the presumption. Other people don't see it; I can see it but not touch it; I haven't slept for three days: all these may cause me to believe that the oasis I see is not real, overriding my presumption of its externality. In the same vein, a man may overhear his girl-friend practising the sounds and movements of passion in front of a mirror, thereby overriding his assumption that she feels it (but not his presumption that she has subjective states of some sort—in fact, if he did catch her doing that, he would be forced to believe all the more in her having very private experiences and would wonder far more strongly about their content).

Morality and the perception of mental states

There are doubtless a number of major reasons for our ubiquitous presumption of subjective experiences in others. Most obvious is the fact that without this presumption, we could not as readily predict or understand the actions of others. Second, in so far as we are taught moral concern for others, such concern is cogent only on the presumption that others have subjective experiences, that we can more or less know them, that their subjective states matter to them more or less as mine matter to me, and that my actions have major effects on what matters to them and on what they subjectively experience. If we genuinely didn't believe that others felt pain, pleasure, fear, joy, and so on, there would be little point to moral locutions or moral exhortations. Morality presupposes that the objects of our moral concern have feelings. And, what is logically equivalent, if there is no presumption of the possibility of feeling in an entity, there is little reason to speak of it 'in the moral tone of voice'.

Of course, the presumption of feeling is only a necessary condition for moral concern, not a sufficient one. One must also believe that the feelings of others warrant our attention. For most people, the mere realization that others experience negative feelings in the same way that they themselves do is enough to generate a stance of moral concern; whether it be out of rational self-interest (Hobbes), innate sympathy (Hume), or a sense of a rational requirement of universalizability (Kant) is irrelevant. To counter this attitude, one must invoke special reasons: that others are not entitled to moral concern because they have sinned in this or in previous lives, because they killed Christ, because their feelings are so primitive that we need not worry about them, or because they cannot feel at all—that is, have no inner experiences. (The latter two claims are often invoked by privileged classes to explain their lack of moral concern for the unprivileged.)

In any case, the attribution of mental states, especially those associated with pleasure and pain, joy and misery, is connected irrevocably with the possibility of morality. For this reason, only a science with blinders to the moral universe, and, most especially, only a *psychology* provided with such blinders could ever deny not only the legitimacy of talking about mental states, but their fundamental place in the world, which a genuine psychology must seek to explain. For some fifty years, millions of college students have shared my experience as an undergraduate, of enrolling in an introductory psychology course wanting to learn about the mind—meaning consciousness, feeling, thinking, imagining, and so on—and being disappointed to discover that these were non-facts to psychology, and that if they were mentioned at all, it was in the same tone of voice as phlogiston and demons. Thus many bright people left psychology when they realized that their desire to understand the mind would not be gratified by running rats through mazes or even by memorizing word lists. By the same token, one can understand the lure of Freud, Jung, and the rest; whatever their scientific deficiencies, at least they seemed to be aware of the same facts as most of the rest of us. In any case, regardless of whether psychology denied the legitimacy of consciousness, people—even psychologists—went on using mentalistic locutions about one another, and continued to base their moral behaviour upon the presupposition that other people had subjective experiences and that what they experienced mattered to them.

Application of this theory to animals

But what of animals? Clearly, common sense and ordinary language have traditionally extended the presumption of mentation to animals. (In some cultures, this was explained by viewing animals as reincarnated humans.) Probably the major reason for doing so was that it works. By assuming that animals feel and have other subjective experiences, we can explain and predict their behaviour (and control it as well). Why beat a dog if it doesn't hurt him? Why does a lion hunt if it isn't hungry? Why does a dog drool and beg for scraps from the table if they don't taste good? Why does a cat in heat rub up against the furniture if it doesn't feel good? Why do animals scratch if they don't itch? Again, common sense continued to think in this way regardless of what scientific ideology dictated, and scientists continued to think in this way in their ordinary moments.

The moral reason for presuming consciousness and mental states in animals does not loom nearly so large. It is an interesting fact that although most cultures in most times and places have attributed mentation to animals, few have clearly set out moral rules for their treatment, and for many, animals do not enter the moral arena. For some philosophers, granting moral status to animals is highly problematic, which is why so many of them have been concerned, like Descartes, to prove that animals are really automata. For others, like Hume, who deem it absurd to deny the full range of mental experience to animals, and who are highly cognizant of the connection between morality and feeling, the question of moral treatment of animals nevertheless does not arise. For ordinary people, though their presumption of mentation in animals is strong, their application of moral notions to animals is minimal or non-existent. Only thus can one explain why for centuries animals were held morally and legally responsible for their actions, subject to trial, punishment, and death, yet at the same time had no legal protection whatever.[30]

Why is this the case? Why has common sense (and until recently the legal system as well[31]) studiously avoided coming to grips with our moral obligations to other creatures? For that matter, why has philosophy, which has notoriously concerned itself with all sorts of questions and which has explored all aspects of ethical theory, been virtually blind to questions concerning animals? (There are, of course, notable exceptions, such as the Pythagoreans; but here we must recall that their concern grew out of the doctrine of transmigration of human souls.) There is no certain answer to this question. Perhaps part of an answer lies in the influence of our theological traditions, most especially the Christian tradition, which has stressed that the proper study of man is man. More plausibly perhaps, a key part of the answer lies in a remark made to me by one of my veterinary students at the end of an ethics course in which I put great stress on moral questions pertaining to animals. 'If I take your teaching seriously', she said, 'no part of my life is untouched, and all parts are severely shaken. For if I ascribe moral status to animals, I must worry about the food I eat, the clothes I wear, the cosmetics I use, the drugs I take, the pets I keep, the horses I ride, the dogs I castrate and euthanize, and the research I do. The price of morality is too high—I'd rather ignore the issue.' Perhaps in a culture which has no choice but to exploit animals in order to survive, one cannot even begin to think about these questions, or even see them as moral choices rather than pragmatic necessities. Or perhaps because the use of animals for our purposes without consideration of their interests is so pervasive and our dependence upon it so great, it becomes invisible to us, in much the same way that exploitation of women and minorities was invisible for too long. Indeed, it is interesting that moral interest in these long-neglected areas arose at virtually the same historical time and place, a point to which we will return.

If I am at all correct, the traditional common-sense view of animals went something like this. On the one hand, common sense took it for granted that animals were conscious and experienced pain, fear, sadness, joy, and a whole range of mental states. Indeed, at times common sense probably gave too much credit to the mental lives of many animals, falling into a mischievous anthropomorphism. Yet in the same breath, common sense consistently ignored the obvious moral problems growing out of attributing thought and feeling to animals, since it had an unavoidable stake in using them in manners which inevitably caused them pain, suffering, and death, and thus, ordinary common sense had its own compartmentalization in this area. Indeed, such animal use was often directed not only at satisfying basic needs such as food and clothing, but more frivolous ones, such as entertainment, as in bear- and bull-baiting, fox-hunting, falconry, bull-fighting, cock-fighting, dog-fighting, gladiatorial contests, and indiscriminate bird shoots. (If the moral issue *was* ever raised, the 'Nature is red in tooth and claw anyway' and 'Animals kill each other' responses quickly dismissed it.)

In this way, though common sense and Darwinism reinforced one another, in general, neither felt the need to draw out the obvious moral implications of its position (allowing, of course, for a few exceptions like E.P. Evans and Henry Salt).[32] And when what we have called the ideology or common sense of science arose, at about the same time that animal experimentation became a crucial part of scientific activity, it had both an ideological and a vested interest in perpetuating blindness to moral issues. Hence, though it violated both common sense and the evolutionary theory which it continued to accept by its denial of animal consciousness, positivistic-behaviouristic ideology, with its denial of the legitimacy of asking moral questions in and about science, buttressed, reinforced, and even to some extent justified and grounded common sense and Darwinism's systematic disregard of moral issues surrounding animal use and exploitation. Though common sense might balk at science's denial of consciousness to animals, it had no problem at all with science's rejection of moral concern for animals, since scientific use of animals was, after all, like agricultural and other uses of animals, one more area of human benefit. Even lurid, periodic newspaper accounts of 'vivisection' aroused in most people not so much moral indignation as aesthetic revulsion—'I don't want to know about that.' (Most people still react that way to slaughterhouses and packing plants.) Thus, for a long time, there was little social moral opposition to scientific denial or ignoring of animal pain, suffering, and mentation, even as common sense might have objected to the strangeness of quantum theory, but certainly had no moral qualms about it. So for science, convenience and ideology went hand in hand, and both were unchecked by common sense, which mostly didn't care much about what scientists did. Though it thought their activities odd, it certainly didn't worry about them morally, most especially in biomedical areas, which promised—and delivered—many glittering advances of direct benefit to all of us. And, as we have seen, common sense's tolerant attitude towards biomedical research was not limited to animal subjects; it extended for a long time to human subjects as well, especially when the subjects were not 'us', but 'them'—prisoners, indigents, primitives, lunatics, retarded persons, and the like.

7 Morality and animal pain: the reappropriation of common sense

The rise of social concern about the morality of animal use

And so a naïve, blind faith on the part of the public that science knew what it was doing, and an ignorance of what it was in fact doing, coupled with common sense's traditional moral disregard for animals, all converged to allow scientists to continue to ignore animal pain, even as they continued to employ animals as pain models. This persisted until the mid-1970s, when, for a variety of reasons, society began to take an interest in moral questions concerning the use of animals in society. Suddenly, common sense seems to have grasped the link between seeing animals as conscious beings capable of a full range of feeling and the fact that our uses of them are morally questionable. In 1984, seventy-five bills pertaining to the use of animals were introduced into the United States Congress. During the decade 1975–85, a well-supported referendum was introduced in Switzerland to ban the use of animals in biomedical research, which garnered much popular support, although it failed to gain a majority. In Britain, the battle over the legitimacy of using animals in biomedical research became front-page news, with those opposing the use of animals becoming increasingly militant and violent. Massive demonstrations, street riots, protests, letter-writing campaigns, and media coverage across the United States and Europe all attested to a soaring concern with animal use of all kinds. When ABC television's news magazine *20/20* aired a documentary on intensive agriculture, or 'factory farming' as it is called, it received more letters, phone calls, and telegrams than ever before in its history. And when the United States Department of Defense announced plans in 1983 to set up a wound lab to teach combat surgeons how to deal with bullet wounds by shooting anaesthetized dogs, they too received more negative correspondence than ever before. Shortly thereafter, the department, which is the largest consumer of laboratory animals in the world, both in its military research and in its biomedical work through the Veterans' Administration, announced that it would no longer use companion animals—dogs and cats—in any of its research programmes, but would switch to such farm animals as pigs, sheep, and goats, animals which it assumed the public did not much care about. Scientists working for the military on animal projects involving cats and dogs were thus barred from their laboratories in mid-project. I was told by one research administrator that they were not allowed to enter even to remove electrodes. Across the United States, traditionally divided animal welfare groups united to oppose 'pound seizure', the use of unwanted pet animals in biomedical research. So powerful, in fact, is public sentiment in the United States against the use of pound animals that, in 1984, the state of Massachusetts, a major stronghold of biomedical research, site of Harvard Medical School, Harvard University, MIT, the Charles River laboratory animal breeding labs, largest in the world, and a variety of other biomedical interest groups, passed a law banning pound seizure.

An increasing number of individuals skilled in civil disobedience, trained in the peace and labour movements, have entered the animal rights arena. Such a person, for example, is Henry Spira, a former union organizer, who has been able to galvanize animal rights groups, regulatory bodies, and industry, into a coalition against the LD50 test. In a remarkable coup, Spira also managed to extract from Revlòn a 'donation' of $750,000 to fund the study of alternatives to the Draize eye-irritancy test in rabbits, pointing out quite reasonably that cosmetics manufacturers would be adversely affected by a campaign which showed that they tortured rabbits to create their product.

These examples indicate that animal welfare concerns—indeed, animal rights concerns (animal welfare being increasingly perceived as too weak and patronizing a notion to capture the sense of our moral obligations to animals)—are becoming more and more pervasive, and represent a powerful social force. More than once, I have heard scientists refer to the animal rights movement as the 'Vietnam of the 80s'; and a 1982 Harvard University report assessing the potential impact of animal rights on American science took it very seriously indeed. Why this sudden upsurge of public interest in the moral status of animals arose internationally after literally thousands of years of neglect is difficult to explain fully. But certain reasons do seem fairly evident, and it is worth pausing briefly to chronicle them, in light of their importance for unsettling the ideology of science.

One can view the growing concern for animals as an inevitable consequence of the liberation movements which arose in the 1960s. One animal rights leader told me only half facetiously that, by the mid-1970s, 'we had liberated everything else!' Unquestionably, a generation which had witnessed the rise of moral concern for the oppressed—blacks, other ethnic minorities, women, homosexuals, people in the Third World, and so on—would inevitably turn the searchlight on animals. It is not surprising, therefore, that the rise in moral concern for animals has been roughly contemporaneous with a rise in widespread moral concern for abused children, battered women, and foetuses. In addition, we live in an era of increasing environmental awareness and, on a superficial level at least, environmental concerns are related to animal rights concerns. (On a deeper level, of course, they tend to be inimical, since most environmentalists are concerned with the preservation of species or ecosystems, whereas animal rights advocates focus on the individual animal.[1])

Further, concern for animal rights was vitalized considerably by the ingression of moral philosophers into what had for a long time been the province of emotion and sentiment. During the 1970s and 1980s, an increasing number of philosophers entered the debate about the moral status of animals, many of them reacting against the aridity and sterility of mainstream philosophical concerns. The works of Peter Singer, Tom Regan, Stephen Clark, and myself have been widely read and reviewed, and have been taken seriously as a basis for lawsuits, legislative efforts, guide-lines, the founding of new organizations, and so on. In addition, animal rights ideas have been promulgated by many philosophers in readily accessible vehicles widely read and heard by the public and by those who use animals. I myself have spoken to and written articles for organizations of attorneys, veterinarians, ranchers, psychologists, biomedical researchers, animal scientists, farmers, government officials, toxicologists, geneticists, and so on. Professional organizations have been formed to press animal rights notions in such fields as law, psychology, and veterinary medicine.

The above factors in turn dovetailed with what appears to be growing public disenchantment with the sciences in general, and growing public demand for accountability in all areas. The boundless enthusiasm for science and technology—the two are rarely distinguished in the public mind—which pervaded common sense in the 1950s and early 1960s has been cooled by increasing awareness of their limitations. The war on cancer has not been won; outer space has not been colonized.

Instead, we face outlandish and ever-rising health-care costs, toxic waste, energy crises, environmental despoliation, iatrogenic disease, acid rain, and so on. Pure science is thus confronted by scepticism and cynicism; and applied science seems to create more problems than it solves. At the same time, as I was told in Britain by social psychologists attempting to explain the powerful hold of animal rights questions on the public imagination, people feel increasingly powerless in a high-technology *Umwelt* the vagaries of whose economy are incomprehensible even to economists. As a result, they identify more and more with animals, viewing them as symbols of helplessness. Striking out on behalf of animals thus becomes a way of asserting one's power against forces one does not understand, but by which one feels victimized. This hypothesis goes a long way towards explaining why, in Britain at least, many of the most militant advocates of animal rights are working-class people, not the middle-class intellectuals one finds espousing environmental concerns.

Be that as it may, even if our admittedly speculative etiology is wide of the mark, it is indubitable that moral concern for animals has seized the popular imagination. And, since the overwhelming bulk of biomedical research—indeed, virtually all scientific research in the United States—is funded by public money, the scientific community cannot remain aloof from public concerns. Moral assumptions made by the community at large, or by large segments thereof, must inevitably shape the form that science will take, even as those assumptions have shaped research on human subjects, and have closed the door on certain areas of research as taboo. Thus, when the public began to ask for assurance that animals were being properly treated, cared for, and used in biomedical research, and whether all possible alternatives had been exhausted before an animal was utilized invasively, the scientific community was ill served by an ideology which said that animals didn't really have feelings, or by the implicit value system which suggested that, even if they did, it didn't matter. For the first time, at least in the twentieth century, scientists had to think about these issues, to realize that they are moral issues, and to engage them.

Moral concern and its impact on science

Many scientists, of course, are unable to shift gears, given the ideology in which they are steeped. Thus we find, in response to moral challenges to animal use, either in general or regarding specific cases, many scientists trotting out the same old saws which worked to allay public concern about animal treatment and use when the vast majority of the public was unconcerned: 'Nothing questionable is ever done to an animal in research.' 'We must do these things to cure human disease.' 'We are fully accountable under the law for what we do with animals.' And so on. But the above are demonstrably false, as increasing numbers of scientists are now admitting. We also find members of the research community utilizing such tactics as wheeling transplant recipients into legislative hearings to mouth statements like 'I love animals, but if it weren't for research I would be dead'—surely a *non sequitur* where the major issue is improving the care and use of laboratory animals, not terminating their use in research altogether.

None the less, more and more scientists are rethinking the issues, the tremors they pick up around them helping to shake loose the ideological blinders they acquired with their training. Inevitably, as moral questions about animal use become pervasive in the culture generally, scientists are helped to recollect many of their common-sense intuitions about animal consciousness, and reappropriate some of the reservations and moral questions about invasive animal use which they buried when learning the common sense of science years before. Whereas ten years ago many researchers would respond to animal rights queries with official ideological pronouncements, one now finds scientists willing to express moral reservations about certain lines of animal research, about certain accepted practices, about failure to use anaesthesia and analgesia. In a climate in which moral concern for animals is expected, its expression becomes permissible, and the hold of ideology begins to weaken. I have seen this dramatically illustrated time and time again, and it never ceases to surprise (and move) me. And the sorts of anecdotes I can relate in this regard appear to mirror, in miniature, the changes taking place in the scientific community at large.

Ten years ago, when I first began teaching ethics at Colorado State University (CSU) veterinary school, and attempting to introduce a moral component into the discussion of the uses of animals in teaching and research at that institution, I confronted the practice, endemic at the time to most veterinary and medical schools in the United States, though illegal in Britain, of multiple, serial, survival use of dogs to teach surgery. For economic reasons, it was deemed expedient to use the same dog repeatedly, sometimes for as long as a semester, with little or no after-care. I raised the obvious moral issues which this practice entailed, and was able to convince the administration that it was unacceptable. The school then moved to single survival surgery, in which an animal was used only once, was then allowed to recover for a day so that students could observe recovery from anaesthesia, and was euthanized, a change which gradually spread to other veterinary colleges, as well as to other institutions, as more and more social concern was addressed to animal use in science. But at the time, some of the surgeons at CSU felt that the animals were theirs, and that they consequently had the right to do as they saw fit with them, without some 'goddamn philosopher who isn't a surgeon telling us what to do'. One man in particular so resented the new policy that he actually threatened to resign. Our relationship naturally became strained, and some years later he left CSU, and I lost track of him. In early 1985, I was invited to lecture at another veterinary college. I was told on my arrival that this same man had become chief of surgery just the year before, had spoken well of me, and wished to see me. Not only that, but upon his arrival, he had immediately abolished the five surgeries which were done on dogs, and had moved to single survival. I went to see him, and congratulated him. 'Oh yes,' he said, 'it was good to be able to do the same thing *we* did at CSU.' Not only did he not recall our conflict, but he saw himself as part of the pioneering reform at CSU, and was proud of it!

It is, I suppose, conceivable that this man would have become more enlightened in this area simply by reflection, but I doubt it. Without social pressure to flag these issues, what reason would he have had to consider them further? His ideology, his training, his peer group, all militated in favour of perpetuating the status quo. Only a strong socio-ethical enlightenment regarding the use of animals could have penetrated the ideological blinders, and made it impossible to ignore my moral critique of standard practice. Without such widespread moral concern, my position would have been dismissed as idiosyncratic, as just the opinion of an 'oddball, non-veterinarian, philosopher', rather than as articulating an idea whose time has come. With such widespread moral concern, scientists go beyond their ideology. Thus, in the summer of 1985, our surgery faculty at CSU voted to abandon even single survival surgery, and for moral reasons went over to terminal surgeries. As in the case of our earlier move to single survival surgery, I expect that this will spread to other veterinary colleges, with both the animals and the moral awareness of students and faculty benefiting from the change.

I am not, of course, suggesting that such terminal surgery—or any terminal procedure—is not itself morally problematic, or that pain is the only issue in animal use. Certainly, in terms of the ethical arguments I have developed elsewhere,[2] animal life is of value in and of itself, and killing animals even without pain is morally questionable. Nor am I happy with the inevitable social tendency, reflected all across society, to see animal life as cheap and expendable, and to see animals as tools whose lives may be taken freely, especially if this is done painlessly. None the less, terminal surgery is clearly an advance over painful, repetitive surgery after which the animal is killed. And I further believe that sensitivity to the moral issue of taking animal life, even painlessly, is beginning to develop. There are a number of scientists challenging the idea that as long as no pain is involved, any scientific project is worth an animal's life, though such a stance is literally incomprehensible to many of their peers. By the same token, a number of medical and veterinary students are challenging terminal laboratory exercises on the grounds that animals are not tools for our use, and in some cases have won the right to pursue alternatives. Further, various legislative initiatives have been forthcoming supporting students' rights to opt out of laboratory exercises on grounds of conscience.

For that matter, we know that certain interests are of higher priority to animals than pain. It is well known that animals will chew off limbs to escape from traps, suggesting that concern with freedom—or life—trumps concern with pain. Animals will also choose sexual contact at the cost of pain. Biologically, as well as morally, it makes sense to suggest that life is more important to an animal than anything else, a point I have argued elsewhere.[3] Even if, as some philosophers have suggested, loss of life is incomprehensible to an animal, being too abstract a notion to impute to it, it does not follow that we do not harm animals by killing them when they are not living a life of misery. By taking life, we take away the possibility of actualizing those interests which would be experienced as good by an animal.

A point very similar to this has been beautifully argued by Sapontzis in his *Morals, Reason and Animals*, a book which attempts to extend our ordinary common-sense, socially pervasive moral notions to animals, rather than to construct an ethic *de novo*. Sapontzis puts it this way in a chapter devoted to the question of animal killings:

> [L]ife is necessary for enjoyment and fulfillment, [and] this generates an interest for sentient beings in remaining alive for as long as life holds (the prospect or even just hope of) sufficient enjoyment or fulfillment for them, and that our common moral goals include a concern with insuring that all those capable of enjoyment and fulfillment have a fair chance at achieving an enjoyable fulfilling life.[4]

The same sort of thing as I described above concerning the surgeon happened to me, even more dramatically, at a seminar in Canada, where I was asked to address the faculty of all the Canadian veterinary colleges at their annual meeting, on the ethics of animal use in science. Appearing on the panel with me, I was told, was an internationally famous surgeon, a legend in experimental surgery, whose work had been enormously influential, but who, in his major book, had tended to equate any questioning of animal use with know-nothing Luddism. When the day of the panel arrived, I was pumping enough adrenalin to supply a football team. After all, I was challenging a legendary veterinary surgeon in front of his peers, on an issue which, to them, was at best flaky, and at worst, a real threat to their autonomy and academic freedom. I delivered my speech on the moral status of animals, and was critical of the traditional failure of the scientific community to engage the moral issues in this area. When I was finished, the surgeon got up to give his speech. 'My topic', he began, 'is fairly mundane, having to do with the need for surgery animals in the coming decades. But before I approach it, I should like to comment on what Dr Rollin said.' I braced myself as he continued. 'I have read some of his and other people's material on this issue, and have been led to reflect on my own use of animals. And I wish to say, in front of my peers, that I realize that my career has been marred and blackened by sins of omission and commission affecting the animals I have used to accomplish my goal, and for this I am deeply sorry, and wish I could undo them.'

As I sat there, like a deflated dirigible, I realized that there was no need for him to have said such a thing. Obviously, the increasing social and moral concern with research animals had drawn his attention, and had made him look at the issue in a new light; this, despite the fact that so doing at the end of a career requires great courage and must surely be very threatening.

The key point is that the factors operative on these individuals have also introduced a moral factor into the scientific community at large. Thus, a counterforce which undercuts the ideological dictates governing animal pain has begun to wax. As a greater number of people come to see animal use as *morally* problematic, they must inevitably begin to ask new questions about it. Twenty years ago, with the exception of small groups of antivivisectionists who were out of the mainstream, no one raised questions about the legitimacy of animal use in science. The *use* of animals was not a moral problem; as long as the animals received proper *care* and as long as they were not anyone's pets, the assumption was that science was important and knew what it was doing. The United States Animal Welfare Act of 1966, enacted mainly to assuage fears of pet-owners that their animals might be kidnapped and sold to research laboratories, as happened to a family dog in a case which received dramatic press coverage, set standards for research animal dealers, animal transport and housing, and so on, but specifically disavowed any concern with the actual conduct of research, and the types of experiment in which animals were used.[5] Furthermore, even the limited provisions of the act covered only those animals which the public felt inspired to protect: dogs, cats, primates, guinea-pigs, and rabbits, and exempted rats, mice, and farm animals, some 85 per cent of the total number of animals used in research! For purposes of the act, a dead dog is an animal, while a live mouse is not.[6] All this reflects the aesthetic, sentimental, and emotional basis for social concern about research animals at that time, a perspective which could not yet be called moral, or how could it justify treating rodents as non-animals?

Indeed, the Animal Welfare Act of 1966, together with its amendments of 1970 and 1976, attests to the traditional power of the research community to block legislation which would have any significant impact on the way animals are used in research. Provisions of the act, and the regulations promulgated under it, dealt mainly with standards of feeding, caging, housing, and housekeeping. Concern with the actual conduct and design of research were statutorily disavowed, and although the act did stipulate proper use of anaesthetics and analgesics, this stipulation could be met simply by institutions stating in an animal report that such use was not deemed necessary. Even a cursory perusal of records of citations under the act by inspectors shows that many of the infractions pertain to matters which do not matter to animals, such as peeling paint, improper storage of material, and other housekeeping items.

The only other kind of regulation of animal research in the United States was even more ineffectual until the mid-1980s. Beginning in the 1960s, NIH, which funds around 85 per cent of biomedical research in the United States, began to issue guide-lines for the care and use of laboratory animals. Although not enjoying the force of law, these regulations, which were in many areas good and sensible, were allegedly binding by contractual agreement on all NIH grantees, with NIH theoretically able to seize all federal funding to any institution not in compliance. But in actual fact, before 1984, NIH had never enforced its own guide-lines. Beginning in the late 1970s, as I was called on to lecture at universities around the United States, I found that although NIH regulations forbade repeated use of animals in teaching surgery, for example, this rule was ignored virtually everywhere. When I called NIH to protest, I was informed, 'We are not in the enforcement business.'

The extent to which animal pain had not been adequately regulated by either the Animal Welfare Act or NIH guide-lines as of 1986 was dramatically documented by a recent submission to the USDA, the agency responsible for enforcing the Animal Welfare Act, by the Animal Legal Defense Fund (formerly Attorneys for Animal Rights). This report documents the inadequacy of regulation in four categories:

1. Painful experiments performed on fully conscious animals where the phenomena examined can be observed in anaesthetized animals.
2. Experiments causing disease or post-procedural conditions for which analgesics or other support therapy is routinely prescribed for humans.
3. Painful procedures routinely done without anaesthetic, analgesic, or tranquillizer drugs, 'which should no longer be permitted because no reasonable person could consider them necessary for human or animal health in the manner intended by Congress'.
4. Painful experiments performed on pigs, chickens, goats, sheep, cattle and other animals not considered animals for purposes of the law.[7]

The questions asked in the last decade are quite different from those asked in the earlier era, and demonstrate public disaffection with such flimsy regulation. They bespeak far deeper moral concern, doubtless growing out of the general moral sensitization of the public in the 1960s and 1970s. The old questions have been replaced by questions like: Do we have the right to hurt animals for our benefit? Does the knowledge gained justify the animal suffering? Do we really need yet another cosmetic at the expense of animal pain? (Amazingly, the readership of *Glamour Magazine*, in response to a survey conducted by the magazine, gave a resounding no to this question.) What assurance do we have that animals are properly cared for? Is all unnecessary suffering of animals in research being eliminated? Is the search for alternatives to animals in research being implemented with sufficient vigour? One need only look at Sunday newspaper supplements, local news programmes, trade magazines, national magazines, and so on to see that these questions are now being asked.[8]

A similar thing appears to have happened in the past forty years regarding human medical ethics. The public and the mass media became fascinated with questions like abortion and euthanasia, and the medical community was forced to explore and to deal with them, as well as to expose nascent doctors to them in medical schools. By the same token, the research community has been galvanized during the past decade into dealing with ethical issues governing research animal use. The responses of researchers have been quite mixed, and many have held seminars on 'how to deal with this new threat'. Others have put out films darkly suggesting that any intrusion into what is done with animals will compromise the lives of sick children—for example, 'Will I be all right, Doctor?' put out by the Association for Biomedical Research. But many have been motivated to start seriously considering the issues, and to articulate their own moral positions (something that scientific ideology hitherto discouraged).

On the other hand, one should avoid totally equating scientists' greater responsiveness to animal ethical issues with a genuine recognition that these issues are real moral problems and that science is not value-free. The hold of scientific ideology is too powerful for such a precipitous change to occur overnight. In many cases, scientists are responding less to the ethics of the issues than to society's perceptions of the ethics of the issues, for, while they may still believe that ethical questions are scientifically meaningless, the fact that society is concerned with them is empirically verifiable, and society must be placated, since it controls the purse-strings of scientific funding.

Law and local project review as a factor in moral change in the United States

One of the major factors giving rise to researcher concern with moral issues surrounding animal use, thereby helping scientists see through the 'science is value-free' ideology, has been the thrust in the United States during the last decade towards local committee review of animal projects, modelled on the committees which are currently mandated for human subjects. The concept of such review was initially championed in the United States largely by a group of people in Colorado, including myself, a laboratory animal veterinarian, a prominent researcher, and an attorney and animal welfare advocate. We realized early in our discussions in 1976 that the chief barriers to moral consideration of laboratory animals were the lack of a consensus ethic in society (or in common sense) about animals and the hold of the self-serving belief that science is value-free, which led to conceptualizing animal use solely in scientific terms, and to viewing any criticism thereof as an 'attack on science'. The problem, we felt, was to a considerable extent an educational one, requiring that researchers acknowledge the moral dimensions of what they do. It was impossible, not to say inappropriate, to have a policeman in every laboratory, and thus researchers needed to police themselves. But given the prevailing ideology which denied legitimacy to moral questions in science in general and to moral questions about animals in particular, this wasn't likely to happen without legislation. So we drafted legislation, at first for Colorado and then for the country as a whole, which in essence would require local review of animal projects by researchers, animal advocates, and others, and which, unlike the Animal Welfare Act, would cover *all* research animals, and would look not only at care and husbandry, but at use, in terms of control of unnecessary pain, cost and benefit, and so on.[9]

By and large, the initial response of the scientific community to any meaningful regulation of animal research was extremely negative and hostile. Various versions of the legislation were put before Congress and regularly defeated during the 1970s and early 1980s. But social pressure and concern regarding animal research was mounting, fuelled in part by a number of revelations of undeniable atrocities in animal treatment. Perhaps most influential were the 1982 Taub case[10] and the release in 1984 of thirty minutes of film distilled from videotapes seized by animal activists in a raid on a University of Pennsylvania head-injury laboratory. These tapes, made by the researchers themselves, documented a welter of abuse and callousness, and made the public aware of the need for regulation. In both the above cases, NIH was forced to stop funding the researchers involved, and eventually to develop mechanisms for enforcing its own rules.

In 1985, local review of projects became law. NIH announced new requirements for grantees, the heart of which is the mandating of institutional animal care committees, which must consist of at least five members, including a veterinarian with a background in laboratory animal medicine, a scientist experienced in animal research, someone from outside the sciences, and someone not affiliated with the institution. The committee must monitor animal care and use, meet regularly, and review, critique, and approve or disapprove research protocols before they are funded. The committees are empowered to stop any activity not in compliance with policy, and to demand control of pain and suffering not essential to the research.[11] In June 1985, these guide-lines were made law by Congress in the Health Research Extension Act of 1985.[12] Furthermore, NIH allocated funding for unannounced inspections to enforce its policies.

At the same time, our goal of amending the Animal Welfare Act was at least partially successful. By 1982, when I was called on to testify on behalf of the so-called Walgren version of our bill, I carried the endorsement of the American Physiological Society, the traditional opponent of any intrusion into the research process, for review committees. And in 1985, Congress passed the Improved Standards for Laboratory Animals Act as an amendment to the Animal Welfare Act which took effect 1 January 1987. Although the full regulations implementing the act have not (as of 1988) yet been promulgated by the USDA, the enforcing agency, the basic features are as follows:

1. Establishment of an institutional animal care committee to monitor animal care and inspect facilities. Members must include a veterinarian and a person not affiliated with the research facility.
2. Standards for exercise of dogs are to be promulgated by the USDA.
3. Standards for a physical environment which promotes 'the psychological well-being of primates' are to be promulgated.
4. Standards for adequate veterinarian care to alleviate 'pain and distress', including use of anaesthetics, analgesics, and tranquillizers, are to be promulgated.
5. No paralytic drugs are to be used without anaesthetics.
6. Alternatives to painful or 'distressful' procedures must be considered by the investigator.
7. Multiple surgery is prohibited except in cases of 'scientific necessity'.
8. The animal care committee must inspect all facilities semi-annually, review practices involving pain, review the conditions of animals, and file an inspection report detailing violations and deficiencies. Minority reports must also be filed.
9. The USDA is directed to establish an information service at the National Agricultural Library which provides information aimed at eliminating duplication of animal experiments and at reducing or replacing animal use, minimizing animal pain and suffering, and aiding in training of animal-users.
10. The research facility must provide training for all animal-users and caretakers on humane practice and experimentation, research methods that limit pain, use of the information service of the National Agricultural Library, and methods of reporting deficiencies in animal care and treatment.
11. A significant penalty is established for any animal care committee member who reveals trade secrets.

12. The USDA is directed to consult with the Department of Health and Human Services (which has responsibility for funding biomedical research through NIH) in establishing the standards described.
13. New civil penalties are provided for violation of the Act.[13]

There are undeniable deficiencies in all this legislation. The Amendment to the Animal Welfare Act still officially disavows concern with the actual conduct of research, and does not mandate, as did our bill, that rats, mice, farm animals, and other legal non-animals be protected by law. But NIH policy covers these animals, and a recent agreement between USDA and NIH empowers USDA to report infractions of NIH policy to NIH.[14] On the positive side, the law and the activities of animal care committees now focus on control of animal pain and suffering; this is arguably the chief concern of local committees. And in the face of laws requiring concern with pain, distress, and suffering, scientific ideology which declares such pain and suffering to be unknowable or unreal or trivial can hardly be sustained. Indeed, the first draft of the USDA regulations interpreting the amendments stipulates clearly that pain-relieving drugs must be used in 'any procedure that would reasonably be expected to cause pain or distress in a human subject were that procedure applied to a human being'—a clear reappropriation of common-sense anthropomorphism.[15]

Nevertheless, scientific ideology dies hard. All this is nicely illustrated by an anecdote related at a national meeting of laboratory animal scientists by the chief administrator of the USDA division charged with enforcing the Animal Welfare Act amendments. He related that the USDA was perplexed at the congressional mandate requiring 'a physical environment . . . which promotes the psychological well-being of primates'. Reasonably, he approached the United States psychological community for expert assistance. Not surprisingly, given the ideology permeating psychology, he was told that there was no such thing. 'There will be after January 1987,' he tellingly replied.[16]

Not only do the new laws force reappropriation of common-sense talk about pain; they force scientists to concern themselves with some of the moral issues involved in the use of animals in research. And although, in theory, committees are supposed to look only at control of animal pain and suffering, not at matters of scientific merit, in practice it is difficult to draw a clear line.[17] In fact, it is impossible to avoid looking at scientific merit when committees are charged with making sure that the proper number and species of animals are used. Even more clearly, NIH policy requires assurance from institutions that 'discomfort and injury to animals will be limited to that which is unavoidable in the conduct of scientifically valuable research'.[18] This implies that committees may judge a piece of research to be, in essence, not valuable. Thus I believe that committees will inevitably move towards judgements of cost benefit in animal research, analogous to what is done in the case of human subjects.[19]

Many of these committees are suffering growing pains, of course, not the least of which stems from the fact that since we as a society lack a coherent consensus ethic on animals, they are a bit like jurors being asked to adjudicate cases without being given the law. None the less, decisions do emerge, and eventually a consensus ethic will emerge too. (At my own institution, there exists a shared sense of obligation to find the point at which we maximize animal benefit while maximizing human benefit as well. This was the basis for our abandonment of multiple and, later, even single survival surgery.) But the key point is that research animal issues are being considered morally, not merely scientifically and pragmatically, and this change in perspective must be salubrious.

British legislation

The history of serious concern with the moral status of animals in general and laboratory animal pain in particular is far more extensive and progressive in Britain than the United States. The first anti-cruelty law was passed in Britain in 1822, and the law was expanded in 1835, 1849, 1861, 1863, 1864, 1876, 1894, and most comprehensively, in 1911.[20] Philosophically, opposition to the Cartesian notions of animals as machines was most eloquently expressed by Hume, as we saw earlier, whose view of animal mind as continuous with human mind may be seen as expressive and paradigmatic of British thought, which of course culminated in Darwinism.[21]

The 1860s and 1870s witnessed the rise in Britain of a powerful antivivisection movement, amid other major currents of social reform, including feminism.[22] (In many ways, the situation presages and parallels similar activity beginning in Britain and the United States in the late 1960s and covering similar issues. Then, as now, concern with animals was strongly linked with women's rights.) Antivivisectionist activism soon galvanized British public opinion, which was especially affected by reports of cruel and unnecessary experiments on animals in Europe.[23] A Royal Commission was appointed in 1875 to examine the issue and to make recommendations. In the wake of testimony by physiologists revealing a cavalier attitude towards pain in animals, legislation was inevitable. Most influential, perhaps, was the testimony of physiologist Emmanuel Klein, whose testimony bespoke an ideological disregard for animal pain, and who indicated that he rarely used anaesthetics, and then only for convenience:

> . . . just as little as a sportsman or a cook goes inquiring into the detail of the whole business while the sportsman is hunting or the cook putting a lobster into boiling water, just as little as one may expect these persons to go inquiring into the detail of the feeling of the animal, just as little can the physiologist or the investigator be expected to devote time and thought to inquiring what this animal will feel while he is doing the experiment. His whole attention is only directed to the making [of] the experiment, how to do it quickly, and to learn the most that he can from it.[24]

Out of the commission's inquiry came the pioneering act of 1876 designed to regulate animal research. At the time, most invasive animal research was physiological, and involved surgical operations designed to increase understanding of bodily function in health and disease. Understandably, the major provisions of the act reflect this interest.

In essence, the major provisions of the act were as follows. No one may perform a painful experiment on a vertebrate without a licence issued by the Secretary of State. Experiments may be done under licence only 'with a view to the advancement by new discovery of physiological knowledge or of knowledge which will be useful for saving or prolonging life or alleviating suffering'. This was interpreted in practice to allow pure research where no benefit was manifest other than the growth of scientific knowledge. Experiments may not be done simply to acquire manual skill or as a 'public exhibition'. Licensure and the latter two requirements are absolute. Other conditions of the law could be dispensed with by acquiring a certificate attached to the licence. These conditions include:

1. The 'pain condition', which requires that
   (a) if an animal is suffering severe pain or pain which is likely to endure during an experiment, and if the object of the experiment has been attained, the animal must be painlessly killed;
   (b) if an animal at any time during an experiment is found to be suffering severe pain which is likely to endure, such animal shall forthwith be painlessly killed; and

   (c) if any animal appears to an inspector to be suffering considerable pain, and the inspector directs that the animal be killed, it must forthwith be killed.
2. The 'anaesthesia condition', which requires
   (a) the use of anaesthetics to prevent the animal feeling pain; and
   (b) the requirement that the animal be killed before recovery if the anaesthesia fails or if the animal is seriously injured.
3. A condition prohibiting the use of paralytic drugs except on decerebrate animals without special permission from the Secretary of State.
4. A condition prohibiting the use of animals in experiments for demonstration or illustration of lectures in medical schools, colleges, and so on.
5. Prohibition of using dogs, cats, horses, asses, or mules in experiments.
6. A requirement that licensees keep records of experiments and report annually to the Secretary of State.

Exemptions from these conditions may be granted by certificates attached to a licence, and are made when insensibility or non-recovery from anaesthesia would frustrate the object of an experiment; when an experiment or demonstration is necessary for the training of students, provided the animals are anaesthetized and are not allowed to recover; when the experiment requires cats, dogs, horses, asses, or mules. There are six such certificates, granting exemptions. The law was administered by inspectors responsible to the Secretary of State.

The law was examined closely by a Royal Commission which sat between 1906 and 1912 and by the Littlewood Commission in 1965, both of which recommended administrative changes in application of the law, but the law itself remained unchanged.[25] Beginning in the 1960s and 1970s, as we saw earlier, social concern with animal experimentation increased dramatically. In Britain, antivivisectionist activism became increasingly militant, and animal experimentation once again became a major social issue. This resulted in growing demands for new legislation. The dramatic changes which had occurred in scientific research in the century after 1876, together with clear evidence that the 1876 act was not adequate to control pain, distress, and suffering in all its forms and in all types of research, laid the groundwork for new proposals.[26] Demands were also made that legislation look more at the purposes behind research, strengthen the termination condition, increase public accountability, reduce the number of animals involved, and accelerate the development of alternatives to animal use. A decade of intense political activity among scientists, animal welfare groups, and veterinarians resulted in the passage of the

Animals (Scientific Procedures) Act of 1986, which went into effect on 1 January 1987 and is projected to replace the 1876 act by 1990.[27]

The major provisions of the new law are as follows: coverage provided by the old law is extended to all vertebrate animals and their embryonic forms from the half-way point in gestation, or when the animals are capable of independent feeding in the case of fish and amphibians. In addition, the act covers not only experiments, but also the production of vaccines and sera, the growth of tumours, and other activities which can produce 'pain, suffering, distress or lasting harm'.

A dual licensing system replaces the old system. Investigators must obtain both a personal licence and a project licence. In granting a project licence, the Secretary of State can not only set limits on the degree and extent of pain allowed in a particular piece of research, but can also, in principle, decide if certain types of research should be done at all. He is directed to weigh the cost to the animal as against the possible benefit to society. This opens the door to the possibility of banning certain types of animal use: for example, for cosmetic testing and behavioural research, both widely perceived, even by biomedical researchers, as odious.[28] In addition the licensee must give evidence that there are no non-animal alternatives available.

To implement the provisions, the inspectorate has been increased, and it has been authorized to consult experts in relevant fields. In addition, the law mandates a national Animal Procedures Committee, consisting of at least thirteen members, no more than half of whom can be licensees. This committee is to advise the Secretary, and must pass on all applications for cosmetic tests, tobacco research, and procedures aimed at attaining manual skill in micro-surgery. To ensure compliance at the local level, each laboratory must have two 'named persons', the first a veterinarian who oversees general health and welfare, anaesthesia, and analgesia, the second a person responsible for the day-to-day care of the animals, usually a senior animal technician.

One major advance over the 1876 act is a significant strengthening of the termination condition, now inviolable. Moreover, whereas under the 1876 act an animal suffering intractable pain or distress had to be killed after the research objective was achieved, the new law requires 'immediate' euthanasia, regardless of whether there is evidence that the suffering is likely to endure. In both this provision and in the cost-benefit directive, it is clear that animal suffering is not always outweighed by human benefit—a significant moral advance over traditional attitudes.[29] As of spring 1987, the Code of Practice (analogous to the USDA regulations) defining the law in greater detail had not been issued by the Home Office, but is scheduled to appear.

As yet, we cannot, of course, assess the direct effects of the new law, but it is unquestionable that the social dimensions of concern which led to it have already had their effect on British science. There is now some pressure for local review of projects even among scientists,[30] but such a policy has been resisted (at least as far as legislation is concerned) by powerful conservative elements in the British scientific community. In my view, the lack of any mechanism for such review in the new law will severely hamper both its effectiveness and its potential for changing the ideology of scientists, since it does not compel scientists to think and deliberate about animal research and pain and suffering in moral terms, but merely to obey rules and jump through what they may well see as bureaucratic hoops.

Ironically, both the 1876 and the 1986 acts were characterized as 'vivisectionist charters' by some radical antivivisectionists, while also being seen as imposing unnecessary restrictions on freedom of inquiry by many researchers. Such disaffection is often the outcome of attempts to introduce incremental change; any change is too much for those who were previously unrestricted, whereas no change short of total overthrow suffices for those who find the status quo morally repugnant. What is forgotten by both parties is that change in a democracy is by nature incremental, a resolution of forces pulling in different directions.

Law and policy in Canada and Australia

Canada has been subject to virtually the same rise of social concern regarding animal experimentation as has the United States, and indeed, has seen social consciousness further raised on animal issues by international criticism of the Newfoundland seal hunt, recently abolished as a result of an official boycott of seal products by the European Economic Community. As of 1988, Canada has no federal legislation governing animal research. On the other hand, since 1970, Canadian research has been monitored by a voluntary system of animal care committees administered by the Canadian Council on Animal Care (CCAC), a body chartered by the Canadian universities. This monitoring does not enjoy the force of law, and although the CCAC may in principle withhold grant money from violators of its guidelines, in practice this is not done.

It is fairly clear that more than voluntary control is necessary, and that animals are entitled to the sort of moral concern which can only be embodied in law.[31] In 1986 the Canadian Medical Research Council chartered a group to explore ways to increase the protection of research animals, with legislation as one such option. More dramatically, the Canadian Law Reform Commission, which reviews the criminal code every twenty-five years and makes recommendations for change, in 1985 released a report entitled 'Offences Against Animals'.[32] In this penetrating analysis, the author critically discusses the traditional view of animals as devoid of consciousness and feeling and their status as property in the law. He also points out that recent philosophical and social thought has been persuasive as regards raising the moral status of animals, which notion should be reflected in law.[33] His discussion of animal experimentation concludes with the following extraordinary paragraph:

> For the present, animal experimentation is likely to remain as a necessary evil; its continuance is strictly an ethical compromise. In these circumstances, every effort must be made to eliminate needless suffering and waste, to reduce the numbers of animals used and to reduce the use of higher animal species whose members are more susceptible to different forms of suffering. Laboratory animals must be maintained in conditions which respect their nature and fundamental interests and all measures must be taken to prevent unnecessary pain. Concerted efforts are also needed to develop alternatives to animal research. These objectives have been succinctly summarized as 'reduction, replacement and refinement'. Although supervision in Canada has traditionally occurred through voluntary standards and funding control it may be appropriate to explore the possibility of bringing experimental procedures within statutory regulation. Whether or not other legislative controls are used, there is a place in a revised Criminal Code for a prohibition on unjustifiable experimentation.[34]

The situation in Australia is also in flux. There is currently no federal legislation governing animal research. On the other hand, the issue of the moral status of animals has become a major one in society, not only as regards laboratory animals, but with reference to the indiscriminate killing of kangaroos, the live sheep trade, and certain agricultural practices. The situation there is not yet as polarized as in Britain or North America, with channels for dialogue still open between the research community and animal advocates. A national senate committee chaired by Senator Georges, a strong animal advocate, is in the process of gathering data and testimony relevant to federal legislation. In late 1985, the state of New South Wales passed legislation mandating a central animal review committee and local institutional animal care and review committees, similar to what has been chartered in the United States. As in Britain and the United States, there is a powerful group of conservative medical researchers who are intent on preserving the status quo, though many other members of the Australian research community are amenable to change.

The new moral attitudes and the study of pain

All these changes in social-ethical valuation of animals and their codification in law have clearly led to changes in scientific attitudes towards animal pain. Obviously, the very first question which comes up in looking at the scientific use of animals from a moral standpoint is whether they are undergoing any pain or suffering at our hands which could be eliminated, avoided, or mitigated. In such a context, the traditional scientific official denial either of animal pain or of its significance has no place. As a result, scientists are paying increasing attention to methods of reducing animal pain—for example, by replacing periorbital (behind the eyeball) extraction of blood from rats, a painful procedure, with tapping a tail vein. In addition, control of animal pain by anaesthesia and analgesia is becoming more common. In circulatory shock research, for example, there is an increasing tendency to use anaesthetized animals in experiments, even though, traditionally, unanaesthetized animals were used for fear of skewing relevant physiological variables.

By the same token, the American Psychological Association, in its new guide-lines for animal research promulgated before the new laws, rejected the traditional, convenient use of paralytic, curariform drugs, which paralyse without anaesthetizing, as a method of restraint in stereotaxic procedures. The International Association for the Study of Pain and its journal *Pain* pioneered in developing moral guide-lines for painful research on animals, guide-lines which require that a researcher try a procedure on himself before calling it minor, and which also prohibit the infliction of inescapable pain on animals. Other scientific journals have set their own moral standards regarding animal use, to which papers submitted must conform; though these are usually fairly tepid, they do represent some advance. Many researchers on pain now permit the animals access to analgesia, and have devised methods for its self-administration. A number of institutions began to use analgesia post-surgically even before the new laws were passed.

There have been more articles on animal pain and its control in scientific journals from 1985 on than can be found in the whole of the previous century. Before 1985, no conferences were held on recognizing, assessing, and alleviating pain in animals. The first was convened by the British Veterinary Association in 1985. In the United States the first such conference which fully addressed these issues was held in Chicago in spring 1987. Moreover, a prominent laboratory animal veterinarian has informed me that recently people in the laboratory animal veterinary community have turned to the voluminous data base which has been acquired in the course of screening analgesics on rodents for human use, with an eye to abstracting information which can be utilized to control pain in rodents.

It is valuable to look at the aforementioned guide-lines developed for *Pain* in 1980,[35] for they elegantly exemplify a paradigm case of the growing pains (no pun intended) which the common sense of science must inevitably suffer as it moves from a positivistic-behaviouristic denial of both the moral value of animals and consciousness in animals towards reluctant and tentative acceptance of both. (Some of the more provocative remarks made in this document were removed three years later in a new, briefer, weakened version. Thus, for example, what were called 'standards' in the 1980 document became 'guide-lines' in 1983).[36] The 1980 guide-lines begin by listing various methods which are used in chronic pain research on animals, including:

> Continuous electrical stimulation of nerves or tooth pulp with implanted electrodes, or repetitive foot shock through the cage floor; experimental inflammation by subcutaneous injection of turpentine, yeast or carrageenan; the formalin test: necrosis and inflammation by subcutaneous injection of formalin; continuous infusion of pain-producing substances (e.g. bradykinin); experimental arthritis induced by certain toxins; fracturing bones; experimental neuroma induced by nerve transection; central deafferentation by dorsal root transection; central nervous system lesions by coagulation or injection of alumina gel; induction of central nervous hyperexcitability by injection of convulsive drugs or toxins.[37]

The report goes on to state (with the ideology of science clearly peering over its shoulder) that the members of the Committee for Research and Ethical Issues of the society felt that 'the methods used in these investigations *might* have induced states in the animals which correspond to pain and suffering in humans'.[38] (Surely the investigators were more confident than that; if they were not reasonably sure that such states were induced, they would hardly bother to do the research!)

The report continues:

> Such experiments deserve careful planning to avoid or at least minimize undue suffering in the animals. Investigators of animal models (for chronic pain) as well as those applying acute painful stimuli to animals, should be aware of the problems pertinent to such studies, and should make every effort to minimize pain and suffering. They should accept a general attitude in which the animal is regarded not as an object for exploitation, but as a living individual.[39]

Without an ethic governing our treatment of other 'living individuals', it is difficult to extract much content from this declaration. It appears to be a lubricant, greasing the path from mechanism and positivism to the beginnings of moral awareness.

The actual guide-lines are then presented:[40]

> 1. It is essential that the necessity of the particular kind of experiment and its benefit for mankind has been scrutinized by colleagues in this field of research, by ethologists, and by lay persons. The investigator should be aware of the ethical need for a continuing justification of his investigations.

Implicit in this guide-line is an embryonic realization that science is inseparable from valuational and moral presuppositions, as well as the explicit appeal to human benefit as a basis of moral justification of invasive animal use. It further acknowledges a role for local review committees which include lay persons.

> 2. If possible, the investigator should try the pain stimulus on himself; this principle applies for most non-invasive stimuli causing acute pain.

In actual practice, local committees have used this criterion to challenge investigators who claim that the pain of a given procedure is negligible.

> 3. To make possible the evaluation of the levels of pain and suffering, the investigator should give a careful assessment of the animal's deviation from normal behaviour. To this end, the largest possible number of physiological and behavioural parameters should be measured, such as EEG, sleeping and sleeping/waking cycle, feeding and drinking behaviour, body weight, distinct classes of motor behaviour (e.g. grooming, exploration), performance in learning or discrimination tasks, copulatory behaviour, rank order in a society. The outcome of this assessment should be included in the manuscript when it is submitted for publication.

This rule is noteworthy in that it treats subjective states of pain and suffering as knowable from physiology and behaviour, and hence as not unscientific notions.

4. The animal should be able to control the effects of acute experimental pain (e.g. by escape or avoidance behaviour).

Note that this rule implicitly imputes choice to the animal.

5. An animal presumably experiencing chronic pain should be treated for relief of pain, or should be allowed to self-administer analgesic agents or procedures, as long as this will not interfere with the aim of the investigation.

Note the salubrious rejection of the traditional disregard for laboratory animal analgesia.

6. The duration of the experiment must be as short as possible, and the number of animals involved kept to a minimum.

7. The investigator should choose a species which is as low as possible in the phylogenetic order, compatible with the aim of the investigation. This recommendation infers that the degree of suffering is smaller in lower than in higher animals, although this assumption cannot be taken as proven.

This last guide-line is extremely interesting, in its explicit and more or less unabashed acknowledgement of a valuational assumption lying at the root of scientific activity. Since there is no evidence that lower on the phylogenetic scale means less suffering, and indeed, there exists the sort of contrary argument developed by Kitchell, discussed earlier, that lower on the scale may mean greater suffering (because less understanding), this assumption must be rooted in the valuational assumption that lower animals are worth less or matter less. This accords with common sense and is a reasonable, if debatable, moral position to take if one is going to look at animal use in moral terms.

Moral concern for animals in society, before and after legislation, must inevitably soak into the presuppositions of science, and in so doing, help erode the ideology of science rooted in positivism, behaviourism, and the denial of moral issues admixed into science. Our focus has been pain, yet it is obvious that the same kind of thinking pertains to all modes of animal consciousness, as long as moral concern for animals remains a social concern. Clearly, other modes of suffering, such as fear, anxiety, boredom, loneliness, and so on, must be of concern if pain is, especially as their physiological and behavioural correlates are identified. Recent work indicates, as we have already mentioned, that most vertebrates possess central nervous system receptors for benzodiazepine, chemicals which serve to alleviate anxiety, just as endorphins and morphine relieve pain. This would surely suggest the presence of anxiety in animals. And articles on animal boredom have begun to appear.[41] It appears that the new regulations tentatively address these states, since they talk of 'distress' as well as pain. And one hopes that, as the 1985 amendment to the Animal Welfare Act suggests in mandating exercise for dogs and proper environments for primates' psychological well-being, society will demand not only controls over noxious experiences, but also conditions which make animals *happy*, a notion we shall define shortly.

One of the major social forces pushing in this direction of acknowledging and studying other noxious modes of experience or consciousness, as well as their positive correlates (modes of happiness or pleasure or satisfaction in animals), is growing unease with confinement or intensive agricultural practices, so-called factory farming. There has been much concern over whether farm animals suffer under the conditions imposed by the latter, and this has sparked a great deal of research and deliberation, research which is also helping to erode the old ideology of science, which we will discuss when we address the general return of consciousness and its study to the scientific arena.

Ideology crumbling: the subjective acknowledged

Before concluding our discussion of pain, one further point must be made. The major thrust of our discussion has been that science has tended to ignore the reality (and moral significance) of the subjective, experiential side of animal pain, for both ideological and conscience-salving reasons. This neglect has been manifest in research protocols, the training of scientists, and standard textbooks. Given our account of changing social-valuational presuppositions and their impact on science, one would expect eventually to find new textbooks mirroring the change in attitudes towards animal awareness and pain. And one is not disappointed. When one looks at a traditional textbook, such as Guyton's *Medical Physiology*, one finds the traditional avoidance of a commitment to explicitly acknowledging animal pain. Instead, in discussing the limbic system and 'reward and punishment centers' in the brain, Guyton points out that reward and punishment are crucial motivators, even as behaviourism presupposed in all its extensive work on learning and reinforcement. None the less, though reward and punishment centres make no sense without implicit appeal to awareness, Guyton, like the behaviourists, cavils at explicitly admitting conscious states in animals. 'Stimulation in these [punishment] areas', he tells us, 'causes the animal to show *all the signs of* displeasure, fear, terror and punishment.'[42] 'All the signs of' is clearly a bow to the ideology of science. Contrast this with the extraordinary statement at the opening of the chapter on pain in Kandel and Schwartz's recent (and definitive) textbook *Principles of Neural Science* (1984). The chapter begins with the claim that 'pain is a primitive, protective *experience* [not mechanism], which we share with all living organisms'.[43] Aside from the fact that this is most certainly false, since plants surely don't feel pain, what is extraordinary about this claim is that it is simply asserted, with no argument, defence, or elaboration, as if it were common knowledge, and as if there had been no seventy-five year period in science during which such claims were anathema! Clearly this passage bespeaks a considerable softening of the old common sense of science, probably as a result of the new moral climate. It goes a good deal further than the sort of 'even if we can't know that animals feel pain, let's give them the benefit of the doubt' stance we saw articulated earlier by Flecknell, which itself was a great change from the 'Since we can't know what animals experience, let's assume they don't' position of positivism and behaviourism. Yet the fact that the attribution of pain is made without any argument whatever is worrisome, for clearly such a claim, in order to be tethered and stable, should be more than simply a response to a change in moral climate. What is needed is a philosophical basis for the claim, of the sort we have been developing. It is not enough that admission of animal pain is back in style. Only when the new philosophical arguments germane to this issue have driven out the traditional positivistic-behaviouristic ones in scientific ideology, can we even begin to feel confident that science will not slip back into denying or ignoring the pain of animals if social conditions again change. Providing such arguments to scientists is a major purpose of this book and a major part of my activities. As I have argued before, self-conscious philosophical thinking must enter into science education.[44]

Kandel and Schwartz's textbook is by no means an isolated case. Beginning in 1984, animal pain and its control had become a major topic in veterinary and laboratory animal journals, after years of silence.[45] We have already mentioned Flecknell's 1984 article, which took a stand in favour of using analgesia with laboratory animals, despite the author's acknowledgement of widespread scientific scepticism about animal pain.[46] In June 1985, an article entitled 'Animal pain: evaluation and control' appeared in the American trade journal *Lab Animal*. Interestingly enough, it begins by making the very point cited earlier from Lloyd Davis: namely, that the same clinicians who routinely give antibiotics without documenting the presence of infection will withhold analgesia because they are not certain of the presence of pain. Yet, say the authors, pain and suffering 'may actually constitute the only situation in which one should go ahead and treat, even if in doubt'.[47] To the standard ideology of science's old saw that animals often do not show pain behaviour—for example, 'the clinician's inappropriate opinion that since [ruminants] do not show pain, they probably do not feel pain, and therefore they do not need pain control'—the authors reply as we did earlier in our discussion that evolutionary pressures militate against an animal flagging itself as injured. Whether in pet animals or in research animals, it is the duty of the veterinarian to alleviate pain whenever it occurs. And contrary to the mainstream view which tends to look at animal pain in terms of the machinery involved, the authors emphasize that 'to understand the nature of pain, we must define, identify, and correctly interpret *its meaning to the animal involved*'.[48] The authors have no difficulty in using 'clinical impressions', rather than 'objective' data, to make such judgements as that 'there are breed differences in pain response . . ., with the sporting and working breeds generally showing the highest pain tolerance among dogs'.[49] They then go on to list the signs of acute and chronic pain, and to acknowledge the existence of anxiety in animals. Even more remarkably, they acknowledge a psychological dimension to pain.

> Simply talking to some animals can relieve their pain or alleviate their anxiety. Stroking or petting some animals in pain often reduces the vocalization, restlessness, rapid breathing, or increased heart rate that were noted before such attention. . . . Removing environmental stresses helps reduce anxiety and lessens some forms of pain. . . . Providing a stall companion (goat, sheep) frequently enables horses or cattle to better handle painful conditions.[50]

Unquestionably, this article reflects, and indeed grows out of, the rising new moral awareness of animals. As in Flecknells's article, there is a sense that the authors are aware of being involved in a revolutionary shift in *Gestalt*. In their conclusion, they point out that

> It is possible that as the veterinary community becomes more adept at recognizing pain in animals *and more concerned with its appropriate control*, newer [analgesic] agents will be developed, and drugs already in use will be investigated in a greater variety of species and situations. The greatest potential for such advancement is in the research area. . . . The ability to recognize and evaluate animal pain and the application of means to alleviate or control it is a

substantial challenge for veterinarians. *Doing so may require changes in attitude as well as additional education*.[51]

Two other articles which appeared in 1985 eloquently buttress the points we have been making. One, appearing in a British journal devoted to veterinary practice, takes much the same tack as Flecknell, and Wright and Marcella, though the author is still sufficiently imbued with the common sense of science to deny that statements about animal pain can be 'proved'.

The experience of pain is subjective and it is therefore impossible to prove that animals actually feel pain. There can be no doubt, though, in the mind of both clinician and physiologist that animals can and do experience pain; the relief of pain must be considered a duty of the veterinarian.

In contrast to human medical care, analgesia is often neglected in veterinary clinical practice. This is partly because of the fact that many animals are stoical and only display behavioural changes which do not immediately suggest that they are in pain. For instance, a cat with a fractured limb will tend to hide away rather than howl in agony. A dog badly bruised following a road traffic accident may simply be quieter than usual. Of course overt signs of pain are also seen, but the veterinarian must suspect the presence of pain when the conditions, and personal experience, suggest that the animal could be expected to feel pain.

There is no doubt that, with the judicious use of analgesics, patient morbidity is reduced and quality of life improved. There is no reason for an animal to tolerate pain simply because it does not complain.[52]

The other, entitled 'Guidelines on the recognition of pain, distress and discomfort in experimental animals and an hypothesis for assessment', appeared in the *Veterinary Record* and is designed to provide a quantitative measure not only of pain, but also of the distress and discomfort suffered by laboratory animals. The article is very interesting for a variety of reasons. First, it specifically acknowledges moral and social pressure as instrumental in pushing questions of animal pain into the foreground. Second, it clearly rejects the obfuscatory point raised by some scientists who wish to avoid dealing with pain—namely, that pain isn't 'defined'. The authors point out, quite correctly, that the onus on the researcher who uses animals is to recognize pain and alleviate it, not to define it. And in a direct swipe at the élitism of scientific common sense, the authors return to ordinary common sense as a basis for enumerating signs of pain. Their criteria for identifying pain are drawn from the clinical observations of animal technicians, animal nurses, veterinarians, and researchers who have looked after animals for a number of years, rather than from controlled research by pain experts. They point out that the existence of conflicting signs of pain in research animals should not be viewed with scepticism, but should be seen as a reflection of 'the types of physiological abnormality that exist in a broad spectrum of progressive debilitation in an animal'.[53]

Contrary to the common sense of science, the authors are not afraid of anthropomorphism, pointing out that the use of animals in research is based, as we have argued, on such anthropomorphism. Furthermore, they see a need to recognize not only chronic and acute pain, but, as the article's title indicates, distress and discomfort as well, such as those occasioned by learned helplessness or maternal deprivation studies. They write:

Pain, suffering and distress are subjective phenomena and with the present state of knowledge it is often possible to recognize these states but less easy to define them. Suffering and distress describe unpleasant emotions which people would normally prefer to avoid. Such feelings may or may not be a consequence of severe pain or lasting harm. Pain may be more easily appreciated when it has an obvious cause such as injury to somatic or visceral tissue and in humans the perception of such injury may influence the pain experienced.[54]

In the past, the very existence of unpleasant emotions in animals was disputed, and if noxious states other than pain were recognized at all, their subjective nature was de-emphasized, and they were assimilated to stress, which was then described mechanistically. Indeed, such a liberal attitude as that expressed in the above paper is still the exception, not the rule.

The primarily moral motivation behind it may be seen in the following remarks:

It has proved difficult in humans to quantify pain where physical factors are complicated by psychological factors. Analgesia is not, however, refused to human patients purely because quantification is unreliable and subjective. Similarly, in animals the responsibility to relieve pain should not be avoided because it cannot be quantified. The animal may not be able to say that it is in pain but there are many clinical signs to guide the observer which, with diligent and intelligent use, are unlikely to be misleading.

There are many similarities between animals and humans in anatomical and chemical pathways of pain perception. These similarities are used to justify the validity of animal research for the benefit of man but the reverse may also be true. Therefore, conditions which are painful in humans should be assumed to be painful in animals until behavioural or clinical signs prove otherwise.[55]

Gratifyingly, the work of Morton and Griffiths has been taken up by the Home Office in Britain as a guide for licensees in assessing pain.[56]

All the above papers, like the quotation discussed earlier from the neurology textbook, portray the general shift in attitude regarding animal pain, and notably on the same data as their predecessors had. No new facts to prove the reality of animal pain are adduced; the old facts are simply viewed differently. Ordinary common sense has begun to be reappropriated. The change is at root philosophical and valuational, not empirical, though it has great implications for empirical work—indeed, for what counts as data! These papers, moreover, clearly show the moral basis for this shift. It is important to note, however, that most of these authors are still working within the umbra of the common sense of science (as one would expect them to be). There is still some suspicion, for example, that when one talks about subjective states, one is somehow doing something slightly illicit. But the papers surely betoken a major change in admitting that moral considerations enter into science, in admitting that animal pain is knowable, in recognizing the need for a vocabulary of noxious experiences other than pain, and so on. Such brave, pioneering efforts should be acknowledged and applauded. However, only when the new philosophy of science which acknowledges values and ethics and subjective experience as legitimate aspects of science has replaced the old combination of positivism and behaviourism will deep change occur. Science textbooks must update their philosophy, as well as their facts, and must teach young scientists to critically examine their assumptions, to be sceptical about the existence of *the* scientific method, admit that subjective experiences are like other theoretical notions, and not be afraid to deal with value issues. Until such a conceptual revolution occurs, all change regarding animal pain and the moral status of animals in research will very likely arise only in response to external pressures.

Subjective mental states as explanatory: the case of stress

Neither common sense nor science can proceed without implicit—hence, why not explicit?—commitment to states of awareness in animals. There is no difference in principle between postulating quarks, genes, quanta, and black holes, and postulating mental states. None of the above are directly observable; on the other hand, they are at least indirectly observable, given certain theoretical commitments and presuppositions—for example, to electron microscopy, radio telegraphy, and evolutionary continuity. Why should one make such commitments? Because they provide the best explanation: they allow us to understand, predict, and explain features of the world which we couldn't otherwise deal with.

An excellent example of how attributing mental states to animals can provide the best explanation of many phenomena comes from some fairly recent work on stress. As we saw earlier, talk about stress, defined purely as a mechanical response involving either fight or flight catecholamine release (for short-term stress) or the general adaptation syndrome or activation of the pituitary adrenal axis for long-term stress, was used in large measure as a way of getting around talking about unpleasant mental states or experiences in animals. It was believed that the same mechanical physiological response occurred in an animal regardless of the nature of the noxious stimulus or the context in which it arose. But research done in the past two decades has made it clear that the common-sense notions discussed earlier are valid—namely, that one cannot avoid talking about the psychological, emotional, or mental states of animals being subjected to stressors, and thus that such states are essential to an adequate explanation of the phenomena in question.

Interestingly enough, Hans Selye, the father of the GAS theory, noted such psychological dimensions, but tended to ignore them. In 1950, he reported, *en passant*, that 'even *mere* emotional stress, for instance that caused by immobilizing animals on a board (taking great care to avoid any physical injury), proved to be a suitable routine procedure for the production of severe alarm reaction'.[57] Twenty years later, more than two hundred publications had demonstrated that purely psychological stimuli can elicit a substantial pituitary-adrenal response. In fact, said one pioneer in such research, 'what has emerged is the conclusion that psychological stimuli are among the most potent of all stimuli affecting the pituitary adrenal cortical response'.[58] Such results did little to change the generally accepted, purely physicalistic view of stress, for reasons of the sort we have discussed throughout this book. First of all, as Mason points out:

> Research on physical stimuli in endocrine regulation is generally confined to endocrinological and physiological laboratories quite far removed from the scene of psychoendocrine studies proceeding in departments of psychiatry and psychology. The territorial separation of the two research areas is almost complete, even to the point of extremely limited communication of published findings between the fields.[59]

Second, all science, including psychology, was dominated during this period by the common sense of science which, as we have seen, rules out a priori states of awareness, and thus was conceptually incapable of assimilating data about the effects of such states, or at least incapable of appreciating just how radically they undercut the standard ideology. As we have said more than once, much research was done presuming mentation, but not realizing or officially accepting this presumption.

Once again, in a theoretical nexus in which mental states were anathema, scientists were forced to reinvent the wheel, to make a major and daring leap in order to reintroduce what common sense had always found obvious. Thus Mason asserts:

> Unless one is extremely mindful of the exquisite sensitivity of the pituitary adrenal cortical system to psychological influences, one is likely to subject animals or humans to stimuli such as exercise, fasting, cold, heat, etc. [i.e., allegedly purely physical stressors] under conditions in which either considerable pain, discomfort or emotional reaction is simultaneously elicited.[60]

That scientists should ignore this is, of course, not surprising, when they have been trained to believe that animals don't experience pain, discomfort, or emotional reaction, and that such attributions are at best anthropomorphic and probably scientifically meaningless.

Yet, as Mason discovered, postulating such mental states is necessary in order to explain the results of experiments on animals. He performed experiments in which, although the physical stressors to which the animals were subjected were the same, the emotional, psychological, or cognitive states of the animals or attitudes of the animals towards the stressor were different, leading to radically different physiological signs of stress—that is, radically different levels of secretion of the urinary steroid 17-OHCS. This shows that the animal does not respond to a noxious stimulus merely mechanically and uniformly, but that what makes a stimulus noxious or stressful to an animal is the animal's 'reading' of what is happening to it, its cognitive state, its emotional attitude, all of which require the imputation of mentation to the animal. In Mason's own words:

> In our first effort to study fasting, we simply stopped feeding 2 monkeys out of a group of 8 housed within the same room. The fasting monkeys were, thus, suddenly deprived of their daily food for 3 days while their 6 neighbors were eating as usual. The familiar animal caretaker was coming into the room at the usual times and the sounds, sights, and odors associated with normal feeding were present as usual, but these 2 monkeys found themselves unaccountably

being ignored in the passing out of food. As might be expected, under these conditions the deprived animals often vocalized in apparent protest. In addition, after fasting is underway there is the factor of the discomfort of an empty gastro-intestinal tract as another possible basis for psychological reaction. It is not surprising, then, that 'fasting' under these conditions was associated with a marked increase in urinary 17-hydroxycorticosteroid (17-OHCS) levels in the first monkeys so studied.

We next made an effort to minimize these possible psychological variables in two ways. First, we placed the fasting monkeys in a small, private cubicle where they were protected from non-involved monkeys and from other extraneous influences. Second, they were given fruit-flavoured but non-nutritive pellets which were similar in appearance and taste to their normal diet. Although the monkeys ate fewer of these non-nutritive cellulose pellets than they did of their normal diet pellets, they did eat enough to provide some bulk within the gastrointestinal tract. Under these conditions designed to minimize psychological reaction, we found no significant 17-OHCS response to fasting, although other hormonal changes (epinephrine, norepinephrine, insulin, growth hormone) were observed. Fasting *per se*, then, appears to elicit little, if any adrenal cortical response, but the fasting *situation* as a whole may indeed evoke a marked 17-OHCS response if it includes factors which elicit psychological reaction or arousal of the subject.

This experience proved to be only one of a steadily lengthening series of observations emphasizing the need for painstaking efforts to isolate physical from psychological stimuli in laboratory experiments. Muscular exercise proved to be another case in point. We found it virtually impossible to force monkeys to perform strenuous muscular exercise, such as walking a treadmill or lifting weights for food, without eliciting signs of displeasure or other emotional reactions that could well, quite by themselves, have explained the marked 17-OHCS responses recorded in these experiments. Perhaps the closest we came to minimizing emotional reaction was in the case of a climbing task in which little, if any, urinary 17-OHCS change was observed in some experiments although a considerable amount of work was performed. More recently we have found that in human subjects exercising under relatively pleasant non-competitive conditions, there is little if any change in either plasma or urinary 17-OHCS levels, even at relatively substantial workloads. . . .

Perhaps the most compelling data we have obtained thus far along these lines have come from studies of heat as a physical stimulus. In our initial heat experiments, we suddenly elevated the temperature in a chamber housing a chair-restrained monkey from 70 degrees to 85 degrees F within a matter of a few minutes. In this experiment a urinary 17-OHCS elevation was observed, very similar to the response recorded in earlier experiments in which monkeys were abruptly exposed to a cold environment. By this time we had become very alert to the problem of interfering psychological variables and were careful to avoid any unusual handling or change of location or strange activities in the laboratory. The only variable which we might have overlooked as a possible source of spurious psychological reaction appeared to be the rapidity of the temperature change, the sudden 'thrusting', as it were of an animal into a hotter environment. The experiment was then repeated with a more gradual, but equally large temperature change, at the rate of a 1 degree F elevation per hr. Under these conditions, there was not only no 17-OHCS elevation in response to heat, but an actual *suppression* of 17-OHCS levels was observed. Furthermore, this suppression of 17-OHCS levels persisted over a period of several weeks during which the temperature was maintained at 85 degrees F, and was terminated only when the temperature was again lowered to 70 degrees F. We have since observed the suppressing effect of heat on 17-OHCS levels in a number of additional experiments.[61]

Neither Mason nor I would wish to rule out the possibility of the psychological features in stress coming to be correlated with some as yet unknown neurophysiological mechanism. I have little doubt that every mental state is correlated with some brain state. Obviously, this issue plunges us headlong into the mind–body problem; but for our purposes, we can skirt it. For whatever one's view—whether one feels that mental states and physical states are correlated (Descartes), run parallel (Leibniz), are aspects of the same thing (Spinoza), or that body is a subset of mind (Berkeley), or mind a subset of body (materialism), the conclusion is the same. One must admit on the strength of Mason's work that what we call 'mind', cannot now describe physically, and speak of in terms of awareness, cognition, and so forth is as much present in animals as in humans, and has explanatory and predictive power. Thus our point is made: mental states in animals do function as theoretical entities, generating the best available explanations.

It is unfortunate and tragic that such a reappropriation of common sense should come to science at the expense of the animals used in Mason's research and in other similar research. Any ordinary person not steeped in scientific common sense would be surprised not at Mason's results, but that anyone could have doubted the conclusion drawn from them. One can but hope that such work will be widely disseminated, so that there will be no need constantly to prove that the emperor is not wearing clothes. Further, one can attempt to affect the direction of research, as we shall see in the concluding chapter, so tha the science which studies animal consciousness does so with full mora respect for animals as objects of moral concern.

I have argued that what makes something stressful is that it is experienced as such, a point ignored by the mechanistic ideology of science. Mason makes much the same point in the article we have been discussing, and in fact goes beyond it. Not only does ignoring animal awareness lead to inexplicable differences in physiological responses in the presence of identical stressor stimuli; it casts doubt on much physiological stress research which has failed to control for psychological influences by defining them out of existence. Furthermore, the failure to recognize just how delicately balanced an animal's bodily functions are relative to its mental and emotional states puts a great deal of biomedical research in jeopardy. How can one be sure, for example, in studying some carcinogen, disease, or drug that an animal's physical state is not as much or more influenced by its fear of being handled or its boredom as by the physical variable?

How ignoring mental states may jeopardize research

In a recent address and a book chapter dealing with the use of swine in biomedical research, I argued a point relevant to our discussion here.[62] There is currently great interest in increasing the use of swine in biomedical research, both because swine are closer to humans than mice, rats, and dogs in their cardio-vascular and digestive systems, and, more ominously, because there are no legal constraints on what can be done to swine and how they are to be cared for or housed. Thus, a recent article frankly admits that in the author's university, research dogs will be replaced by swine. 'Since existing standards within the swine industry clearly demonstrate a lack of public concern about the fate of a pig, this will minimize the opportunities for those opposed to use of animals for research'.[63] The Animal Welfare Act simply exempts farm animals (as well as rats and mice) from its limited coverage, and NIH guide-lines for laboratory animals, mandatory for all grantees of federal research funding, aside from space standards, require only that the care and housing of farm animals be comparable to those on 'a well-run farm'. The latter is surely absurd, since a well-run farm is one which is economically profitable, and there is a great deal of literature showing that intensive agricultural management techniques, while generating profit, involve levels of stress on animals which, if they existed in a research colony, would badly skew data based on animals therefrom. Thus, from a scientific point of view, without strict standards for care and husbandry, the data from different laboratories are likely to be wildly incommensurable. (As of 1987, NIH shows signs of moving towards more detailed standards for farm animals used in biomedical research.)

Pigs are acknowledged to be highly sensitive animals, and are thus very susceptible to being upset. A vast literature is extant chronicling the sorts of stressors which affect a variety of physiological and metabolic variables. Social isolation and confinement lead to reduced libido, poor copulatory performance, low sperm count, and poor fertility and fecundity.[64] Early weaning of piglets leads to abnormal behaviour, increased adrenal weight, and pathological changes in the small intestine.[65] Variation in environmental conditions in the neonatal period can change brain chemistry in piglets. Piglets removed from a sow at 12–14 days of age and kept in separate cages were found to have a lower concentration of catecholamine metabolites and of the enzyme tyrosine hydroxylase in the caudate nucleus than did their siblings reared by the sow.[66] As suggested in our earlier remarks, purely psychological factors can be as significant physiologically as more obvious physical deprivations or assaults. In pigs, exposure to a new environment increases their corticosteroid level more dramatically than does electric shock.[67] Kilgour has shown that an awareness of social facilitation among pigs can cut post-weaning weight loss and eliminate growth check. He points out that since, under natural conditions, weaning is gradual, this should be approximated under confinement conditions.[68] It is also known that play behaviour, and therefore room to play, is of great significance in the development of piglets.[69]

Another, related area of concern is handling. Few laboratory animal facilities control for the personality of handlers. Dairymen have always known, and research has shown, just how significant this variable is to animal performance, health, and welfare.[70] Barnett and co-workers studied the effects on pigs of being gently stroked versus slapped or shocked. The pigs which were treated badly showed significantly less growth and significantly lower feed conversion. The pregnancy rate for badly treated gilts was 33.3 per cent, whereas that for well-treated gilts was 87.5 per cent. Boars handled roughly had smaller testicles at 160 days of age and reached behavioural puberty later.[71] Acquired fear of humans has been shown to reduce reproductive performance of sows.[72] In the face of these sorts of considerations, as well as common sense and common decency, the traditional 'macho muscling' of pigs used in research is indefensible, especially given the existence of a comfortable minimum-stress method of restraint.[73]

In the face of such data, it is self-defeating scientifically to use swine for research. Though researchers may be drawn to use swine because there are few constraints on their treatment, this very lack militates against standard, trustworthy, reliable data which can be compared from laboratory to laboratory. Thus, aside from moral reasons for prescribing standards of care and use, there are scientific reasons as well. Such standards, however, cannot be based merely on seeing animals as machines which have certain needs for fuel and space. Given all that we have said about stress, standards should control not only for physical needs, but for psychological ones too. Thus conditions like boredom, anxiety, and fear should be avoided.

In my view, standards of care, husbandry, and use of all laboratory animals should be based on what makes an animal *happy*, not merely on avoiding pain and distress. Happiness is the theoretical notion which best captures what we are after, both in wanting to avoid noxious experiences for the animal and in wanting to maximize its well-being. It is plausible to suggest that happiness resides in the satisfaction of the unique set of needs and interests, physical and psychological, which make up what I have called the *telos*, or nature, of the animal in question.[74] Each animal has a nature which is genetically and environmentally constrained, from which flow certain interests and needs, whose fulfillment or lack of it *matter* to the animal. Psychological dimensions are most important; physical needs like hunger and thirst are processed psychologically, and are experienced as either deficiency or satisfaction. In the case of the domestic pig, contrary to claims made by many confinement agriculturalists, its *telos* is not that far removed from that of its wild relatives. Indeed, Stolba has shown that domestic pigs placed in an ecologically rich environment display all the behaviour patterns of wild pigs.[75] Obviously, extreme confinement, isolation, roughness in handling, failure to habituate the animal to invasive procedures, and so on are likely to make the animal unhappy, and thus introduce unknown, uncontrolled stress variables. As mentioned earlier, new regulations seem to be beginning to move towards concern for making animals happy, not merely avoiding pain and suffering. Keeping animals under conditions which are convenient for us—for example, keeping rats in stainless steel cages—must give way to conditions oriented towards the animals' happiness—rats, for example, are nocturnal burrowing creatures, and some provision needs to be made for burrowing.

The work of Mason and others has shown unmistakably that the degree to which an invasive stressor affects an animal, physically and psychologically, has a strong relationship to the animal's mental state, emotional and cognitive. Weiss demonstrated that rats which could predict an electric shock (because it was consistently preceded by an audible tone) developed virtually no gastric ulcers, whereas other rats receiving the same shock, with the tone sounded randomly with respect to the shock, developed a great deal of ulceration. Similarly, rats which could control the shock, and thus cope with the noxious stimulus, developed far fewer ulcers.[76] Again, these experiments bespeak the explanatory power of attributing mentation to animals; indeed, they cannot be explained without such postulation. This is consonant with the currently popular, but barbarous, work done in the United States on learned helplessness, which shows that animals receiving inescapable shocks inevitably develop a state of pathological withdrawal and inaction. This work is morally defended by researchers on the grounds that it models human depression—another eloquent attestation to the theoretical relevance of imputing mental states to animals, surely.

The major point is clear: analysis of the concept of stress in animals is impossible in purely mechanical-physical terms. Stress responses depend on an irreducible element of mentation, emotional and cognitive. The same stimulus can produce highly different physiological responses, depending on the mental state of an animal, whether it knows the stimulus is transient, can predict it, control it, is frightened during administration of the stimulus, is isolated from its fellows, can perceive other animals eating when it is being starved, and so forth. All this suggests that there is nothing mystical about attributing mental states to animals. Mental states are just like any other theoretical entity, not directly observable but having explanatory and predictive power. Indeed, clinical evidence supports the application of the principles discussed above to pain as well as to stress. As we saw earlier, Wright *et al.* (1985) point out that stroking or petting an animal mitigates its pain, as do providing stall-mates for some animals, and reducing environmental stresses. In other words, in animals as in people, one's mental state determines one's experience of pain. Our discussion also shows the power of the ideology of science. While common sense has always seen stress as inseparable from mentation, physiology artificially rent them asunder. And, more bizarrely, even though psychology studied the major effects of psychological states, it refused to acknowledge them as mental states, because it was ideologically committed to their non-existence. Thus, throughout the reign of behaviourism, psychology studied what can only be called mental states and states of consciousness, but refused to admit that it was doing so until the mid-1960s in the case of humans and the mid-1970s (if at all) in the case of animals.

We have seen throughout our discussion that there is no good reason to adhere to the ideology of science regarding animal pain—indeed, such adherence is abhorrent to common sense, inimical to moral thinking, and harmful to science itself. Open admission of animal mental states will surely cause many procedures in science to be perceived as more morally problematic than they have been in the past, but this is salutary. For science is not value-free, and only by exposing its values to strong light can one assure their coherence and defensibility. Moral theory and a science of all animal consciousness, not merely pain, will inevitably stand in a dialectical relationship, for the burgeoning questions regarding the moral status of animals lead to a variety of questions about animal awareness which science must try to answer, and the study of animal awareness in turn generates its own moral questions.

8 Consciousness lost

Consciousness under behaviourism

That behaviourism was the dominant approach to animal psychology for fifty or so years is unquestionable. As D.E. Broadbent, a leading British psychologist wrote in 1961, behaviourism was dominant among 'those people in the English-speaking countries who engage in pure academic research in psychology'.[1] And though there were many variations on the Watsonian themes we discussed (even in their treatment by Watson, who apparently viewed consistency as a hobgoblin of little minds[2]), the tendency to deny the existence of mental experience in humans or animals, or less radically, its relevance to a scientific psychology, tended to be a major theme of virtually all mainstream behaviourists, at least in their stated ideology, with the very limited exception of Tolman. Clark Hull's stimulus–response psychology, dominant in the 1940s, does not depart in this regard from the Watsonian themes we have discussed. In his effort to make psychology scientific, Hull felt compelled to banish consciousness from its purview:

> After many centuries the physical sciences have largely banished the subjective from their fields, but for various reasons this is far less easy of accomplishment and is far less well advanced in the field of behavior. The only known cure for this unfortunate tendency to which all men are more or less subject is a grim and inflexible insistence that all deductions take place according to the explicitly formulated rules stating the functional relationships of A to X and of X to B. This latter is the essence of the scientifically objective. A genuinely scientific theory no more needs the anthropomorphic intuitions of the theorist to eke out the deduction of its implications than an automatic calculating machine needs the intuitions of the operator in the determination of a quotient, once the keys representing the dividend and the divisor have been depressed.[3]

To guard against the danger of 'anthropomorphic subjectivism', Hull suggests that we 'regard . . . the behaving organism as a completely self-maintaining robot, constructed of materials as unlike ourselves as may be'.[4] Interestingly enough, the words 'consciousness' and 'awareness' do not even appear in the index to his *Principles of Behavior*.

Arguably, B.F. Skinner is the last major behaviourist figure, and his writings sound essentially the same themes as those first articulated by Watson: the irrelevance of subjective experience to psychology (although, like Watson, Skinner is inconsistent on this[5]), the restriction of science to what is observable, the emphasis on control of behaviour as the chief justification for psychological research, and the vision of a utopian society moulded and shaped by behavioural technology and objective science. In short, behaviourism continued to adhere to the valuational basis developed early in its history and, at least in Skinner's hands, became scientism as much as science—an ideological creed formed from positivism and a fervent belief in science as curer of all social ills. Common sense has nothing to contribute; it is 'science or nothing'.[6]

Although behaviourists were not interested in animal consciousness, they were extremely interested in animals. The bulk of behaviouristic psychological work has been animal work, primarily with rats, pigeons, and cats, with an occasional non-human primate. (Indeed, emphasis on this sort of research earned behaviourists the sobriquet 'rat runners'.) And though they were non-Darwinian in their failure to trace subjective experience through the phylogenetic scale as Darwin and Romanes had done, they were consistent in their methodological, if not ontological, rejection of consciousness at all levels. Their tendency to focus on relatively few species of animals, which they studied under laboratory conditions, reflected their own version of Darwinism—namely, the belief that the key concept for psychology was learning, and that learning was pretty much the same all along the evolutionary scale, differing only in degree or complexity.[7] Thus principles of learning arrived at by studying one sort of organism gave conclusions which were believed to have universal validity.

In his 1939 presidential address to the American Psychological Association, Gordon Allport alludes to this tendency, remarking that whereas in the late nineteenth century animals were studied for the purpose of understanding animals, 'today animals are not studied for their own sakes, but rather for what Fabre, the naturalist, called the "universal psychology revealed by all animals from insects to *Homo sapiens*" '.[8]

According to Allport, admittedly a critic of behaviourism, in its efforts to wear the trappings of real science, behaviourism has ruled out the traditional and interesting questions of psychology:

The body-mind problem, never solved, has been declared popularly null and void. Dualism evokes rejection responses of considerable vehemence. (Indeed, of all philosophical problems, psychologists seem most allergic to this.) Higher mental processes as exhibited in the speech of human beings are relatively neglected, marked preferences being shown for studies of non-verbal behavior and for animal subjects.[9]

At the 1939 American Psychological Association meetings, Allport informs us, fully 25 per cent of the papers were based on animal research. (The full significance of this becomes apparent only when we realize that these meetings are a forum for all psychologists, not merely experimentalists, and that social psychologists, clinical psychologists, and psychotherapists constitute the vast majority of those attending.)

The degree of blind faith in animal research as the real key to psychological progress is vividly depicted by Allport when he tells his listeners that

A colleague, a good friend of mine, recently challenged me to name a single psychological problem not referable to rats for its solution. Considerably startled, I murmured something, I think, about the psychology of reading disability. But to my mind came flooding the historic problems of the aesthetic, humorous, religious, and cultural behavior of men. I thought how men build clavichords and cathedrals, how they write books, and how they laugh uproariously at Mickey Mouse; how they plan their lives five, ten, or twenty years ahead; how, by an elaborate metaphysic of their own contrivance, they deny the utility of their own experience, including the utility of the metaphysic that led them to this denial. I thought of poetry and puns, of propaganda and revolution, of stock markets and suicide, and of man's despairing hope for peace. I thought, too, of the elementary fact that human problem-solving, unlike that of the rat, is saturated through and through with verbal function, so that we have no way of knowing whether the delay, the volition, the symbolizing and categorizing typical of human learning are even faintly adumbrated by findings in animal learning.[10]

And in a number of passages which must surely have made the behaviourists in his audiences squirm, Allport reminds them that the mark of a mature science is its ability to predict and control, and that thus far, animal research shows all the trappings of scientism, but has given us no insight into, or ability to control, any aspect of human behaviour.

At the same time, behaviourists' commitment to learning as *the* mark of the mental, as well as their valuational commitment to the perfectibility of humans, led to a deep hostility towards instinct, biological bases of behaviour, or notions of innateness. In the eyes of behaviourists, innateness was suspect on many levels. For one thing, to them as to all empiricists and positivists, claims about innateness of mind smacked of mysticism, being both unverifiable and subject to misuse as an explanatory *deus ex machina*. Second, and equally important, innate constraints suggested that behaviour was not infinitely malleable, and therefore that one could not eliminate social 'evil' through scientific behavioural technology. Innateness was essentially 'un-American', in that it constrained human possibility, mocked the behaviourist (and democratic) dream of creating kings out of cabbage-heads. Ironically, genetic determination was as much anathema in freedom-loving America as it was in Stalinist Russia; for both countries it represented a stumbling-block to rational social planning. Stalin, of course, embraced Pavlov and Lysenko, and simply forbade Darwinism, even in biology. American behaviourism simply ignored genetic determination of behaviour, and focused exclusively on learning.

European approaches to animal psychology: ethology

As one might expect, such a move appeared problematic to many, and in Europe, new approaches to animal psychology arose, designed primarily to bridge the glaring gap between Darwinism, the accepted basis of biological science, and a psychology which ignored the existence of heritable constraints. Behaviourism, while rejecting traditional Cartesian dualism, had created a new dualism of mind and body, with body subject to genetic laws and determinations, but mind (or behaviour) somehow exempted from anything but environmental control.

The rise of ethology, or the science of animal behaviour, in Europe was, in many ways, a return to Darwinism. The primary figures, Lorenz and Tinbergen (later awarded the Nobel prize for their work), were primarily biologists and zoologists, rather than psychologists, and thus found the study of animal behaviour conceptually inseparable from its evolutionary, and hence genetic, base. Furthermore, and consonant with an evolutionary, biological perspective, they were interested in animal behaviour in its natural environment, since, from a Darwinian point of view, behaviour evolves under environmental pressures and constraints; so it is essentially meaningless to study it purely or even primarily under laboratory conditions. Thus a natural-historical perspective was restored to the study of animal behaviour, and, since the establishment of controlled conditions for natural observation is impossible, so too was an element of anecdote, though of course the reports considered valid are by 'trained observers'.

Furthermore, Lorenz, at least, did not hold as a value the perfectibility of humans through learning and the shaping of environment and behaviour. On the contrary, he was strongly oriented towards genetics, a position which later led him to find Nazism not uncongenial. Interestingly enough, yet predictably, this political and moral aspect of Lorenz's work turned out to be a significant factor in retarding the acceptance of ethology in the United States, and in making it suspect. Even now, it is the subject of scholarly articles in scientific journals,[11] once again showing that the rejection and acceptance of scientific views is hardly the value-free process that the common sense of science would have us believe it is.

All this was sufficient to put ethology at loggerheads with mainstream behaviourism. Whereas behaviourism placed a premium on learning, ethologists, while certainly not ignoring learning, emphasized its biological context. Whereas behaviourists emphasized the unity of learning behaviour across species, ethologists were interested in behavioural diversity and explaining each species' behaviour in terms of its adaptive value.

For much of both of their histories, behaviourism did not engage ethology so much as ignore it, though at any rate some of their development was roughly contemporaneous. When the two approaches did interact, the differences were readily apparent. Karl Lashley, for example, a highly respected, thoroughgoing, but physiologically oriented, behaviourist whose work we alluded to earlier, succinctly outlined behaviouristic objections to ethology.[12] American psychology, said Lashley, is not interested in animal behaviour *per se*, but in learning. As a result, it has no need to study a great variety of species. Furthermore, ethology is natural history, not science; it is fundamentally descriptive, not based on controlled study. Third, behaviourism for the most part rejects the notions of instinct and innate tendencies, as well as explanation of behaviour by appeal to biological wiring; whereas such conceptual baggage is essential to ethological investigation. On the other hand, such biological concepts are not explored in neurophysiological terms in ethology, but rather in terms of life history and ecology—a major debit, in Lashley's view.

As a result, Lashley points out, ethological work, written primarily in German, was essentially unknown to American psychologists until the publication in 1957 of *Instinctive Behavior*, edited by Claire Schiller, a volume which collected and translated for the first time many of the classic papers of Lorenz and Tinbergen. Lashley is also troubled by at least some of the earlier figures in ethology, notably Jakob von Uexküll, one of Lorenz's teachers, who was openly interested in the mental life of animals, though again within the context of their natural environments, and whose work we shall consider shortly. Lashley views this element as an unfortunate continuation of the Romanes tradition, one from which later ethologists never escape, citing in his indictment their major concern with 'sensory components of behavior'.[13]

Not surprisingly, Tinbergen, one of the fathers of ethology, writing in the same volume as Lashley, also marks major differences between behaviourism and ethology. Remarking that 'whenever I meet American behaviourists I am struck by the very great difference in approach between them and us',[14] he proceeds to outline some of those differences. The overriding one, he suggests, is that ethology is rooted in biology, behaviourism in psychology. Thus ethology is interested in evolutionary and adaptive explanations of behaviour, whereas psychology has essentially ignored these, a glaring omission for a life science at a time when everyone realizes that all such science must operate within an evolutionary framework. (Elsewhere in the volume, Lorenz makes the same point, asserting that behaviourists 'proceed as if there were no such thing as evolution and phylogenetic adaptedness of form and function to the ubiquitous necessity of survival'.[15])

Second, whereas behaviourists are interested in learning, ethologists are interested in the fact that 'learning is a change in something, and that it must pay to study the something before a change occurs'.[16] At the same time, he admits that ethologists have erred in the other direction, paying insufficient attention to learning. Finally, ethology is interested in describing the full, natural behavioural repertoire of organisms, and is closer to natural history; whereas behaviourism is interested in discrete units of behaviour, and is thus more closely aligned with 'analytical and experimental procedure'.[17]

Thus, behaviourism and ethology represent quite different, incommensurable approaches. Is it therefore the case that ethologists eluded the strait-jacket imposed by the common sense of science, and were 'softer' than behaviourists on consciousness and subjective experience in animals? On the whole, the answer is no. Ethologists were just as susceptible to scientific ideology and just as anxious to demonstrate their fealty to it, which meant either denying or ignoring questions of subjective experience. Thus, in the first substantive paragraph of his preface to the aforementioned volume designed to introduce his and other ethologists' work to the American scientific world, Tinbergen declares that behaviourism and ethology

> are fundamentally the same in that both have as their object an observable phenomenon, in that they are both 'objective'. Both schools study behavior, and not a mysterious 'psyche', nor unobservable subjective phenomena. Both aim at descriptions, and at formulations of problems, concepts, and conclusions that are as objective as any other natural science. It is true that both often use words borrowed from everyday language, words which, stemming from a period in which subjective and objective thinking were not yet separated, are charged naturally with blended subjective and objective meaning. But both try to overcome the dangers of this by giving new, purely objective definitions. This fundamental identity of approach justifies the expectation that both fields will ultimately merge into one.[18]

Thus Tinbergen enunciates the same scientific ideological notions which characterize the common sense of science, according to which subjectivity and consciousness are not legitimate objects of study. And the subsequent history of mainstream ethology, in both its research publications and its textbooks, did not deviate from these norms until the 1970s.[19] Like students who took psychology classes to learn about the mind and consciousness, students who enrolled in ethology classes to study the minds of animals were sorely disappointed. They encountered nary a mention of consciousness, and, indeed, the term simply cannot be found in mainstream animal behaviour textbooks, at least until the late 1970s. Neither Manning's *Introduction to Animal Behavior* (1979) nor Lehner's *Handbook of Ethological Methods* (1979) even make passing reference to consciousness or awareness in animals. Instead, they chronicle or press methodologies which stress the 'objective facts' of animal behaviour characteristic of the genetic heritage of particular species—aggressive behaviour, courtship behaviour, and so on. Moreover, the desire for genuine scientific status still further blunted the inherently interesting nature of the subject-matter with the development of 'quantitative ethology', an attempt to mathematize and quantify the data collected. Unfortunately, the mathematization is typically forced and artificial, substituting formulas for insight, as a great deal of sociology does, and concerning itself with such issues as 'Cleaning behavior in centraichid fish', 'Thermal limitations to escape responses in desert grasshoppers', 'Sexual behavior in unisexual lizards', 'Influence of male plumage on mate selection in the female mallard'—all titles chosen at random from a variety of volumes of the journal *Animal Behaviour*. I could find virtually no explicit references to consciousness in any issue of the journal.

Lorenz and consciousness

The key point is that most ethologists treated consciousness in animals the way behaviourists did, as either non-existent or not amenable to study. Interestingly enough, although Lorenz's scientific work does not deal with consciousness and awareness, he specifically addresses the question of animal consciousness in a number of places. This is not surprising, since Lorenz was strongly grounded in German philosophy, interested in philosophical questions, and wrote frequently on epistemological questions—indeed, the second volume of his *Studies in Animal and Human Behavior* (1971), the first extensive collection of his works to be published in English, 'is intended to convey to the interested reader some understanding of the philosophy as well as of the theory of knowledge on which ethological approach is based'.[20]

From a philosophical point of view, Lorenz is an identity theorist. Like Spinoza and Lloyd Morgan, he viewed mind and body as two aspects of the same thing:

> Of the three attitudes it is, in theory, possible to take towards the body-soul problem, one assuming identity, the second interaction, and the third parallelism between both, I find myself compelled to profess the one first mentioned. For the practical purposes of psychophysiological research it is irrelevant which of these three positions one assumes, as long as one keeps conscious that none of them release us from the obligation to keep constantly in mind the existence and the nature of the impenetrable partition between our two fields of knowledge.[21]

For Lorenz, our knowledge of other human beings as 'a unity of body and soul (or mind)' is not based on inferences from analogies between their behaviour and mine. Rather, reflecting his Kantian commitments, Lorenz sees this knowledge as presuppositional or categorical, 'which the cognitive functions I am endowed with make it impossible for me to doubt', a point similar to one we argued earlier.[22] Lorenz does not go on to say quite the same thing about animals; but he hints at it when he asserts that 'my knowledge about the subjective experiences of my fellow men and my *conviction* that higher animals (such as a dog) also experience things are two quite closely related phenomena'.[23]

It appears that Lorenz's position comes down to suggesting that we can't help attributing consciousness to either humans or higher animals because of the way our minds are constituted, but that we can have only belief about animal minds, while we can have *knowledge* of human minds. 'In human beings, of course, psychology proper, that is to say, the investigation of Man's subjective phenomena, can be pursued by the method of true induction.'[24]

This is presumably the case because of the possibility of first-person reports. But regarding animal consciousness, one has no such access, and thus animal consciousness cannot be part of science proper:

> The student of animal behaviour is in a highly conflicting situation regarding the problem of the relationship between objective, physiological, and subjective, psychological processes. While fully aware that it is only the former that are accessible to the methods of inductive research, he cannot help believing in the reality of the latter. No normal man is able to avoid that type of 'empathy', and it is simply not honest not to confess to it. After all, we profess it by refraining from inflicting torment on animals. But, at the same time one must keep strictly aware that empathy does not convey to us even the slightest degree of knowledge. Though I 'know' that the tame greylag goose accompanying me is 'frightened' when it stretches its neck upwards, flattens its feathers against its body, and utters a certain note, it is not any empathic flow of communication flowing from the bird's soul to mine which gives me information about its behaviour. It is the exact opposite of this process that has taken place. From experience, repeated literally thousands of times, I have learned that in a goose performing just those expression movements all thresholds of escape responses are extremely low and that, in all probability the goose will fly off the next moment. If I admit, as I don't hesitate to do, my own belief that the goose does experience, at that moment and in correlation with that behaviour, something like fright, this belief has nothing to do with scientific knowledge. But it is based on scientific knowledge if I predict that the goose will fly off, for science is, in F. Fremont-Smith's aphoristic definition, 'Everything that makes things predictable'.[25]

This passage is repeated almost verbatim in Lorenz's 1963 essay 'Do animals undergo subjective experience?' The essay essentially replicates the points we have made so far. We inevitably believe in animal subjective experience; such belief is not scientific knowledge; analogical arguments for subjective experience weaken progressively as we move away from fellow humans. Though mind and body are identical, he believes, the mind–body problem and all questions relating to it are fundamentally unanswerable, and thus one can only speculate.[26] Additional points are also made. Especially interesting is Lorenz's warning that one not equate subjective experiences only with highly sophisticated and evolutionarily advanced new physiological processes.[27] Some very basic physiological mechanisms are there to let us know that something is wrong without telling us what or where—feeling ill, feeling sea-sick; others, like pain, tell us where.

> Such very simple nervous processes accompanied by experience almost always have the plus or minus sign of pleasure or displeasure and they have a positive or negative conditioning effect on the behaviour which produced them. 'That was good, do it again soon!' is the encouragement of pleasure; 'Leave that alone in future!' is the warning of displeasure. This is the subjective side of the process which Ivan Petrovitsch Pavlov has termed the 'conditioned reflex'. Such positive or negative conditioning mechanisms, prior to any individual learning, already incorporate phylogenetically acquired information. They 'know' in advance what is good and what is damaging for the continued existence of the organism.[28]

On the other hand, some very complex neurophysiological processes are performed without any 'emotional participation' or conscious experience accessible to introspection. An excellent example is all aspects of our perceptual system, which constructs objects out of sense-data—perceptual constancy effects are a paradigm case.

We are most conscious of ourselves in three kinds of situation: when we must make important decisions, when something goes wrong, and when we have pleasant experiences.

> The message of displeasure ('that was wrong') and of pleasure ('do it like that again') are probably the most powerful generalized and abbreviated forms of information contained in our ego. The 'ability to experience pleasure and sorrow' . . . is doubtless the primeval form of experience. And this is exactly what I feel can be attributed to higher animals.[29]

This attribution is based somewhat on analogy with the conditioning effects of pleasure and pain in humans. But more convincing for Lorenz is the fact that animals have generalized behaviour not betokening a particular sort of negative experience, which could be thought to be mechanically connected with the experience or event in question, general indicators of pleasure and displeasure which span a variety of different experiences.

> . . . very many higher animals exhibit motor and vocal display patterns which do not express a special kind of pleasurable or unpleasant experience, but indicate pleasure or displeasure in general. One can at once see that a dog is sad, but not why this is the case. With a young greylag goose, the discontentment

call (which we usually refer to briefly as 'crying', and which alarms the mother goose) can be heard in exactly the same form when it has lost its parents, when it is hungry, when it is cold, or when it wishes to crawl beneath the parent and go to sleep. In brief, the call is given in all unpleasant situations. It will, for example, be uttered even after fledging by a goose which is not yet independent of its parents, when it has fallen through thin ice on the lake and cannot get out. The goose does not readily hit upon the solution of flying out of its predicament. On one occasion, an amusing experience with a very tame young snow goose provided me with some extremely convincing evidence of the 'you' that such a bird also possesses the generalizing negative experience of displeasure. The goose had been greatly spoilt in order to ensure that it would be as closely attached to me as possible. Every day, as I went down to the lake, I used to take it a handful of wheat. On one occasion, the wheat was exhausted, and I had taken along oats as a substitute. The goose—whose name, characteristically enough, was 'Little Princess'—delightedly flew up to me from some distance away. It was just about to pick greedily at the grain when it noticed that it was not wheat, but oats. 'Little Princess' began to cry loudly and heartbreakingly, just like a small child whose doll has been taken away. In such a situation, one automatically feels that an animal really experiences things. My teacher Heinroth, the grandfather of objectivized behavioural research, was often reproached with the criticism that he treated an animal like a machine with respect to its physiological processes. He often used to reply: 'On the contrary, animals are emotional people with very little ability to reason.'[30]

Thus, although Lorenz is willing to admit the existence of animal subjective states, he, like methodological behaviourists, excludes them from science. Indeed, he goes so far as to correct McDougall's use of a term like 'appetite', which smacks of subjectivism, and substitute the more objective 'appetitive behaviour', an approach which dominates ethological work.[31]

Vestiges of consciousness: America and Britain

So ultimately ethology, like behaviourism, eschewed the study of consciousness for much the same sort of common sense of science reasons. It should not be assumed, however, that absolutely all psychologists and ethologists for the fifty or so years in which consciousness was anathema to the mainstream accepted this verdict. Psychology (and ethology) are not able to command full international allegiance of everyone in the field to one doctrine, as physics and chemistry can. There always have been, and probably will be for the foreseeable future, divergent schools of psychology and ethology which, while not dominant, are not wholly devoid of respected adherents. To use

Kuhnian terms, these fields have never fully operated under a dominant paradigm which precludes any deviance. And it is thus no surprise that one can find examples of thinkers who did not reject the legitimacy of studying animal mind and consciousness even in the heyday of behaviourism and objective ethology.

Tolman's purposive behaviourism

It is sometimes suggested that Edward Tolman's 'purposive behaviorism', as discussed in his 1932 work *Purposive Behavior in Animals and Men*, which recognizes the need to talk about purposes and goals in animal behaviour, is an example of such a heterodoxy. Yet, if one examines his work, one finds that Tolman takes pains to separate himself from 'mentalism'. His interest is in finding observable ways of cashing out notions like 'purpose', without resorting to unknowable contents of consciousness. His work, which he characterized as 'molar', contrary to Watson's 'molecular' interests, still deals with behaviour; but it is more interested in wholes than parts. In other words, as his autobiographical essay attests, he was interested in going beyond mere responses, or 'muscle twitchism', as he characterized Watson's thinking, to concepts which defined relationships between organism and environment, such as purpose, emotion, and consciousness.[32] On the other hand, he describes these concepts in purely behavioural terms. Thus, for him, consciousness in a rat is defined as 'sampling behavior' or 'running back and forth behavior'.

> Our doctrine, now, is that such a running, or looking, back and forth—such a moment of 'sampling' defines, or rather *constitutes*, a conscious awareness on the part of the rat—an awareness of the white as contrasted with the black, or of the black as contrasted with the white, or of both together in a 'step-up-ness' or 'step-down-ness' whole. The running back and forth constitutes, that is, an awareness of this differential aspect of the field which, previous to this instant of consciousness, becomes decisively more determinative.[33]

Like Sommerhof and Braithwaite later, he defined purposive behaviour functionally, as behaviour which persists and is plastic until a given goal is achieved:

> *Purpose (purposive)*. Is a generic term for one of the two classes of immanent determinants (q.v.) of behavior. A purpose is a demand (q.v.) to get to or from a given type of goal-object. Such a purpose is testified to objectively by the fact that behavior tends to persist to or from and to show docility (q.v.) relative to getting to or from specific types of goal-object (q.v.) or goal-situation.[34]

Whatever Tolman is saying, he does not seem to think that he is affirming privileged access to subjective experiences. Like Thorndike before him and Ryle after, he seems to be cashing out words like 'consciousness' in terms of behaviour in an environmental context which is as accessible to others as to the subject. Thus he asserts that 'our psychology is a purposivism, but it is an objective behaviorist purposivism, not a mentalistic one'.[35] Tolman also asserted that 'immediate experience . . . may be left to art and metaphysics'.[36]

McDougall's unabashed mentalism

Tolman was seen as a mentalist by behaviourists and as a behaviourist by mentalists. Much more frankly mentalistic in his willingness to attribute subjective experiences to animals is William McDougall, whose differences from Tolman were characterized by the latter as follows:

. . . whereas, for McDougall, the hormic drives, and their dependent purposes and cognitions, which are resident in the instincts, seem to be in the last analysis, mentalistic, introspectively defined affairs, for us, they are, as has been emphasized perhaps ad nauseam, but functionally defined entities—quite objective variables invented to be inserted into the objectively definable equations which exist between stimuli on the one side and responses on the other. Thus, whereas for McDougall the objective behavior facts of purpose and cognition are a mere external testimony—a testimony to a probably ultimate dualism in nature—a testimony to the fact that mind is somehow, in some degree, metaphysically other than body—for us, these same facts of purpose and cognition are but an expression of certain very complex activities in organic bodies.[37]

McDougall, as his 1920 volume *Body and Mind: A History and A Defense of Animism* shows, is frankly psycho-physical, 'a dualist'.[38] Neither animal nor human behaviour can be explained without postulating the existence of mental substance. The facts that animals must synthesize disparate sensations into unified experience even as Kant argued we must do, that animals experience different portions of the world in terms of their meaning or significance for the purposes they have, and that behaviour is purposive, and purposiveness implies foresight, something not explicable mechanistically, all imply the need for postulating mental reality.[39]

Purposive action is, then, action that seems to be governed or directed in some degree by prevision of its effects, by prevision of that which still lies in the future, of events which have not yet happened, but which are likely to happen, and to the happening of which the action itself may contribute. Purposiveness in this sense seems to be of the essence of mental activity; and it is because all actions which have the marks of behavior seem to be purposive, in however lowly and vague a degree, that we regard them as expressions of Mind.[40]

The following passage, applying this notion even to an insect, succinctly summarizes McDougall's points in this regard:

A solitary wasp, after digging a hole in the ground to serve as a nest for her eggs, sets out in search of prey to be stored in the nest as food for her grubs; having found a caterpillar at any point within a radius of some hundreds of feet of her nest, she drags it over the rough ground and between the many obstacles that obscure for her all vision of the nest or its immediate surroundings; in spite of these obstacles, she takes approximately and on the whole the shortest possible course to her nest, and arrives there with her prey in virtue of a long-sustained series of varied movements all directed towards the one end, every deviation from the direct path necessitated by obstacles being rectified as soon as possible.

At every step of this prolonged journey the wasp is guided by visual impressions of the surroundings, which by many explorations she has made familiar to herself. How totally different from a series of reflexes are the movements by which she maintains and regains her true direction! A mere familiarity with, or power of recognizing, a certain number, even a very large number, of the objects that she encounters would by no means suffice to account for her behaviour. In order to guide herself she must not merely recognize objects previously seen; she must recognize objects (or the parts of the landscape immediately presented to her vision) as related in some determinate manner to the whole field of her explorations, and especially to that point of it at which her nest is situated; that is to say, each visual perception that guides her course not only involves a synthesis of a large number of details of the field of view to a unitary whole (or a synthesis of the effects of a manifold of sense-stimuli), but also must be related in a determinate fashion to a larger whole, namely, the scheme of the whole region which in some sense and manner she carries with her. Nor is this all. Her reactions to the complex visual impressions by which her course is maintained are determined also by the nature of the task in hand at the moment; for her reactions to each part of the landscape are different according as she is looking for a spot suitable for her nest, is seeking her prey, or is carrying it back to her nest; in psychological terms, each part of the landscape has for her a meaning or significance which is dependent upon her dominant purpose at the moment she perceives it; and this meaning is a decisive factor in determining the nature of her reaction.[41]

The chief argument advanced by McDougall seems to be that explaining the apprehension of meaning, whether it be conventional meaning as embodied in language or natural meaning by animals, requires going beyond mere mechanical explanation. Some mental activity is required. Indeed, this is the key argument of the Cartesian and Kantian tradition for equating language and mind—that language can deal with and express meaning which goes beyond the here and now, can deal with the past, the future, the universal, and the non-existent. Even if language or linguistic expression can be correlated with a particular set of brain events, its meaning (for example, allusions to the past or the future) cannot be so correlated, and is not exhausted by a description of the physiological processes. It is for this reason that Descartes sees language as definite proof of the presence of mind in others; it cannot be exhaustively translated into features of *res extensa* (matter), but remains inextricably tied to *res cogitans*.

McDougall's point, of course, is that the ability to read even non-linguistic or natural meaning in perception entails something not definable exhaustively in physical terms. For even natural meaning—for example, the meaning of a spoor or a track—carries the interpreter of the sign to the past or elsewhere. As I and others have argued at length elsewhere, the same sort of ability to transcend the confines of the momentary or immediate is entailed by an ability to apprehend signs, conventional or natural; for, in the final analysis, much of sign interpretation involves being carried from what is present to what is not.[42] Furthermore, as Peirce pointed out, all sign cognition is triadic, involving sign, what is signified, and some interpretation in the mind of the subject. Thus, even the ability to respond to natural signs bespeaks the presence of mind. And in so far as animals, even insects, interpret the world, perceive objects and events as signs of food, shelter, egg-laying space, or whatever, they must be said to have some subjective mental experience.

Thus McDougall, who spent most of his life in England but his last years in a United States teeming with behaviourists, is a truly anomalous figure who appears not to have minded being a gadfly, as shown both by his ill-fated attempts to prove the inheritance of acquired characteristics and by statements like the following:

The commandments of natural science must, of course, be obeyed. But the behaviorists forget to mention one such commandment, in fact, the one which should be given first rank. If in a given field we make certain observations, and particularly if we make them all the time, such observations must be accepted under all circumstances, whatever may happen elsewhere. Otherwise, why should activities in this field be honored by the name of empirical science? The behaviourists, however, not only fail to mention this commandment, they disobey it consistently. What is the reason for this strange conduct? There can be only one reason: In this fashion they are enabled to choose such facts as fit their particular philosophy, and to ignore all those which do not fit. Now there is something that clearly does not fit. Watch a rat in a maze. The most outstanding characteristic of his behavior is striving. Always the creature is after something, or tries to get away from something else. It is, of course, the same with man, including the behaviorist who, for a certain purpose, selects some empirical data, and refuses to recognize others. Striving is the very essence of mental life. Really, behaviorists ought to learn about the more elementary facts of life before they advertise their junior-size science.[43]

Vestiges of consciousness: Europe

In Europe, one finds far more cases of deviation from the philosophical orthodoxy of scientific common sense. In part, this seems to be because European scientists are generally better educated philosophically than their American counterparts, and are less ready to see philosophical questions as irrelevant to scientific activity. Thus, the second volume of Lorenz's papers, as we mentioned earlier, is specifically intended to discuss the philosophical basis for ethology. Similarly, German-speaking physicists like Heisenberg, Schrödinger, Einstein, and Bohr were always very interested in the philosophical underpinnings of their scientific activity. Not surprisingly, a number of German psychologists and ethologists had their philosophical roots in traditions other than positivism and behaviourism.

Köhler and Gestalt *psychology*

The *Gestalt* psychologists, for example, were highly philosophical in their approach. Köhler even wrote a book unabashedly entitled *The Place of Value in a World of Facts*. In essence, and in part under the influence of the Kantian rather than British empiricist tradition, *Gestalt* psychology rejected the reductionist approach to psychological phenomena which characterized both behaviourism and the ideology of science. Köhler, for example, cites with disdain Ebbinghaus's reductionist credo: 'I am not sure whether psychological facts are merely aggregates of psychological atoms; but being scientists, we must proceed as though this were true'. Says Köhler: 'What a sad statement. It seems to tell us that certain alleged necessities of scientific procedure are more important than the nature of the facts we investigate, with the consequence that we may ignore such facts as seem to be at odds with these "scientific necessities".'[44]

There is hardly a better statement of the point I have been trying to make. The common sense of science has, in my view, excluded certain data by fiat, and is the poorer for it. In their desire to avoid such impoverishment, *Gestalt* psychologists were extremely catholic in what they allowed to count as data for creating a psychological science. Köhler makes this point specifically in a 1953 essay:

In the first place, experimentation need not be a quantitative procedure. First attempts at clarification in a new field will often be greatly facilitated if the investigator observes in a qualitative way what happens under one condition and another. Outstanding work of this kind has sometimes been done even in modern physics. Secondly, observations of the very greatest significance may also be made without experimentation in any sense. Naturally, psychologists from Europe are more likely to admit that this is true than are their American colleagues, because psychological observations of a very simple kind had been part of the cultural tradition in Europe long before psychology learned to use experimental methods. Aristotle knew about the association of ideas. Both he and the Arabs were familiar with the moon illusion. Leonardo studied the curious system formed by the various colors. Contrast, afterimages, and dark adaptation were matters of fairly general interest around 1800. When McDougall pointed to the virtual omnipresence of purpose in mental life he also simply reported what seemed to him a plain fact. As such a fact, it then became a matter of experimentation. Let us not forget that other subjects of present technical investigation in psychology have come to us from the same humble source. When Ebbinghaus invented methods by which certain forms of memory could be studied with great precision, his first questions in this field were clearly derived from a knowledge of memory which was acquired without the help of any experimentation. Everybody can gain similar knowledge in one part of psychology or another. No more is needed for this purpose than that he be interested in facts, and that he watch. Since in this fashion so much has been gained for psychology in the past, contempt for equally direct observation now and in the future would clearly be a dangerous attitude. Is further evidence to be accepted only if it is already polished with the special tools of science? I cannot believe that anybody will seriously recommend such a policy. It would deprive us of our best chances to extend our work beyond its present scope.[45]

Controlled experiments are cited along with simple observations in Köhler's classic work *The Mentality of Apes*. As Carroll Pratt tells us in the introduction to Köhler's *The Task of Gestalt Psychology*, for *Gestalt* psychologists, 'the data of introspection and behavior are both part of the phenomenal world, and thus suitable material for scientific study'.[46] Like Kant, these psychologists realized that just because experience, subjective and objective, could be reduced to atoms of sensation did not mean that it was nothing but these atoms. Again, like Kant, they realized that there are principles of perceptual and cognitive organization which are logically prior to learning, and that learning depends on them. These are the innate *Gestalt* principles. Indeed, contrary to what behaviourists would have us believe, in psychological (subjective) phenomena like 'insight' we can find clear evidence of causal relations, and far more definitively than by induction in physics. Thus, 'When a young man sees a lovely figure with a lovely face on top, and then finds himself moving in the direction of so much perfection, does he need the indirect techniques of induction for discovering that those lovely conditions and the displacement of himself as a system are causally related?'[47]

Given such a liberal and, ultimately, commonsensical approach to the data which psychology ought to study, it is no surprise to find Köhler taking for granted the existence of mental experiences in 'higher animals' which parallel our own, and reporting his data in what, to a behaviourist, must appear to be hopeless anthropomorphic terms, yet which to an ordinary person are simply natural explanatory, if not descriptive, categories:

I will give another striking example, to show how the momentary practical advantage can quite recede into the background, when some emotional state has first to be expressed. One night, when it was raining unusually heavily and continuously, I heard two animals, who were kept alone in a special place, complaining bitterly. Rushing out, I found that the keeper had left them out in the open, because he had broken the key to the door of that place. I tried to force open the lock. It gave way, and I stood aside to let the two chimpanzees run in as quickly as possible, into their warm, dry sleeping-den. But although the cold water was streaming down their shivering bodies on all sides, and although they had just shown the greatest misery and impatience, and I myself was standing in the middle of the pouring torrent, before slipping into their den they turned to me and put their arms round me, one round my body, the other round my knees, in a frenzy of joy. And it was not until they were satisfied in this way that they threw themselves into the warm straw of their sleeping apartment.

The incidents just described are characteristic of the behaviour of chimpanzees towards familiar adult human beings. With little children and infants there is no need for these apes to know them at all, in order to be taken with them. When a young creature like this was brought near the railings of the animals' play-ground, one or other of the animals regularly came up and looked at the apparition for a long time with interest, a good-natured, satisfied expression on its face, tried to look under the clothes encasing the infant and nodded from time to time pleasantly in the direction of the child in front of him. This behaviour was most noticeable in the eldest of the females. As she had only come to the station when fully-grown, it may be that she had already had dealings with baby chimpanzees. But much younger animals behaved in a similar way, and I found it very clearly marked in a female orang, long before sexual maturity.[48]

And later:

the following incident shows how closely some of the perceptions of the higher animal species must resemble ours, though in this instance one might at first tend to regard human perception as the product of very complicated 'higher processes.' I was riding down a mountain path on a cold moonlight night in winter, when a faint mist made everything seem important and fantastic in outline. My guide was walking behind me, and his very lively little dog trotted wearily about one hundred metres in front. A little ahead the path dipped into a dark cleft in the hills; facing us in the moonlight several pine stumps stood out upon the opposite slope. As I rode on I perceived, to my surprise, that an old man was squatting in an attitude eloquent of misery and dejection on the top of one of these stumps, alone and motionless in the cold night among the mountain solitudes; a pitiful and, at the same time, a somewhat uncanny spectacle. As our pathway led close by the old man's perch, I waited in silence till we should reach him. The guide seemed to have noticed nothing—at any rate, he, too, was silent. When the dog had got near to the crouching form, he sprang with an excited bark towards him, and circled round for some time, still barking. Then he appeared to regain his composure and went on his way. Meanwhile we had come up, and in a few rapid processes which I cannot describe, the old man had resolved himself into a group of short branches and shoots of one of the pine stumps. I then asked the guide why the dog had barked—they were the first words uttered. 'Oh, he probably thought that there was an old man,' was the answer. 'I thought so, too, at first.' For two human beings, at the same time but quite independently, to have received the same optical delusion the complex of optical stimuli must have had a very compulsory effect at some distance; and now the dog, who trotted heedlessly past dozens of pine stumps—some of them with strongly developed twigs and shoots—was much excited under the same conditions. Even if we admit that he may not have received exactly the same impression as we did, the occurrence remains curious enough.[49]

These examples, quoted in detail as much to give the 'colour' of Köhler's work as its content, eloquently attest to the fact that animal subjective experience was very much alive for *Gestalt* psychology.

The idealistic tradition: Buytendijk and von Uexküll

The ability of thinkers to resist the pull of the positivistic common sense of science correlates directly with the availability of rich, alternative philosophical traditions. In Europe, such traditions have always coexisted with scientific activity, and they continued to do so even at the height of positivism. The Dutch philosopher and physiologist Buytendijk was basically a phenomenologist, and this philosophical commitment led him to buck the tide on animal consciousness. Thus, during a long career, he produced a variety of works decidedly out of the mainstream. In 1935, he published *The Mind of the Dog*, which attempted to reconstruct, from known data, the mental life of the domestic dog. And even as the bulk of the scientific community was treating pain in animals as a purely physiological mechanical question, Buytendijk examined pain from a phenomenological, as well as physiological, perspective in his 1943 book, translated in 1961 as *Pain: Its Modes and Functions*, and stressed the experiential dimensions in both humans and animals. In 1953, he published the classic paper we alluded to earlier, entitled 'Toucher et être touché', in which he elegantly demonstrated that the octopus could distinguish between the experience of being touched and that of actively touching.

Buytendijk devoted an entire chapter of *Pain* to the question of animal pain. While admitting that it is difficult to judge painful sensation in animals, he none the less asserted that 'we cannot say that animal emotional life is a closed book for us or that we have no understanding at all of their sensations of pain. In principle the same factors determine the degree of certainty with regard to animal pain as to the pain of the mentally ill and children.'[50]

In an interesting and idiosyncratic move, Buytendijk argues against the traditional (among Darwinians) view that so-called higher animals have a greater degree of consciousness than lower ones: 'Development of bodily structure does not always correspond to a difference in function and behaviour. An animal's position in the zoological scale is no measure of the clarity of its consciousness.'[51]

The degree of consciousness, he claims, depends rather on the type of life the animal lives. For example, a hunting animal must have a more finely tuned consciousness of its surroundings than a grazing one. The former needs to spot its prey, creep up on it, pursue it, fight, and so on; the latter need only spot food areas and react to others of its species. Thus a hunting invertebrate should have a more developed consciousness than a ruminant vertebrate: 'The consciousness of the octopus might be said to be clearer than that of the carp or even the cow.'[52]

In identifying pain or other emotions in lower animals, according to Buytendijk, we must be careful not to confuse it with reflexes. The best evidence of such feeling is the presence in the animal of 'expressive' reactions not tied to withdrawal response—for example, the change in colour of an octopus's eyes when it approaches prey or is endangered bespeaks emotional affect. In animals phylogenetically closer to us, we can better trust analogy to recognize pain. Buytendijk is almost alone in stressing individual differences among animals, even within a species or subspecies (or breed). Such differences, he tells us, 'confirm our impression that real feelings are expressed and not merely typical reactions of the species'.[53] Notwithstanding such differences, he goes on to distinguish five general types of pain reaction in, for example, the dog: flight reactions, kinetic disorganization, aggressive expression, check on movements, and care and defence of an injured part.[54] Throughout his discussion, the integration of empirical knowledge with phenomenological analysis provides a valuable and unusual perspective on animal pain and awareness.

An equally interesting and equally neglected thinker who resisted positivism was Jakob von Uexküll, one of Lorenz's teachers, whom Lorenz later took somewhat to task for an excessively speculative streak. Like the *Gestalt* psychologists and many other German thinkers, von Uexküll was profoundly influenced by Kant and the idealist tradition. Positivism (and of course behaviourism) tended to see experience as writing on the blank slate which was the subject—what John Herman Randall called the 'dartboard theory'. Behaviourist psychology and empiricist epistemology likewise embraced this fundamentally untenable picture, most brilliantly articulated by Hume. Hume himself was too astute a thinker not to see the failings of such a model, failings of which he became acutely conscious when attempting, in the *Treatise*, to give an account of the 'I'. Kant made the critique of atomistic phenomenalism explicit, by pointing out that, although our experience can be reduced to sensory atoms, it is nevertheless experienced in units by an enduring 'I'; hence his emphasis on the contribution of the experiencing subject to knowledge. It is the organizing capabilities of the subject which bring order to the chaos of ever-changing sensory particles, and which create from these coruscating particles the experience of objects engaged in causal interaction.

Von Uexküll is solidly in, and acknowledges himself to be in, the Kantian tradition. In the late nineteenth century, he had been seduced by Helmholtzian, reductionistic, mechanistic thinking, and, with Bethe and Beer, had proclaimed a programme for eliminating any terminology in science which smacked of subjectivism.[55] But by 1934, when he published his brilliant piece 'A walk through the worlds of animals and men: A picture book of invisible worlds,' von Uexküll had come—pardon the pun—to his senses, and had done a complete turnabout. He now realized that far from subjective notions being in principle eliminable, biology cannot be understood without explicit appeal to subjects.

The mechanists have pieced together the sensory and motor organs of animals, like so many parts of a machine, ignoring their real functions of perceiving and acting, and have even gone on to mechanize man himself. According to the behaviourists, man's own sensations and will are mere appearance, to be considered, if at all, only as disturbing static. But we who still hold that our sense organs serve our perceptions, and our motor organs our actions, see in animals as well not only the mechanical structure, but also the operator, who is built into their organs, as we are into our bodies. We no longer regard animals as mere machines, but as subjects whose essential activity consists of perceiving and acting. We thus unlock the gates that lead to other realms, for all that a subject perceives becomes his perceptual world and all that he does, his effector world. Perceptual and effector worlds together form a closed unit, the *Umwelt*. These different worlds, which are as manifold as the animals themselves, present to all nature lovers new lands of such wealth and beauty that a walk through them is well worth while.[56]

Von Uexküll's fundamental insight is rooted in philosophical idealism, going back to Leibniz and, later, Lotze. The world, for these thinkers, is a world of subjects, each mirroring the world from its own perspective. There is, ultimately, no objective world, though it is of course convenient to speak of our human scientific perspective as objective. There are only the myriad points of view of different subjects. It is this metaphysical perspective which he cashes out in his biology. We must, he graphically informs us, 'blow a soap bubble around each creature to represent its own world, filled with perceptions which it alone knows'.[57]

Each animal, however lowly, has access to the world through a variety of sense- and effector-organs. To use his famous example, the tick, though blind, has access to prey via its senses of smell, temperature, and touch. When it senses prey, it is then led to burrow and to nourish itself on the blood of its prey. The range of the three sense-organs constitutes the *Umwelt* of the tick, its experienced universe. But even at this level of organization, we are dealing with a subject, with a unity of experience, for the information from the three organs must be brought together and synthesized, and a single subject must then act. There are indeed organisms which are merely bundles of mechanical reflexes, each operating independently of the other with no central unity. Such a 'reflex republic' can be found in a sea-urchin, which is essentially a group of independent reflex perceptions and actions. The tick, however, seems to be higher than that, and certainly, animals above the tick are unified subjects:

The tick's life manifestations, as we have seen, consist substantially of three reflexes. Even this represents a superior type of organism, since the functional cycles, instead of using these isolated reflex arcs, have a common receptor organ. In the tick's world the prey may therefore possibly exist as an entity, even though consisting only of butyric acid stimulus, tactile stimulus and heat stimulus. For the sea urchin, this possibility does not exist. Its receptor cues, which are composed of graduated pressure stimuli and chemical stimuli, constitute totally isolated magnitudes.[58]

By the time one gets to a snail, there is unquestionably a unified subject. To use von Uexküll's example:

A vineyard snail is placed on a rubber ball which, carried by water, slides under it without friction. The snail's shell is held in place by a bracket. Thus the snail, unhampered in its crawling movements, remains in the same place. If a small stick is then moved up to its foot, the snail will climb up on it. If the snail is given one to three taps with the stick each second, it will turn away, but if four or more taps are administered per second it will begin to climb onto the stick. In the snail's world a rod that oscillates four times per second has become stationary. We may infer from this that the snail's receptor time moves at a tempo of three to four moments per second. As a result, all motor processes in the snail's world occur much faster than in ours. Nor do its own motions seem slower to the snail than ours do to us.[59]

In this example, we see von Uexküll's philosophical base cashed out in a scientific context, for he has inferred part of what a snail's experienced world must be like. And if one takes seriously the presence of perception in animals, as betokened by the existence of sense-organs, one should not find such a move problematic. There is certainly an element of speculative imagination involved in going from sense-organs and form of life to what an animal actually experiences, but, as we have emphasized all along, such speculation is no more pernicious than that involved in other theoretical constructs or in evolutionary explanations in biology, which are accepted as legitimate by all but diehard mechanists like Loeb. Such explanations of animal subjectivity should be grounded in a strong, detailed understanding of the animal's physiology and anatomy, of the way these work in the animal's environment, and of the variables which the animal must deal with. In his discussions of the tick and the snail, von Uexküll provides fine exemplars of such synthetic, factually based thought. Far from being unscientific, they are science at its best, for the alternative is essentially to say that we can know nothing of perception in other creatures, save to describe its machinery. Indeed, the study of perception has always been an embarrassment to behaviourists, since perception, by its very nature, involves not behaviour, but awareness. Later we shall discuss a possible method of using perception as a key to the experiences of animals.

Thus, we have seen that, despite a few deviants, like McDougall in psychology and von Uexküll in ethology, both fields were pretty much dominated by the ideology of science and by the exclusion of subjective states as legitimate objects of enquiry. That such a position was fundamentally at loggerheads with ordinary common sense bothered no one. British psychologist R. B. Joynson has documented the extent to which behaviourism self-consciously prided itself on its rejection of ordinary common sense and its replacement by the common sense of science.[60] Skinner, for example, wrote in 1972: 'What, after all, have we to show for nonscientific or prescientific good judgement, or common sense, or the insights gained through personal experience? . . . [There is] a tremendous weight of traditional "knowledge" which must be corrected or displaced by scientific analysis.'[61]

By the same token, as Lashley and Tinbergen remarked, psychology ignored for a long time the biological constraints on animal behaviour and learning, while ethology ignored the dimension of learning. And behaviouristic psychology and ethology in turn ignored each other. All this occurred not primarily for what the common sense of science considers 'scientific' reasons, but for philosophical and valuational reasons. Thus Lorenz and Tinbergen often asserted that their work was rejected in the United States because it was too biologically deterministic for Americans. Lorenz remarked: 'Liberal and intellectual Americans . . . accepted [behaviourism] almost without exception, particularly because here was a doctrine proclaiming itself as a liberating and democratic principle.'[62]

On the other hand, Lorenz clearly had his own valuational agenda for rejecting behaviourism, an agenda which Loren Graham has argued was 'highly reminiscent of right-wing political views in Germany and Central Europe in the first half of the century, the time when Lorenz' world view was formed'.[63]

During the Nazi period, Lorenz was wont to draw political capital from his science, arguing that the Nazi state, with its policies of racial purification, represented the proper approach to the improvement of humans and society.[64] This political and moral version of ethology in turn served as a further barrier to the acceptance of ethology by the American community, which was inclined to reject ethology as fascistic, and to look for scientific reasons to back this rejection. Once again we see the direct relevance of philosophical and valuational notions to scientific ideas, despite their official dismissal by the ideology of science.

The conflict between ethology and behaviourism was not in itself a major factor contributing to a softening of scientific attitudes on consciousness, since, as we saw, the one thing the two approaches shared was the common sense of science's rejection of the legitimacy of talking about consciousness! It appears that the tentative return of consciousness to psychology and ethology was precipitated by a variety of factors, social, scientific, and valuational.

9 Consciousness regained: psychology

Internal critiques of behaviourism

In so far as it has begun to be legitimate to talk about consciousness again, it is, we have seen, less a matter of profound new ideas or experiments or conceptual analyses overthrowing and refuting the rejection of consciousness than of a change in intellectual climate.

Perhaps the major factor in creating a climate in which it was possible to talk about consciousness in psychology was the rise of criticisms, starting in the late 1940s, of a number of standard behaviourist beliefs and practices. Although not directed at consciousness, these criticisms helped weaken the edifice of behaviourism enough to allow all sorts of new ideas to enter, even that of consciousness. Behaviourism was subjected to a series of devastating critiques, many of them 'in house', which led to a reappropriation of what was, to a non-behaviourist, common sense and obvious. Thus things which were obvious to countless generations of ordinary people, when published by respected individuals who had worked in the behaviourist tradition and/or when backed by laboratory data, became major crises for behaviourism, thereby weakening its doctrinaire hold.

One of these criticisms can be found in a famous article by Frank Beach, based on his presidential address to the Division of Experimental Psychology of the American Psychological Association in 1949, and later published in *The American Psychologist*, the major psychological journal in the United States, as 'The snark was a boojum'. In this article, Beach vigorously attacks American experimental psychologists since the advent of behaviourism for doing more and more research on fewer and fewer animal species, and thus neglecting to generate a truly comparative psychology. Between World War I and 1948, over 50 per cent of animal research in psychology was done on the Norwegian rat, 0.001 per cent of the known species! This is dramatically illustrated in Figure 9.1.

By the same token, of seven possible categories of animal behaviour research, including conditioning and learning, sensory capacity,

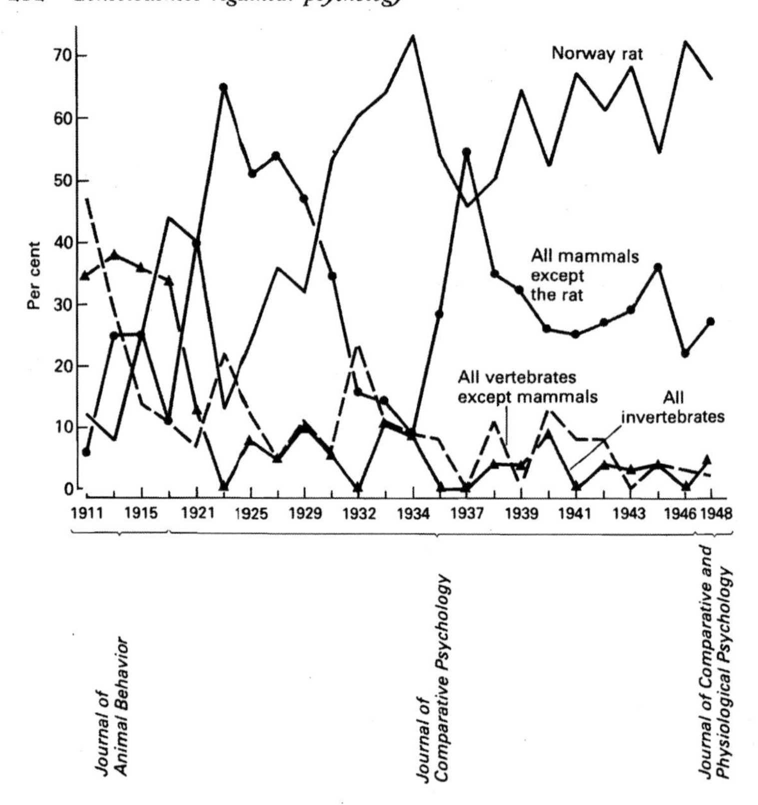

Fig. 9.1. Beach's graph of the percentage of all articles devoted to various phyla classes, or species between 1911 and 1948

general habits, reproductive behaviour, feeding behaviour, emotional behaviour, and social behaviour, over 50 per cent of the research done since World War I fell into the category of conditioning and learning, as Figure 9.2 illustrates.

Other data show that 70 per cent of the papers published in the 1940s dealing with animal learning mentioned Hull's stimulus–response approach to behaviourism.[1] Beach wittily proposes having one psychological journal called the *Journal of Rat Learning*, but points out that most psychologists who study the rat think they are studying all learning.[2] Indeed, Tolman's book, *Purposive Behavior of Organisms*, is based exclusively on rat behaviour in bar-pressing situations![3]

Beach goes on to argue that the experimental emphasis on the white rat is a matter of tradition and has no sound basis. No systematic study

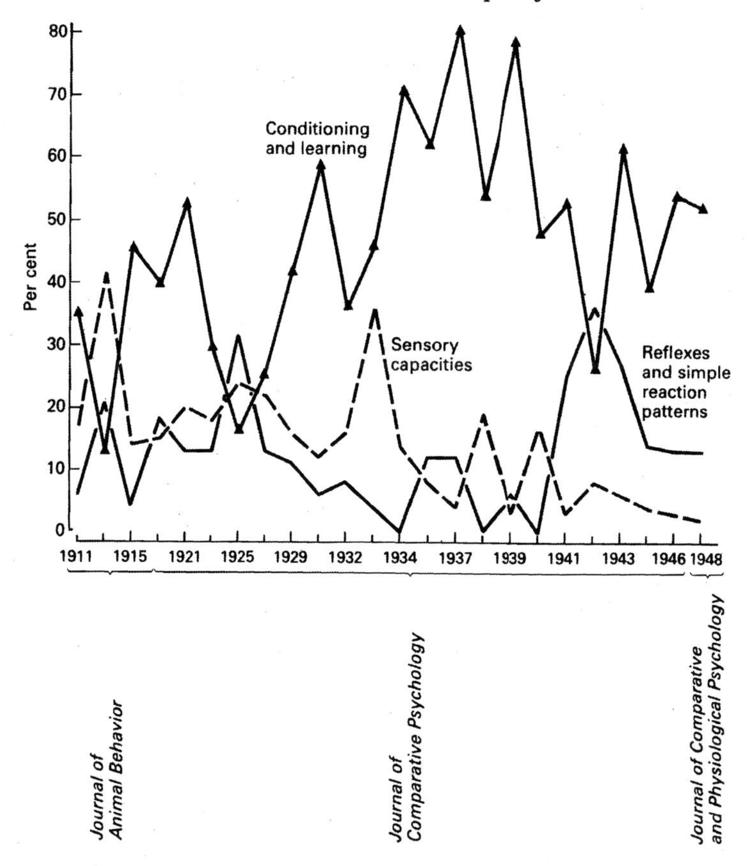

Fig. 9.2. Beach's graph of the percentage of all articles concerned with various psychological functions between 1911 and 1948

has been done to justify the rat as the best species, nor studies demonstrating that the principles derived from rat learning are at all applicable to any other species. Furthermore, given the ubiquitous stress on learning, 'psychologists are led to neglect many complex patterns of response that stand in urgent need of systematic analysis'.[4] As a key example, Beach cites so-called instinctive or, as behaviourists call it, 'unlearned behavior'. This area, as we saw ethologists claiming, was very much neglected by behaviourists. Beach's critique is withering:

Any pattern of response that does not fit into the category of learned behavior as currently defined is usually classified as 'unlearned' even though it has not been analysed directly. Please note that the classification is made in strictly negative terms in spite of the fact that the positive side of the implied dichotomy is very poorly defined. Specialists in learning are not in accord as to the nature of the processes involved, nor can they agree concerning the number and kinds of learning that may occur. But in spite of this uncertainty most 'learning psychologists' confidently identify a number of complex behaviour patterns as 'unlearned.' Now the obvious question arises: Unless we know what learning is—unless we can recognize it in all of its manifestations—how in the name of common sense can we identify any reaction as 'unlearned'?

The fact of the matter is that none of the responses generally classified as 'instinctive' have been studied as extensively or intensively as maze learning or problem-solving behavior. Data relevant to all but a few 'unlearned' reactions are too scanty to permit any definite conclusion concerning the role of experience in the shaping of the response. And those few cases in which an exhaustive analysis has been attempted show that the development of the behaviour under scrutiny is usually more complicated than a superficial examination could possibly indicate.[5]

Beach goes on to show that types of learning which do not fit the behaviourist approach have been ignored, citing Lorenz's work on imprinting and Tinbergen's on single-trial learning in wasps. (Single-trial learning is, of course, a great embarrassment to conditioning theorists.) By the same token, ethological work has shown that, contrary to behaviourist dogma, hunger does not necessarily produce maze-learning, as when food-deprived ferrets none the less continue to explore every blind alley in a maze on the way to a goal box containing food.[6] Furthermore, argues Beach, knowing the natural behaviour patterns of a given animal (something ignored in the case of the rat) would surely help us to decide whether the animal is a good model for a given behaviour. Thus it is absurd to study hoarding behaviour in rats, since they aren't natural hoarders.[7]

Finally, Beach suggests that if psychology is to be credible, it must study humans to understand humans, despite the fact that they are not as easily dealt with scientifically as rats. Scientism, he implies, has replaced science, a point tellingly and devastatingly made in his conclusion:

If we as experimental psychologists are missing an opportunity to make significant contributions to natural science—if we are failing to assume leadership in an area of behavior investigation where we might be useful and effective—if these things are true, and I believe that they are, then we have no one but ourselves to blame. We insist that our students become well versed in experimental design. We drill them in objective and quantitative methods. We do everything we can to make them into first-rate experimentalists. And then we give them so narrow a view of the field of behavior that they are satisfied to work on the same kinds of problems and to employ the same methods that have been used for the past quarter of a century. It would be much better if some of our well-trained experimentalists were encouraged to do a little pioneering. We have a great deal to offer in the way of professional preparation that the average biologist lacks. And the field of animal behavior offers rich returns to the psychologist who will devote himself to its exploration.[8]

As Graham has pointed out, Beach's article was enormously influential—not least in preparing Americans to accept ethology.[9] What is especially interesting about Beach's article, though, is that it offers criticisms of behaviourism that any bright student in an introductory psychology class could and often did make: namely, that rats are not the only animals; that not all animals are the same; that rats surely can't model people in every way; that not all learning is like maze-running; that not all animal behaviour is learned, since animals have innate behaviour patterns; that animals can learn without obvious drive motivation; that naturalistic observations of animals under normal conditions are surely logically prior to laboratory manipulations of them; that psychologists don't give good reasons for studying rats; and so forth. Yet when such criticisms are voiced by 'outsiders' or persons without the requisite credentials, they are simply not heard. In other words, contrary to the fable, it takes more than a child to say that the emperor is nude; it takes an expert in anatomy!

Attacks on learning theory

Beach's critique was followed by others, which struck at various tenets of behaviourism. In 1951, Lashley, one of the fathers of behaviourism, struck a major blow at behaviourist conceptual underpinnings when he argued in 'The problem of serial order in behaviour', that the simple model of learning presupposed by behaviourists, which reduced complex behaviour to a series of conditioned responses, was totally inadequate to explain temporal integration. This, he argued, was true with regard to animals as well as humans. In 1960, the Brelands attacked the cherished behaviourist assumption that all responses of an organism are equally capable of being attached to any discriminable stimulus. They showed that instinctive behaviour can override reinforcement; thus a raccoon will wash an object even if so doing interferes with its fulfilling a food-rewarding task.[10] Later, and in the same vein, Seligman showed that certain responses may be unconditionable in a species—for example, yawning in a dog—and that certain stimuli are more readily linked with certain responses than others. Thus animals are more inclined—Seligman's term is 'prepared'—to link intestinal upset and nausea with novel tastes than with flashing lights (to which they are 'contraprepared').[11]

Other experiments conducted in the 1950s[12] challenged the behaviouristic dogma that learning proceeds incrementally by repetition, and reintroduced the notion of one-trial learning, something known to both common sense and associationists like Hume. Any farmer can tell you that it takes only one shock to teach animals not to touch electric fences. Indeed, I was advised by an old rancher not to use electric horse-water-tank defrosters because 'they can short out in hard water, shock the animals, and *they will then never drink from that tank again*!' Once again, we see psychology reappropriating common sense and traditional biological knowledge.

Chomsky's critique of behaviourism

One of the most devastating blows to behaviourism came in Noam Chomsky's 1959 review of Skinner's 1957 book *Verbal Behavior*. The latter was an attempt to explain human language by the traditional behaviouristic notions of stimulus, response, and reinforcement. For Skinner, language is like maze-learning in rats; that is, it is built up from simple reinforcements into an apparently complex entity. Though Chomsky's review is nominally directed at the theory of language, it is really, as Chomsky himself later admitted, an attack on all behaviouristic 'speculation as to the nature of higher mental processes'.[13] In his review, Chomsky took to task the behaviouristic assumption that language can be explained solely in terms of the history of reinforcement, without consideration of innate factors, and asked what grounds there were for excluding innate factors a priori. Furthermore, Chomsky showed that a welter of basic behaviouristic notions are unclear—what, for example, is a 'stimulus' or a 'response'? If everything in the environment is a stimulus, then why do organisms react only to some of them? And if only some things are stimuli, what makes them so? If language is learned by 'reinforcement', how do we construct sentences for which we have never been reinforced, or only negatively reinforced? Furthermore, Skinner cannot explicate mentalistic terms like 'intend', 'desire', 'persuade', and so on in behaviouristic terms; his attempts to do so are either patently false or involve covert appeal to common sense and/or the mentalism he deplores, and so on.

Chomsky drew attention to the fact that even animals work to solve problems for reasons other than reinforcement, a point he attributes to Romanes. (In this context, Chomsky also remarked on the great extent to which adherence to a successful theory can blind scientists to the real insights found in earlier works which antedate the theory in question.)

This review, both in its critical aspect and in its postulation of innate mental capacities to explain such mental processes as language, did much to undercut the hold of behaviourism. Chomsky's major concern was with human psychology, but his works in general are spotted with references to the inadequacy of behaviourism as an approach to animal mind as well. Interestingly enough, Chomsky's rejection of behaviourism was motivated by ethical considerations as well as intellectual ones (even as we saw that behaviourism itself was so motivated). From comments in many of his works, including *Problems of Knowledge and Freedom* (1971) and *Reflections on Language* (1975), it is clear that he sees behaviourism as inimical to human political freedom. Ironically, a philosophy which was liberating in the eighteenth century was seen to have great potential for oppression in the twentieth. Behaviourism is, in some respects, updated empiricism, in that both see the human being as a *tabula rasa*. In the eighteenth century, this metaphysical commitment opposed the paralyzing notions of innate and hereditable superiority which formed the basis of a hierarchical, stratified society. Change the environment and change the person—peasants are potentially equal to kings. But by the twentieth century, this notion of human malleability had become a mainstay of massive totalitarian attempts to shape human beings. It is no accident that Pavlovian behaviourism was the official psychology of the Soviet Union, or that Skinner should write a book about our need to go *Beyond Freedom and Dignity* (1972) to a perfectly planned utopia shaped by behavioural psychological means. In this context, an emphasis on innate constraints is salubrious, in de-emphasizing the malleability of the individual.

Chomsky's work provided a powerful blow to behaviourism and a powerful boost for what has come to be called 'cognitive psychology', a term covering a multitude of ideas but which, for our purposes, can be interpreted in the broadest sense as a more liberal approach to mind than that of behaviourism, one which doesn't cavil at calling a mind a mind, and which looks at other aspects of mind besides learning, especially cognitive abilities. Chomsky self-consciously placed himself in the mentalistic tradition in his *Cartesian Linguistics* of 1966, wherein he assimilated his own thinking on mind to that of Descartes and the

Port-Royalists, and attempted to show that there were historically viable alternatives to empiricism.[14] In later work—for example, *Rules and Representations*[15]—Chomsky made even more explicit his general view of mind, and went well beyond language into various aspects of mentation.

Cyril Burt's defence of consciousness

Thus, by the early 1960s, behaviourism was leaking from many holes. Some psychologists began to abandon orthodoxy and to strike off in different directions. Inevitably, the issue of consciousness was one of the major ones raised, and in an increasingly explicit manner. One of the most articulate voices raised during this period, critically assessing the still-dominant tendency to ignore or deny consciousness, was that of Cyril Burt. In a seminal 1962 article entitled 'The concept of consciousness', Burt made a case for restoring the study of consciousness to centre-stage in psychology. He began by describing the status quo:

> Nearly half a century has passed since Watson (1913) . . . proclaimed his manifesto. Today, . . . the vast majority of psychologists, both in [Britain] and in America, still follow his lead. The result, as a cynical onlooker might be tempted to say, is that psychology, having first bargained away its soul and then gone out of its mind, seems now, as it faces an untimely end, to have lost all consciousness.[16]

The chief arguments advanced against the legitimacy of consciousness and introspection, from Watson to later behaviourists like Broadbent, seem to be that statements about consciousness are not certain, and are private rather than public. Burt's reply was that all sciences which rely upon individual observations, even astronomy, are at best probable. By the same token, statements about private experience are made in a public language, and thus are meaningful, though also only highly probable. Furthermore, as we saw earlier, all sense-experience is private, and thus any statement, whether about public objects like stars or about inner experience, are at root about private experiences.

To behaviourists, statements about private experiences, even perceptions, are scientifically illegitimate. Burt pointed out that to behaviourists, statements like 'I see O' mean that if presented with stimulus *X*, I react by saying, 'I see O.' 'For psychology, the subject thus becomes nothing more than a stimulus-response machine: you put a stimulus in one of the slots, and out comes a packet of reactions.'[17]

But such moves, says Burt, misdirect our attention. When someone reports a hallucination as a result of ingesting a drug like LSD, what they say is not the major object of interest, but rather, what they *experience*. While it is true that words cannot replicate the exact content of that experience, words can't replicate physical objects either.

Indeed, the methods (and vocabulary) developed by the introspective psychologists were far closer to ordinary experience—the subject-matter of psychology—than the explicit and artificial notions coined by behaviourists, like 'sign-*Gestalt*-expectations' or 'demanded type of means-or-goal'. Burt writes: 'Few readers would want their favourite novelist to translate his "subjective nomenclature" into a behaviouristic idiom; and all that the introspective psychologist seeks to do is to apply the scientist's technique to the novelist's material and refine and standardize the novelist's . . . vocabulary.'[18]

Such a move, however, has been 'smeared' by Watson and others as 'unscientific', despite the fact that their notion of science is an outmoded, nineteenth-century mechanistic one, no longer even true, as we saw earlier, of physics, the so-called master science. Furthermore, the related attempt to axiomatize psychology according to the principle of parsimony, as Hull tried to do,[19] if possible at all, is premature, since such axiomatization must take place after a science is well developed, not at its inception. And, we might add, there is nothing to preclude using mentalistic concepts in any such axiomatic treatment of psychology.

In essence, Burt asks for a retreat on the part of psychology from the rarefied heights of scientific common sense back to ordinary common sense. Failure to make this move, as we have remarked throughout our discussion, means ignoring all the interesting questions which ordinary people expect psychology to deal with, those which characterize our ordinary, everyday view of mind. Burt puts his point as follows:

> I am all for a kind of psychological passport officer, who shall scrutinize the credentials of the vast crowd of popular concepts, which, more than in any other science, are apt to intrude from the common discussions of everyday life. But the behaviourist's criteria seem to me as unfair as they are old-fashioned.[20]

Burt goes on to an elegant conceptual discussion of the irreducibility of locutions pertaining to consciousness, and argues convincingly that any such reduction will end up being circular, in that it will presuppose some notion of awareness. Consciousness cannot be eliminated, and thus introspection cannot be eliminated either. The study of perception, logical thinking, and, especially, mental imagery depend on introspection. (Burt cites the phi-phenomenon, where an observer who is fixed on a stationary light eventually reports that it moves, as paradigmatic of work presupposing introspection.)

Turning Watson's valuational polemic back against him, Burt reminds us that part of psychology's role is to control and guide behaviour. Once again, introspection is essential. Thus, mental imagery, for example, is of great practical value in remedial work on reading disability. Burt's conclusion is worth quoting in its entirety:

> The rule that all introspective terms must be expunged from the vocabulary of 'objective psychology' has played havoc with the interpretation of mental tests and mental factors. In the early days of such research no investigator would have claimed that some new test or factor really measured this or that particular ability, until he had first checked his conclusion by securing introspections from his subjects. Today such identifications are commonly made simply on the basis of some hypothetical preconception—often quite differently by different interpreters. And in almost every field of cognitive psychology—in the study of colourblindness, of sensitivity to noise, of artistic creativity and musical appreciation, of the processes of thought and reasoning, to pick but a few of the most recent examples—this obsessive psychophobia leads to the most far-fetched and misleading periphrases merely to express what are essentially psychical characteristics in purely behaviouristic terms. It is, however, on the motivational side rather than the intellectual that the gulf between the introspective method and a strict behaviourist approach is most conspicuous. 'Will', 'purpose', 'conation', these are the concepts which Watson singled out for his most ruthless attacks. Instead we must 'substitute mechanical explanations for the meaningless jargon of affective and conative processes', and hold fast to the view that man is 'an assembled organic machine'.
>
> 'If you think we are waxworks', said Tweedledum to Alice, 'you ought to pay.' And Watson has to pay a heavy price for his strict adherence to mechanistic principles. It makes nonsense of every form of applied psychology. Educational psychology, vocational psychology, criminology, and psychotherapy—all become impossible if we are to look upon men and women, patients and pupils, as mere automata, devoid alike of reason and feeling. Even in laboratory research on animal behaviour it has become necessary to discard the original restrictions; and it is largely around these motivational problems that the controversies between later behaviourists have revolved. Little by little, however, some of the younger members of the school have ventured to reintroduce a few basic concepts reminiscent of the discredited introspectionist psychology, usually disguised by new names or fresh descriptions. The traditional notion of willing, wanting, or striving—'conation', in short—is rechristened 'drive'; a 'purpose' appears as a 'goal'; and instead of the old pleasure-pain principle—the 'algedonic' law—we have a

'law of effect'. We are forbidden to say the 'satisfaction' we feel at the success which has attended a particular action increases the probability that the action will be repeated; but we are allowed to say that 'the nervous process resulting from a "reward" tends to "reinforce" the nervous process which issued in the reward'.

The object of these heroic circumlocutions is, so we are told, to abolish terms implying 'occult notions, such as perception, pleasure, purpose, or conscious state', and to replace them by interpretations that 'imply nothing more than the mechanical operation of physical causes and effects'. But the futility of the attempt is patent. The hypothetical 'nervous processes' or 'excitations' which the interpretation presupposes are far more 'occult' than the conscious states whose mention is taboo'd. After all, feelings of satisfaction and the like are things that everybody can observe; no one has ever observed the alleged nervous processes that are put in their place, nor has anyone the slightest idea of what they are like, how they can be identified, or indeed whether they really exist.[21]

Burt's paper was less a major influence on psychology—he had held these beliefs throughout a long career—than a clarion call heralding a change in *Zeitgeist*. For consciousness did begin to return in all the areas of psychology he listed, most especially cognitive psychology. Quite predictably, it was talk of human consciousness which first became respectable, but animal consciousness followed.

The work of R.B. Joynson

The points made by Burt were carried on and developed in Britain during the early 1970s by R.B. Joynson, in a series of articles and in an excellent little book entitled *Psychology and Common Sense* (1974). In this book, Joynson points up the failure of behaviourism to deal with all that is interesting in psychology, as well as all that a lay person means by 'mind'. The tone of Joynson's book is set by an epigraph taken from Köhler: 'To a deplorable degree, the obvious has disappeared from learned psychology, so that we have to rediscover it.' Joynson's book is an attempt to point the way towards the reappropriation of common sense in human psychology.

One of the most interesting claims made by Joynson is that, despite pronouncements and protestations, some avowed behaviourists were in fact beginning to reintroduce mental locutions and concepts, without acknowledging them as such. These were cases where 'the behaviourist uses the concepts of mental life, and uses them in their proper original meaning as referring to subjective experience, and yet tries to persuade himself that this total defeat may be portrayed as the final victory of Watson's great endeavour'.[22]

Even more interesting—and baffling—are cases of behaviouristic psychologists who quite self-consciously (as it were) reintroduce notions like introspection and consciousness into psychology. He cites J.G. Taylor, described as 'a leading behaviourist', who makes the extraordinary assertion that 'Psychologists, and above all behaviourists, must restore conscious experience to its rightful place as a legitimate area of research . . . a behaviourism that persists in neglecting this important sector of behaviour might as well sink into the limbo of forgotten theories.'[23]

Joynson points out that behaviourism with consciousness restored is hardly behaviourism, for the essence of behaviourism is the denial of consciousness. A behaviourist who accepts consciousness would be an oxymoron, for 'it would leave us all in doubt, when a man called himself a behaviourist, whether he was urging us to reject consciousness or imploring us to restore it'.[24]

Ernest Hilgard and the return of consciousness

Between the 1960s and the 1980s, there was gradual return in psychology of the legitimacy of talk about consciousness. Perhaps the best indicator of this attitudinal softening and rejection of hardline behaviourism may be found in Ernest Hilgard, a mainstream learning theorist and sometime president of the American Psychological Association. In 1975, the sixth edition of Hilgard and co-workers' enormously popular *Introduction to Psychology*, first published in 1953, appeared. What is notable about this edition is Chapter 6, entitled 'States of Consciousness', in that it was the first time that consciousness had been explicitly recognized as a legitimate object of study by a major psychology textbook since the behaviourist house-cleaning. In their introduction to that chapter, the authors offhandedly remark on their revolutionary inclusion:

> An earlier behaviourism, in seeking to make psychology more like other sciences rejected introspective reports (reflecting conscious experience). But the facts of consciousness are too pervasive and too important to be neglected. . . . The question of whether there are facts of 'consciousness' . . . can be answered readily enough: of course there are.[25]

Thus, with no fanfare—and no argument—consciousness as a legitimate object of study returned to mainstream psychology. No new discoveries forced its return; the authors' few remarks are really just a reaffirmation of ordinary common sense against the common sense of science, though by a highly respected psychologist and two colleagues. Consciousness (at least in humans) has simply come back in fashion.

Why did this occur? Clearly, there have been no new empirical discoveries which refute behaviourism or major conceptual analyses which uncover absurdities hitherto unnoticed. Behaviourism simply began to suffer the same fate which it had imposed on the Darwin–Romanes approach. As with ordinary-language philosophy, its stranglehold on creative minds began inexorably to loosen, for a variety of reasons. Adherence to it was exacting too great a price—as the authors of the textbook mentioned above say, it required that obvious facts be ignored and banished. Further, a critical mass of research slowly began to tip the balance away from some of the tenets of behaviourism—for example, the classical disregard of biological constraints on learning. It isn't that *logically* there hadn't been enough data to refute such tenets much earlier; but rather that until such data reached a psychological critical mass, or until cultural factors forced researchers to attend to them, it was easy to persist in ignoring them.

In May 1987, I gave a speech at the American Veterinary Medical Association conference on pain and distress in animals.[26] I also served on the AVMA panel charged with issuing a report on pain, which stressed the validity of anthromorphism and the moral need to control pain. I urged that the panel express the notion that the existence and knowability of *felt* pain in animals had been ignored for bad reasons, and was being restored in response to social pressure. Not so, I was told by one of the scientists. We now have 'enough' neurophysiological data to license the claim that animals feel pain *as we do*. He refused to accept the fact that common sense had always had enough data to make this claim, and that, given the common sense of science, no amount of new data could logically compel this conclusion! Nor was he able to point to new data unavailable ten years earlier which could be said to have turned the tide.

Someone who has not been through a doctoral programme in order to be certified as a professional can probably not fully appreciate the overpowering influence of what Heidegger calls *das Mann*—'them'—on a person's intellectual development. One is simply conditioned to accept certain areas, questions, and philosophies as mainstream or central, others as trivial. Furthermore, one learns to suppress scepticism based on common sense about current orthodoxies. In sciences involving work with animals, one also learns to suppress moral reservations about hurting animals, on pain—forgive the pun—of being declared 'too soft' for the field in question. In philosophy, a generation of fledgling philosophers learned to suppress their scepticism about all philosophical problems being outgrowths of linguistic abuse and about the history of philosophy being, by and large, an irrelevant chronicle of error and folly. In both philosophy and science, one continues to suppress one's reservations until one of two things happens: either one reaches the top of the field where one can set the tune instead of merely dance to it, or cultural pressures once more permit and legitimize one's doubts. Thus, in the case of philosophy twenty years ago, a number of pivotal figures (bored, no doubt, with critical constraints on their curiosity) began to write about the history of philosophy—Hampshire about Spinoza, Ryle about Plato, Danto about Nietzsche; and, all of a sudden, the history of philosophy again became acceptable. By the same token, increasing student vocationalism, stemming in large measure from economic pressures and cultural changes in the 1970s, forced philosophy departments that wanted to survive into concern with more 'relevant' matters.

The same thing clearly happened in psychology. A senior and unimpeachable figure like Hilgard could voice concerns and doubts about orthodox behaviourism which lesser figures would never dare to (or think of) expressing, but which thousands of undergraduates had felt for decades. And interestingly enough, Hilgard himself acknowledged explicitly in an article on 'Consciousness in contemporary psychology', as well as implicitly in his and his colleagues' textbook, that cultural factors in the 1960s influenced the re-entry of talk about consciousness. When society showed a great deal of interest in drug experiences, the expanding of consciousness, meditation, conscious control over normally involuntary bodily processes (biofeedback), it became more implausible than ever to ignore or deny the reality of consciousness—though, needless to say, people had dreams, fantasies, and so on long before the 1960s!

The cognitive turn

The sort of liberalization forthrightly expressed by Hilgard and his co-authors may be taken as symbolic of what has been called the cognitive turn in psychology. Though encompassing a *mélange* of ideas, and sometimes appearing to be little more than a new bottle for the old behaviouristic whine, cognitive psychology does represent a great liberalization of experimental psychology from Watsonian and operationalistic constraints. As Hilgard points out, cognitive psychology is informed by a new metaphor, that of the computer and its processing of information.[27] Using this metaphor as a source of a new perspective on mind seems to have freed psychologists from the constraints of the reductionistic, atomistic explanations associated with behaviourism. Though the metaphor sometimes seems tortured and over-inflated, it has unquestionably served as a lever with which to pry open hitherto closed doors. Thus, for example, mental imagery became a legitimate object of study under the aegis of cognitive psychology, and has generated books and even journals of its own. (The *Journal of Mental Imagery*, begun in 1978, must surely have Watson spinning in his grave!)

As R.R. Holt put it in a 1964 article in the *American Psychologist*, the study of imagery represented a 'return of the ostracized'.[28] Similarly, as Hilgard pointed out, the cognitive turn has re-licensed the study of other forms of introspection. W.K. Estes, one of Skinner's best-known Ph.D.'s, asserted in the introduction to his six-volume *Handbook of Learning and Cognitive Processes* of 1975 that:

> Conceptions of learning and cognition couched in terms of mental processes did not begin to grow to the stature of formal theories until the recent relaxation of the hold of behaviouristic thinking. . . . Only in the last few years have we seen a major release from inhibition and the appearance in the experimental literature on a large scale of studies reporting the introspections of subjects undergoing memory searches, manipulation of images, and the like.[29]

To some extent, cognitive psychology seems to me to be an excuse for looking at things too long anathema for less than defensible reasons. But whether it is a genuinely unified movement or simply a ticket to exploring the hitherto forbidden, it has certainly been liberating for psychology. Hilgard seems to come close to my 'fashion' view of scientific change in his discussion of Piaget's developmental psychology:

> The return of interest in Piaget, after many years of neglect, is a phenomenon to be understood according to the sociology of science. The climate for a cognitive psychology rather than a motivational one had to be great before his genetic epistemology could replace Freud's dynamic psychology as the most influential position in developmental psychology. . . . I am inclined to believe, though I cannot document it, that American psychologists needed Piaget, rather than that they adopted Piaget because they were convinced by reading his books.[30]

With cognitive psychology, whether as cause or effect, has come a liberalization in what psychologists officially allow as a legitimate aspect of their activity, which now includes the study of dreams, hypnosis, drug experiences, stream of consciousness, and so on. And, as we shall see, in so far as cognitive psychology made it possible to talk of consciousness again, this liberalization eventually extended to animal consciousness as well. None the less, the dominance of artificial intelligence as a central metaphor for cognitive psychology is disturbing. For the shadow of mechanism—very sophisticated mechanism, but mechanism in any case—looms large in those who take their inspiration from the computer. Granted that this metaphor allows for talking of conceptual maps, strategies, plans, expectations, as well as of the 'software' of mind, surely an advance over behaviourism's atomistic, *tabula rasa* learning model, still, the place of consciousness is unclear. Contents of consciousness, after all, do not in the final analysis fit well into a computer, and thus it is not surprising that some people who embrace cognitive psychology on the artificial intelligence model are agnostic at the very least with regard to consciousness. Thus, Steven Harnad, editor of *The Behavioral and Brain Sciences*, a leading journal in cognitive science, recently wrote a letter to the *Times Literary Supplement* to correct the claim that behaviourism failed because it did not deal with consciousness. Harnad asserted that

> Behaviourism's contemporary rival is not a bit more successful in its treatment of consciousness. . . . The cognitivist is as powerless as the behaviourist to observe directly what (if anything) is going on in anyone's head but his own. And more important, what cognitive theory attributes to our heads has no reason whatever for being conscious. Cognitive theory, like behaviourist 'theory', will always be equally (and indistinguishably) true of an organism (or machine) that behaves exactly as if it were conscious (but is not) and an organism that actually is conscious. That's why it's still a kind of behaviourism.[31]

Animal consciousness regained: the study of primate communication

Nevertheless, the new cognitivist approach did result in the tentative restoration of animal consciousness as legitimate subject-matter for psychology. This occurred in both general and specific ways. The general way is obvious: once consciousness came back into science, it was difficult for evolutionary reasons to restrict it to human beings. Furthermore, if talk of consciousness was again legitimate, then scientists were much more disposed to see things in terms of that category, and thus the common-sense tendency to impute mental states to animals was likely to resurface. Furthermore, as we saw in discussing pain, and shall see again shortly, social concerns with animal suffering helped to amplify and buttress the cognitivist approach, and are more acceptable to scientists not ideologically conditioned to see such locutions as meaningless.

To this general set of considerations should be added the specific interest in primate language and communication and the research thereon which was instituted in the 1960s and persists to this day. The work of the Gardners, Rumbaughs, Premack and Woodruff, Patterson and Linden, and others with teaching primates various forms of non-verbal signing systems and the resulting insight into the mental processes of these animals,[32] was so impressive and so seized the imagination, that it became increasingly difficult to deny mentation to these animals. The common sense of science has resisted these findings, of course, notably by denying that primates demonstrate true *language* in the fullest sense, or by asserting that the animals' behaviour is a very subtle version of the 'Clever Hans' phenomenon, the animals being subtly cued by the investigators to give the desired response.

None the less, the data are too impressive to ignore. Primates have scored 75 to 85 on standard IQ tests, have put signs together in novel ways to express new ideas, have shown the ability to lie, have taught signing systems to others, and so on. And even if their achievements fall short of true (human) language, they certainly reflect mentation. Chomsky has provocatively suggested that although animals lack true language, primates at least may well possess something like what is operative in humans whose linguistic centres in the brain have been ablated by stroke or injury—namely, a back-up system which, while lacking the full richness of language, none the less allows for expression and communication.[33] Furthermore, it has been pointed out that even if all this evidence was 'simply' of the Clever Hans phenomenon, the animals producing the evidence would have to be possessed of some very high level of mentation just to be subtle enough to unerringly pick up the cues of the experimenter.

Animal consciousness in the psychological literature

At all events, by the 1970s and 1980s, animal consciousness had been re-established to some extent as a legitimate object of study in scientific psychology. Thus, in a 1970 article published in the *Psychological Review* and entitled 'The Problem of Volition', the authors indicate *en passant* that 'the concept of [mental] imagery has become respectable once again in psychology'.[34] This, in itself, as we have seen, is not surprising. What is surprising, and also indicative of the dramatic changes which have taken place by 1970, is the following paragraph:

In the case of human subjects, it is possible to determine something about images from verbal reports. With lower animals the situation is more difficult but not impossible. Konorski (1967), for example, proposes the existence of perceptual images in the dog. He suggests the following experimental demonstration: allow two hungry dogs to eat exactly the same amount of food but, for one dog, the food dish is empty after the meal; for the other, some food remains. Now, remove the two dogs from the situation. If the second dog returns to the food dish and the first does not, it would indicate that the two animals retain an image of the partially full and empty food dishes.[35]

What is remarkable about this paragraph is precisely what is unremarkable—namely, the fact that an ordinary commonsensical response to the question of whether animals have imagery has become a credible, scientific one. Once again, no new discoveries have forced this change. It is the *climate* which has rendered the above example legitimate in 1870, illegitimate in 1940, and legitimate again in 1970, rather than any feature of the example itself. This of course points to the valuational basis of notions of evidence. The common sense of science (at least in psychology) has simply grown closer to ordinary common sense.

Exactly the same point can be found exemplified throughout recent psychological literature. Thus, in 1982, there appeared a book entitled *Mental Images and Their Transformations* by Shepard and Cooper. As its title suggests, the book is about the ability to transform and rotate mental images. The authors point out that the ability to transform mental images of three-dimensional reality is surely indispensable evolutionarily to any creature living in a world of three-dimensional objects:

Biological evolution would seem to have embarked on an uncharacteristically profligate course if the very most enduring and pervasive facts about our world—such as that it is three-dimensional, locally Euclidean, isotropic (except for the unique upright direction conferred by the earth's gravitational field), and populated with objects that therefore have just six degrees of freedom of rigid motion—had not been in any way incorporated into our genetically transmittable perceptual wisdom. Without such incorporation, this wisdom, so crucial for survival in such a world, would have to be learned, laboriously and at appreciable risk, by each and every newborn individual *de novo*.[36]

Thus the authors combine a return to ordinary common sense with a sensible appeal to evolutionary thinking to legitimate their study. Given such a philosophical stance, it is not at all surprising that they apply such an approach to animals.

The need to anticipate the consequences of structure-preserving transformation, though most obvious in the species that has advanced furthest in the use of tools and the construction of shelters and other artifacts, does not appear to be exclusively human. The possible utility of mental rotation for the canine species is suggested by the following behavior that one of us witnessed on the part of a German shepherd (Alsatian). In pursuit of a long stick hurled over a fence, the dog passed through a narrow opening where one of the vertical boards was missing from the fence and, seizing the stick, plunged back toward that same narrow opening—with the long stick now clamped in its teeth. Just as catastrophe appeared certain, the dog stopped short, paused for a (thoughtful?) moment and, rotating its head and, thus, the stick through 90 degrees, proceeded through the opening without mishap.[37]

What is so interesting about this case is again its mundane nature. Shepard and Cooper take it for granted that the dog's behaviour bespeaks the presence of subjective mental experiences. I have little doubt that such cases would have been known to every behaviourist from Watson on, yet none of them would ever have drawn the above conclusion; they would have explained the dog's behaviour in terms of 'random strategy generators' and reinforcement. What has changed is the valuational and ideological stance towards the same data.

It appears, then, that the common sense of science *vis-à-vis* animal mentation has retreated somewhat in psychological circles, allowing for a return of the facts of ordinary common sense. As in our discussion of pain, we may ask to what extent social-ethical concerns motivated this change. Presumably, they were not without influence. In his 1983 book *Animal Thought*, Stephen Walker marshals a variety of evidence from mainstream psychology to show that animals think and enjoy subjective experience against 'the weight of opinion that animals don't think but people do'.[38] Especially interesting is his use of neurophysiological data to show analogies between human and animal thought.

My argument in this book is as follows: human thought is intimately connected with the activities of the human brain; other vertebrate animals apart from ourselves have very complicated brains, and in some cases brains which appear to be physically very much like our own; this suggests that what goes on in animal brains has a good deal in common with what goes on in human brains; and laboratory experiments on animal behaviour provide some measure of support for this suggestion.[39]

Walker, like Griffin, whose work we shall discuss shortly, often goes back to behaviouristic experiments to support his thesis, illustrating once again how changes in ideology change perspective on the same 'facts'. But it is also interesting that however willing Walker is to take on behaviourist ideology on the facts of animal consciousness, he makes it clear on the very first page of his book that he is not willing to depart from it sufficiently to engage moral questions. Thus: 'I have deliberately avoided the question of how theories of animal psychology might impinge on opinions about our moral responsibilities towards animals.'[40]

One can speculate that Walker must have been at least partially motivated to explore animal thinking by the rise of concern about animal ethics, even though he was reluctant to deal with such questions as a scientist. Nor does one need to be an oracle to predict with confidence that the new laws mandating environments conducive to the psychological well-being of primates in laboratories which we discussed earlier will guide and motivate research at least into primate mentation; indeed, such research has already begun.

In August 1985, the *American Psychologist*, the major psychological journal in the United States, published an article by Gordon Burghardt entitled 'Animal awareness: current perceptions and historical perspective', in which the author surveys and defends the return of the legitimacy of discussing animal mentation. He lists a variety of factors in restoring the legitimacy of consciousness: research, Griffin's work, the rise of cognitive psychology, and 'the increasingly activist environmental and animal welfare movements who need and will use, whatever scientific data they can to protect animals—indeed the entire biosphere—from wanton and needless violation'.[41]

Interestingly enough, Burghardt is an ethologist rather than a psychologist, and we shall see that ethology has been more ready to engage the issue of animal awareness (and its connection to ethics) than psychology has. But it is significant that the *American Psychologist* published the paper, and did so in the same issue as an article of mine entitled 'The moral status of research animals in psychology', a topic almost never discussed in psychological journals. Unquestionably, social and moral concern have had their impact on psychological science in terms of preparing the ground for legitimizing talk about animal mentation.

10 Consciousness regained: ethology and beyond

The work of Donald Griffin

The influence of ethical concerns is evident in the recent ethological tradition, a movement still quite separate from psychology. Stephen Walker's 435-page book *Animal Thought*, which appeared in 1983, devotes only one page to Griffin's influential 1976 book *The Question of Animal Awareness*, and contains only three references to ethology. Yet ethologists are just as uncomfortable as psychologists with having to deal with ethics, and take pains to disavow any attempt to engage ethical issues.

Far and away the most influential writings *vis-à-vis* the attempt to restore animal consciousness to scientific discussion and make it legitimate are those of Donald Griffin, most notably his *The Question of Animal Awareness*. Griffin is a mainstream, 'establishment' ethologist, based at prestigious Rockefeller University, who made a solid reputation in orthodox, non-controversial ethology. He was best known for his pioneering work in echolocation in bats, work which in some ways harks back to von Uexküll's notion of understanding the *Umwelt* of other creatures by examining their perceptual apparatuses.[1] In *The Question of Animal Awareness*, subtitled 'Evolutionary Continuity of Mental Experience', Griffin calls into question the behaviouristic, ideology-of-science-based notion that animal consciousness cannot be dealt with in science or even known. Much of his brief (100-page) discussion deals with questioning philosophical and ideological reasons for denying animal mentation, as well as with giving reasons for believing that animals may be conscious. In the course of his discussion, he cites a variety of factors, including neurophysiological data, evolutionary considerations, experimental results, and most important, data on animal communication, notably the ape-signing mentioned earlier and the language of bees.

The important point about Griffin's book is that though it is well argued and well supported, it contains virtually no new empirical evidence; all the evidence that he marshals to show that animals are aware could and probably would have been available to the diehard well-read behaviourist. What Griffin is doing is ultimately philosophical and valuational; he is recommending the replacement of an inadequate philosophy by a better one. Interestingly enough, Griffin's ideas are a return to the logical implications of evolutionary theory in the manner of Romanes—that is, to the likelihood of evolutionary continuity of mentation, and to the perspective of common sense. Indeed, on first reading Griffin, my naïve response was 'This is news from nowhere; any ordinary person knows that animals are conscious.' (At that time, I was not aware of scientific ideology.) As we have seen, this is the case for ordinary persons, but not for every scientist! Had an ordinary person without academic credentials written the identical book, it would very likely not have been published, and would certainly not have been read by scientists. But Griffin had the proper credentials, so he could not be totally ignored; thus he could reinvent the wheel of ordinary common sense.

When Griffin's book appeared, I was just beginning to become involved in the animal ethics issue. Naïvely, I expected him to explore the moral implications of attributing mentation to animals. When I discovered no mention of this issue, I wrote to him asking his view of these implications. His reply bespeaks the influence of the common sense of science. Though he admitted that he was stimulated to write the book by discussions with philosophers whose major concern was the moral status of animals, he felt that it was not the job of a scientist to get involved in such questions. Like Walker, then, Griffin was quite conscious of the moral implications of his scientific work, yet loath to address them *qua* scientist.[2]

Griffin's most influential idea was probably that some animal communication (including the language of bees) bespeaks intentionality, and thus cannot be interpreted purely mechanistically. While there are certainly differences between animal communication and human language, argues Griffin, they are not so vast as to license one to exclude mentation from animals as Cartesians have done, a point I made in a different way in an earlier book.[3] Griffin envisions a 'cognitive ethology' which, using animal communication systems as a basis for experimentation and even for interaction between animal and experimenter, could give us 'a window into the minds of other creatures'.[4]

Though Griffin's work was welcomed by those espousing moral concern for animals, his reception by his peers was a good deal less enthusiastic. Reactions ranged from scepticism to outrage. The review of *The Question of Animal Awareness* by N.K. Humphries which appeared in the major ethological journal *Animal Behaviour* can be taken as paradigmatic of the response of orthodox adherents to the common sense of science. It accuses Griffin of asking behavioural scientists 'to open up an enquiry into the souls, large and small, of animals'.[5] It further caricatures Griffin's programme as follows:

> The enquiry should be conducted in a spirit of radical anthropomorphism by scientists who, refusing to be hobbled by the Behaviourist mafia, have the courage to trust their intuitive judgment about the inner working of animals' minds. Away with critical standards, tight measurements and definitions. If an anthropomorphic explanation feels right, try it and see; if it doesn't feel right, try it anyway.[6]

Once Humphries has worked himself up, the nastiness flows freely: '[For Griffin] . . . the bower-bird's nest is constructed by a cock imbued with aesthetic sensibility who very likely has a "mental image" of the effect his beautiful bower may have on a potential mate.'[7] Again:

> The chief witnesses at Griffin's enquiry are to be disciples of St. Francis of Assisi. Griffin believes that the way forward for behavioural scientists who wish to study the subjective feelings of animals is to get in on the act themselves, to establish two-way communication with their animal subjects. He points out that an anthropologist when studying a foreign tribe does not simply sit on the periphery with a notebook and tape recorder but learns the language of the people and attempts to establish a dialogue with them. In the case of, say, honeybees, the 'participant ethologist' would probably have to work in partnership with a bee-like stooge, a remotely controlled mechanical model which could be placed in the hive; but in the case of the great apes the ethologist could dress up in a furry suit and take the stage himself.[8]

Having accused Griffin of some of the most heinous sins against scientistic orthodoxy—anthropomorphism, mysticism, non-operationalism—Humphries unleashes his major charge of heresy: guilt by association with outmoded theories and discredited thinkers:

> Griffin seems hardly aware how closely this book mirrors the writings of the functional psychologists: James, Angell, Jennings and others, who were active in America around the turn of the last century. They too were interested in the 'evolutionary continuity of mental experience.' They considered it their job to 'discern and portray the typical operations of consciousness under actual life

conditions' (Angell 1907), and they were quite prepared to discuss the possibility of inter-subjectivity between man and an Amoeba (Jennings 1906). *Plus ça change, plus*—unfortunately—*c'est la même chose*. Griffin's own interests, his type of argument, his evangelism all have a 70-year-old freshness to them.[9]

Notably, Humphries fails to mention that Darwin too was interested in evolutionary continuity of mental experience.

Clearly Humphries, solidly immersed in the common sense of science, felt that Griffin was essentially a crank. As such he was not entitled to either empirical refutation or philosophical confutation, only to rhetorical dismissal. (Philosophical confutation in any case, as we have seen, is officially alien to the common sense of science.) Thus the only appropriate response is a patronizing sarcasm.

Writing about animal ethics often elicits the same sort of response as writing about animal consciousness—namely, the spleen reserved for heretics (or traitors). The common sense of science, like ordinary common sense, resists direct challenges to it, and responds to them with the same hostility which new ideas—for example, equality for some disenfranchised group—elicit from ordinary people or established religions. The Church tried Galileo, but it could just as well have been the other way around. All this has been well documented by historians and philosophers of science—for example, Velikovsky was denounced as a charlatan by mainstream scientists who had not even read his books. Once again, we should remember, as in our discussion of fraud in science, that scientists are human, and that one should expect no better and no worse of them than one expects of 'ordinary people'—though obviously the things which irk them, given their particular common sense, are often significantly different from what piques the common man.

Social concern with farm animals as a factor in reinstating consciousness

We have seen that moral concern about animals was causally influential in Griffin's rediscovery of animal consciousness, though one could scarcely deduce this from the book itself. In other areas of ethology, the link with burgeoning moral concern for animals is far more manifest, and indeed shaped the form which the revival of interest in animal consciousness took. Specifically, perhaps the major moral and social factor militating in favour of reintroducing consciousness into science is two decades of concern with farm animal welfare, especially in Britain.

Historically, most social concern for the welfare of animals was directed either towards perceived overt cruelty such as dog-fighting, bear-baiting, and so on, or else towards scientific research on animals, which was seen as frightening, incomprehensible, bizarre, and purposeless—a manifestation of the Frankenstein archetype which I have discussed at length elsewhere.[10] Agricultural uses of animals, with the exception of 'overt cruelties' such as branding or shipping of animals under extreme conditions (historically a very early concern), were largely ignored, since farmers by and large had a positive image with the public. (Children sing of Old MacDonald's farm, but no one sings of Old MacDonald's laboratory.) The lot of animals in traditional agricultural settings was viewed as idyllic—witness the Carnation Company's touting of its 'contented cows'. In the United States the barnyard and pasture view of agriculture meshes well with the ideal of rural America, and was rarely critiqued in terms of animal welfare when it was indeed the dominant mode of animal agriculture. The majority of those concerned with animal welfare historically were, and probably still are, meat-eaters and consumers of agricultural animal products; so agricultural uses of animals were taken for granted.

Some forty years ago, however, traditional agriculture began to change dramatically, and was transformed from the ancient extensive management systems, where animals roamed freely in 'natural' settings—cows on pasture, chickens in the barnyard, pigs in mud—to intensive management, where many animals were kept in close confinement in small spaces, and machinery replaced human labour in managing them. The industrial model was carried over into agriculture, and traditional farming was replaced by factory farming. With fewer and fewer people growing up on farms, increasing numbers of people were naïve regarding the changes in agriculture which had taken place, genuinely believing that animals were still raised under Old MacDonald conditions. Even as recently as a few years ago, one of the largest, most intensive chicken producers in the United States, brazenly ran a series of advertisements showing chickens in a barnyard setting, not in the close confinement in which they are in fact produced, and boasted of raising 'happy chickens'.

In the 1960s public attention was focused on the dramatic changes which had, unnoticed, become a *fait accompli* in agriculture, when Ruth Harrison published her *Animal Machines* in Britain, a dramatic account of the way in which animals had become part of an industrial process. Harrison's work stirred a great deal of coverage of the issue in the popular press and other mass media, which in turn resulted in the chartering of the Brambell Commission by the British government. The Brambell Commission report took it for granted that animals could have noxious subjective experiences, and cited fear, frustration, and exhaustion, as well as pain, as examples. Indeed, the Brambell report defined animal welfare as 'a wide term that embraces both the physical and mental well-being of the animal. Any attempt to evaluate welfare, therefore, must take into account the scientific evidence available concerning the feelings of animals that can be derived from their structure and functions and also from their behaviour.'[11]

Large numbers of people, appalled by the dramatic departure of intensive agriculture from the traditional, comfortable picture of the barnyard, demanded to know whether farm animals were suffering. And thus was born a great deal of scientific effort to establish answers to this question. Though many of the answers offered were (and still are) couched in terms of notions like 'stress' in order to avoid coming to grips with subjective experience in animals, it soon became apparent that this could not be done. An increasing number of scientists began to talk in terms of animal suffering, acknowledging a full range of negative subjective experiences in animals. To the orthodox common sense of science, such talk was unacceptable, and it became necessary to reappropriate ordinary common sense, which took such locutions for granted.

The work of Marian Dawkins and other farm-animal ethologists

Though a great many ethologists and animal scientists took part in this reappropriation, undoubtedly the best known is Marian Dawkins, whose *Animal Suffering* of 1980 is the main apologia for reintroducing these banished notions into science. Significantly, Dawkin's book is subtitled 'The Science of Animal Welfare'. Indeed, her goal is to demonstrate that one can make scientific sense of attributing noxious subjective experiences to animals. Like Morton and Griffiths's 1985 paper,[12] which we discussed earlier in dealing with the evaluation of animal pain, Dawkins essentially plugs the perspectives of common sense into a scientifically acceptable framework. Thus, the core of her book consists of a discussion of the various methods by which one can assess the suffering of animals. Though no single method provides indubitable proof, taken together these methods justify the attribution of suffering or happiness. In her discussion, Dawkins surveys the strengths and weaknesses of each method. Productivity has tended to be used by the intensive agricultural industry as *the* measure of welfare.[13] If animals are productive, it is claimed, they are well off and thus are not suffering. Dawkins, like others, points out that while productivity of the individual animal may sometimes indeed indicate that the animal is not suffering, the concept is ambiguous, since it is equally often applied to an agricultural operation as a whole, as an economic measure.

Thus, it has been demonstrated by poultry researchers that keeping laying hens under conditions where they have relatively more space than the usual confinement operation will increase individual hens' egg production. But the producer who buys an expensively capitalized laying operation will make more money if he crowds many hens into a cage, for even though individual productivity will decline, hens are relatively cheap, and the extra hens crowded into the cages will more than compensate for the loss in individual productivity. Further, even if one looks at the productivity of the individual hen, and it is high, it does not follow that the hen is not suffering. We know that unhappy humans often gain weight; this may well be the case for animals, and seems to be. In the same vein, physical health is not infallible as a sign of absence of suffering; apparently healthy animals (and humans) may still show psychological and behavioural disturbances.

Comparison with alleged wild counterparts of domestic animals is another valuable, but not infallible indicator, of suffering. Aside from the fact that the wild counterpart may be significantly different genetically and environmentally from the domestic animal, differences in behaviour in and of themselves do not necessarily indicate differences in suffering, as the environmental conditions are different. (Nevertheless, Stolba has shown that domestic pigs, when placed in an environment like that of wild pigs, show virtually all the behavioural patterns of the wild pig.[14])

On my view, one can in principle assess the genetic similarity of the wild to the domestic. If they are close, yet the living conditions are significantly different for the domestic animal, then one may have a prima-facie reason to believe that the animals' *telos* is being violated—that a square peg is being forced into a round hole—and that it is not living as it evolved to do. It is, however, conceivable (but unlikely) that even slight genetic differences make the animal congenial to the different environment. So one must, as Dawkins suggests, look at the animals' physiological parameters—for example, cortisol, endorphins, reproductive hormones, some of the traditional measures of stress—and also at the animals' behaviour. There are problems with each of these, of course. As we saw in the case of Gärtner's work with rats, the mere act of measuring some physiological parameter may be distressing enough to the animal to skew the measurement. Moreover, it is sometimes difficult to tell when a piece of behaviour betokens suffering. On the other hand, certain behaviour is clearly pathological since it damages the animal—for example, the self-mutilation often encountered in frustrated or confined animals. In the case of stereotypical behaviour—for example, the pacing of a stallion or a tiger in a cage—this too seems to betoken some suffering, as it both clearly deviates from the animal's behaviour under the conditions in which it evolved, and is often correlated with physical ill health.

One of Dawkins's major contributions was the development and popularization of preference testing. This is a way of 'asking' an animal what it prefers by giving it a choice of food, bedding, housing, and so on and seeing what it picks. One can even obtain a measure of *degree* of preference by 'sweetening the pot' or 'stacking the deck'. Thus, suppose an animal chooses open housing over confinement. One can test how much it prefers the former by giving it *ad libitum* food in confinement but irregular feeding in the open and seeing whether it still chooses the open space. The major problem with preference testing, of course, is that the animals, like us, may prefer and choose things not in their interest! In addition, they may prefer different things at different times.

Finally, Dawkins talks about analogies with ourselves—selective anthropomorphism, or critical anthropomorphism as Burghardt calls it,[15] for which we argued earlier. This is especially useful for generating hypotheses that can be tested by the other methods. To this list, as an adjunct to the method of analogy, I would add evidence from the animal's *telos*. By this I mean simply that if an animal has bones and muscles, for example, and is given no opportunity to use them, this provides a prima-facie reason to postulate suffering. Thwarting an animal's most basic urges and needs must surely be a major cause of unpleasant sensations, once we have admitted that it makes sense to attribute unpleasant sensations to an animal. This is being increasingly recognized socially, most recently by demands for enriched, non-sterile environments for laboratory animals, and for exercise for caged animals. (As we saw, the recently passed 1985 amendments to the United States Animal Welfare Act actually mandate enriched environments for primates and exercise for dogs, but neither of these for other animals, showing both how far we have come and how far we have to go.)

This is only a very brief sketch of Dawkins's book, a very clearly written work which is accessible to scientists in all fields and to lay people. Like Griffin's book, it is a watershed. In some measure, both books capture and articulate the essence of an evolving *Zeitgeist*, while at the same time accelerating and encouraging its evolution. Most important, Dawkins's work gave scientists a framework for legitimizing what they were under social-ethical pressure to do—namely, reappropriate and incorporate into their activities the common-sense notions that animals can suffer a full gamut of noxious experiences.

Other ethological advances in the farm-animal area

Starting roughly at the time of publication of Dawkins's book, research into animal suffering and preference took off, and older research came to be re-examined in a new light, illuminating the possibilities of mentation. Also in 1980, Ian J. H. Duncan, another British scientist and an expert in poultry, delivered a seminal lecture to the Poultry Science Association annual meeting at Purdue University, published in 1981 in *Poultry Science* as 'Animal rights—animal welfare: A scientist's assessment', in which he made some major points similar to those of Dawkins. Duncan begins with a subtle, learned philosophical discussion in which he shows that animal welfare and its assessment are inseparable from presuppositions of animal mentation, as well as from ethical notions—a major deviation from the common sense of science. In Duncan's view, as in mine, talking about animal experience is in principle no more difficult than using other theoretical scientific constructs. Notions like fear, discomfort, and frustration can be tied to 'facts' about which one can be 'objective'. In a passage reminiscent of the tradition of Darwin, Romanes, and Lloyd Morgan, Duncan argues forcefully in a way which surely must have astounded his audience of poultry production scientists, sitting comfortably on their cushion of scientific ideology, and which I quote in its entirety to show the extraordinary changes taking place.

> Of course, what we want to know ultimately is whether or not animals are suffering. The term 'suffering' implies a particular type of mental experience, a subjective consciousness, and that is what the Brambell committee was trying to take account of when they referred to 'mental well-being' and 'feelings of animals'. Subjective feelings are not directly accessible to scientific investigation, but that does not mean that they do not exist. We cannot prove that other human beings have subjective feelings any more than we can for a hen. However, common sense tells us that they probably do; their psychology

and behavior are very similar to our own, and, moreover, they can use language to describe what they feel. If we accept this to be true of human beings without hard proof, why not accept that it may be true in other species? In his book *The Question of Animal Awareness*, Griffin (1976) has argued that there is compelling evidence from animal orientation and navigation studies and from animal communication studies suggesting that animals do have mental images, subjective feelings, and intentions. Even classical ethologists such as Konrad Lorenz have suggested that animals probably do experience subjective phenomena (Lorenz, 1971). Students of phenomenology such as Buytendijk (1958) have made a more comprehensive study of subjective feelings in animals and man. In an elegant little paper entitled 'Toucher et être touché', Buytendijk (1953) showed that when deprived of visual cues, some animals, such as octopi (Octopus), were able to distinguish between touching something actively and being touched passively. Other animals, such as starfish (Asterias), were unable to make the distinction due to insufficient integration between receptor and effector control. Buytendijk interpreted this finding as demonstrating that animals above a certain phylogenetic level have a mental image of their immediate environment, the space occupied by their bodies in that environment and the stability (or otherwise) of stimuli in that environment. Of course this could only be considered as a very primitive 'mental image'. However, recent experiments with chimpanzees have shown that they can demonstrate intentionality (Menzel and Halperin, 1975), that they are capable of self-awareness (Gallup, 1979), and that they can learn to use surprisingly large vocabularies of artificial languages to communicate complex messages (Gardner and Gardner, 1975). This adds up to something much closer to human mental experience. We should probably assume that the hen, lying somewhere between the chimpanzee and the octopus on the phylogenetic scale, has some intermediate capability for experiencing subjective phenomena.

Careful experimentation in the future, therefore, may allow us some insight into what the hen is experiencing, including whether or not it is suffering mentally.[16]

Duncan's substantive discussion is very similar to that of Dawkins, deviating only in his introduction of yet another method of assessing suffering, one which he pioneered. This involves some invasiveness—namely, subjecting the animals experimentally to noxious conditions like fright, deprivation, frustration, and so on, and then comparing the psychological and behavioural results to those under 'normal' conditions and to conditions occurring under the system in which the animal is being raised. This sort of research, while illuminating, does raise moral questions of the sort raised by pain research—namely, that though the study is directed at improving the lot of the type of animals being studied, the particular animals studied are subjected to deliberate noxious stimuli not to their benefit. We shall return to this point shortly.

The pioneering work of people like Duncan, Dawkins, B.O. Hughes, Kilgour, Hemsworth, and others, some of which we alluded to at the end of Chapter 7, has catalysed a great deal of new research into animal suffering or, more accurately, has helped to accelerate the turn of the scientific *Gestalt* towards considerations demanded by both rational ethical principles and growing social concern with the moral status of animals. Thus, scientists have become more and more willing—though this is by no means universal—to admit that animals suffer in various ways, and to try to study such suffering. I have already alluded to Wemelsfelder's 1985 survey article on animal boredom, and numerous factors, both scientific and social, have helped to establish the legitimacy of this concern. Some farmers (not labouring under the common sense of science) have ascertained that providing confined pigs with toys such as hanging chains and bowling balls cuts down significantly on what the common sense of science called 'vices': tail-biting, bar-biting, vacuum-chewing, and so on, which the farmers saw as plainly a result of keeping fairly intelligent animals in a sterile environment, in which they did these things as a form of stimulation. This has in turn penetrated the thinking of scientists. Thus, a number of scientists in the laboratory animal field, most notably Spinelli and Markowitz,[17] have explored the physical and psychological benefits of creating enriched environments for laboratory animals, especially primates.

New methods for studying animal consciousness: Herrnstein and Gallup

Increasing numbers of scientists in psychology, ethology, zoology, animal science, and other fields have developed ingenious new methods for studying animal consciousness in dimensions beyond its relevance to farm and laboratory animal issues and beyond the modalities of suffering. It would be impossible to survey the many approaches recently taken to these questions, as well as pointless, since such surveys have been made, and made well, by Griffin, Walker, and to some extent Dawkins.[18] But a few examples are appropriate to illustrate new developments. The 1964 work of Herrnstein, ironically a leading behaviourist, and his associates on the concepts possessed by pigeons is one such fascinating development. Herrnstein, in a series of experiments, showed that pigeons possessed (or could be taught) a generalized concept of 'human being'. He showed pigeons many photographic slides, some including pictures of human beings, others not. The pigeons were rewarded with food if they pecked at a key in the presence of human beings in the picture, but not if they pecked when there were no humans. The pictures included people of both sexes and all races, ages, postures, nude and clad, and the pigeons were consistently correct in their pecking, showing that they did indeed have the general concept (despite their absence of language, *contra* the Cartesian and Kantian tradition we discussed earlier). In subsequent experiments, Herrnstein and his co-workers showed that pigeons could recognize individual persons, trees, water, fish under water, and so on.[19] Obviously, some of these discriminations—for example, of particular persons or of underwater conditions—could not be simply explained away as genetically determined 'hard wiring', but show, rather, a sophisticated ability to consciously discriminate general features.

The view that what appears to be consciousness is really mechanical was explored in a series of interesting papers by Gordon Gallup.[20] Basically, Gallup's original work involved trying to determine if animals recognize themselves in a mirror. Gallup showed that only chimpanzees and orang-utans behave as if they realize that their mirror image is a likeness of themselves. Gorillas do not. Thus chimps will use the mirror to become aware of and remove painted spots from places on their bodies that they can see only in the mirror, while gorillas will not. On the other hand, gorillas will remove spots they can see on their bodies without a mirror, leading Gallup to conclude that while gorillas do not want spots on their bodies, they fail to recognize that the image in the mirror is theirs. From this, Gallup concludes that gorillas have no sense of self, no self-awareness, and that their behaviour is 'hard-wired'—that is, mechanical, devoid of self-consciousness. For Gallup, an animal unable to recognize itself in a mirror is not truly conscious, because it is not aware that it is aware. Such an animal's apparent conscious behaviour is to be seen as analogous to human blind sight—cases of brain-injured humans who don't think they can see, but who, when asked to guess shapes, do so with accuracy significantly greater than chance.

Recognizing oneself in a mirror is not the only criterion of mentation for Gallup. Ability to deliberately mislead, as witnessed in certain chimpanzees,[21] reflects intentionality, which to Gallup bespeaks mind, as opposed to the hard-wired (mechanical) explanations possible for behaviour like that of nesting birds who act injured to mislead predators. Gallup seems to think that whereas the latter behaviour is most economically explained mindlessly and mechanically, as hard-wired and not requiring the postulation of consciousness, but only evolutionarily implanted mechanisms, the deliberate attempts of chimps to mislead humans when they are in competition for a goal shows that the chimps must 'introspectively anticipate the probable reaction' of humans or other animals. Though Gallup admits that it is not always easy to discriminate genuine mentation 'which entails self-awareness' from 'hard-wired analogs' which do not, the 'hard-wired' species ought to be unable to show 'intentional instances of [such behavior as] gratitude, grudging, sympathy, empathy, reconciliation, attribution, deception, and sorrow'.[22]

While I applaud Gallup for his ingenuity in theory and experiment, I do not agree with his results, since in my view, he has not given an adequate criterion for mentation, but rather, one that is too arbitrary and limited. Why should we take Gallup's criterion of recognizing one's mirror image as definitive for mind? Perhaps gorillas recognize themselves in a mirror, but don't show it. Perhaps there are species-specific reasons for their not showing it—maybe gorillas evolved near crocodile-infested waters, where those who attended to their own reflections were selected out!

I agree with Gallup that there is indeed a connection between awareness and self-awareness; we have already alluded to this in our discussion of Kant. But that is no reason to accept Gallup's restrictive, esoteric mirror criterion. As we saw earlier, even being able to perceive objects (or to make causal connections) requires some unity on the part of of the experiencer—that is, the same 'I' must be able to hold the discrete or progressive units of experience together. Why not simply take the fact that animals perceive or are able to recognize dangers and threats to themselves as evidence of self-awareness? Furthermore, the existence of even rudimentary cross-species communication and mutual adjustment, especially among species not normally found together in nature, bespeaks awareness, self-awareness, awareness of others, and conscious coping, not mechanical hard-wiring.

Other problems with Gallup's account are manifest. Why ignore other traits in gorillas which bespeak mind even on his criteria—like loyalty and affection—for the sake of the mirror criterion? Why accept a rare, flukish oddity like blind sight as a metaphor for the multitude of apparently conscious animal behaviours? There is a huge functional discrepancy between blind sight and a great deal of, if not most, animal behaviour. Gallup also suggests that human babies and humans who sleepwalk are hard-wired, not conscious. But this surely begs the question. After all, it is just as plausible to suggest that in both cases, the human being is conscious, but for some reason does not remember. Gallup suggests that at a certain age, humans go from non-conscious to conscious. Presumably that ought to tie up with acquisition of language, yet the evidence he cites shows that the age at which children allegedly become conscious does not coincide with the age of language acquisition. And even if it did, this could equally show that experience framed linguistically is simply more easily remembered.

Further, it is a bit difficult to believe that chimps, social primates for whom consciousness is presumably an advantage for social interaction, would be conscious, whereas gorillas, also social primates to whom consciousness would be a real advantage, are hard-wired. As we saw earlier, dogs can consciously mislead, and in many ways seem to meet Gallup's criteria. So presumably they are conscious. But this means that consciousness is haphazardly distributed through the animal kingdom, with no relevance to place on the phylogenetic scale, again an implausibility. In my view, Gallup's interesting criteria provide sufficient conditions for attributing consciousness to animals, but not necessary ones. As I have argued all along, it is more plausible to believe that awareness manifests itself in many different ways, from pains and itches to planning and duplicity—and mirror recognition of one's own body. Despite all these reservations, Gallup's work is very important, because it shows how an ingenious scientist can study what is considered 'unstudiable' by many of his peers.

A new approach to studying animal consciousness

In the course of thinking about the material presented in this book, and even more, while faced almost daily with scepticism from those steeped in the ideology of science concerning our ability to 'know what is going on in an animal's head', I have often tried to construct an experimental situation which would answer this question. Finally, I arrived at what I believe is a novel approach, though something similar was done by Blough.[23] There exists a phenomenon in human perception theory known as 'subjective contours'.[24] Thus when one looks at a certain sort of configuration (see Figures 10.1 and 10.2), even though what is

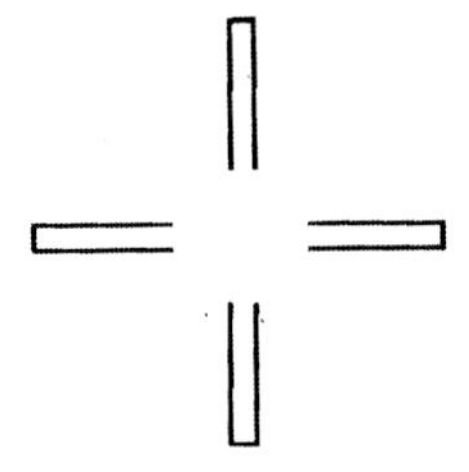

Fig. 10.1. Rectilinear figure which gives rise to the subjective perception of a circle.

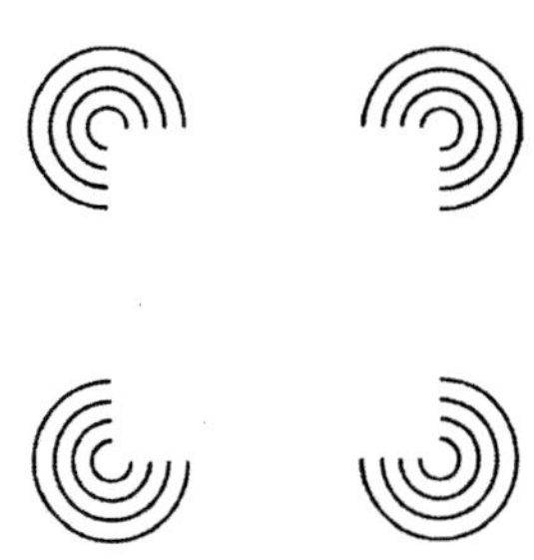

Fig. 10.2. Curved figure which gives rise to the subjective perception of a square.

actually printed on the page is one thing, what the eye sees is something else. In other words, in these sorts of diagrams, humans perceive visible contours which do not actually exist. Figure 10.1 gives rise to the experience of a circle in humans, Figure 10.2 to a square. I propose to divide a group of experimental animals (pigeons) into two. Group 1 will be trained to peck at circles, group 2 at squares. The group that is trained to peck at circles will receive positive reinforcement when they do so. When they are presented with square figures, they will receive nothing if they peck at them. We will then present that group with Figure 10.1 and see if the animals peck at it—in other words, see whether they perceive it as a circle. (Figure 10.1 in fact stacks the deck against the pigeons, because it contains rectilinear elements which might cue the pigeons to discern squareness, and has no objective lines suggesting circularity.) The second group will be trained to peck at squares but not circles, and will receive rewards when they do so. We will then present them with Figure 10.2, which again stacks the deck against them, since it contains curved shapes. If they peck at it, we may assume that they in fact subjectively experience a square as we do.

Similar tests, more refined so as to exclude various possibilities, and using other such figures could also be run. Indeed, and this is crucial, the whole area of illusory phenomena in humans, optical and otherwise, raises a unique set of possibilities for testing the subjective experiences of animals. Thus, for example, a black-and-white wheel spun at a certain speed produces subjective experience of colours in humans, and tests could be devised to see if this is the case in animals who have colour vision.

Unfortunately, I have not yet run the experiment outlined above. The key point, however, is not the results, but the method, for it demonstrates that even the most private of experiences, an animal's experience of something that isn't objectively there, is open to experimental examination.

The study of animal consciousness in mainstream science

All the preceding examples show that people do study animal mentation, and that new avenues for doing so have yet to be tapped. Whereas some people like Gallup and Dawkins are quite aware of the fact that they are studying animal mentation, other scientists just study it, without such philosophical apperception. For example, in February and March 1985, a number of relevant experiments were reported in *Science.* In one article, 'The goldfish as a retinex animal', the author studied the ability of goldfish to identify 'colors as constant under a variety of critical illuminated variations'.[25] The author concluded that 'the experiments indicate that goldfish determine the color of an object, as human observers may, within the context of the background array, rather than by the spectral composition of light reflected from the object'.[26] Though nothing in this experiment refers directly to consciousness, it is difficult to treat constancy and generalization as anything other than a mode of the goldfish's awareness, not simply as a mechanical process.

The next month, a more remarkable article appeared, dealing with how bees remember. A great deal of attention has been focused on the language of bees, first described by von Frisch. Griffin made much of the possibility of our understanding bees' language and of the possibility of even communicating with bees. Philosophers like Jonathan Bennett stressed that bee communication was not really a language,[27] just as we saw was done with ape 'language'. J.L. Gould, in a paper entitled 'How bees remember flower shapes', was concerned with the question of the quality of bees' mental images. In an elegant series of experiments reported in *Science*, Gould showed that bees seem to remember in 'photographic' images. In his words:

> Bees are able to learn to distinguish between flowers with different shapes or patterns. Some studies have suggested that bees remember only isolated features such as spatial frequency and line angles, rather than the photographic search images that are characteristic of vertebrates. New data indicate that this presumptive vertebrate-invertebrate dichotomy is false; bees can store flower patterns as a low-resolution eidetic image or photograph.[28]

We are obviously a long way from Watson here. In Gould's article, we see a topnotch scientist unconstrained by behaviouristic dogma, demonstrating empirically the characteristics of bees' memory images. Consciousness has indeed been regained. And even as Gould in his work turned his back on the tired old debate about whether bee communication is or isn't 'a language' in order to concentrate on the bees' experience, the same sort of thing was reported in *Nature* in May 1985 by Matsuzawa regarding his work with chimpanzees.[29] According to Matsuzawa, an ape was able to identify six arabic numerals by selecting from a keyboard the numeral that fitted the number of objects displayed to the animal. In addition, the ape was able to match the appropriate colour-word and noun for about 300 specimens, demonstrating this by stringing together the symbols—for example, '3 red pens'. Such an ability again speaks volumes about the mental life of apes, specifically about their potential for grasping concepts long thought by many to be the exclusive domain of humans. Another scholar, H. Davis, has looked at counting in other animals in a number of recent papers.[30]

One could cite further examples from diverse sources attesting to the amenability to study of animal consciousness, and its quiet return to scientific legitimacy. But the key point is that animal consciousness is alive and well in science. And, as the social and philosophical or ideological *Zeitgeist* changes, more and more scientists will be studying consciousness in animals, and more important, will be taking it for granted that it can be studied; thus their work will in turn further weaken the hold of the old ideology. Thus the common sense of science has begun to change, and will continue to do so, in response to the changing conditions and constraints under which science operates. Behaviour, automatically and arbitrarily dismissed under the ideology of science as purely mechanical, may now be more sympathetically viewed as evidence of genuine mentation.

Convincing the sceptics: A thought experiment

There are, of course, many true believers in the old ideology, and it is still being taught as gospel in many places. Further, many scientists who have laboured under it for ten or twenty or more years—and were trained in it—still have difficulty shaking it off. Virtually every time I discuss any of these issues with groups of scientists, I find large numbers who are still unwilling to admit that talking about animal consciousness of any sort is scientifically possible or legitimate. 'I believe in it at home,' a veterinary ethologist told me not long ago, 'but I leave it behind when I go to the lab.' 'The rules of science don't allow us to talk about animal consciousness,' said another. 'If you don't like it, get the rules changed.'

How does one deal with this entrenched scepticism? In many cases, when there is time, I attempt to present the arguments of this book. I have done this with many colleagues, and it has been quite effective, since many did not realize that they had uncritically learned a debatable philosophy along with their science. In most cases, however, I am seeing people for only a brief period, and must make my point forcefully and briefly. Under such circumstances, I usually perform a thought experiment. I ask my audience or the person to whom I am talking to imagine that social concern with animal treatment has reached proportions such that legislators, industry, and so on need detailed information on how animals think and feel in order to establish rules and laws for their proper treatment. As a result, the federal government makes $500 million available for research into animal mentation. (NIH has in fact made funds available for validating ways of decreasing animal suffering, and in the wake of the new United States law discussed earlier, researchers are now studying ways of enriching primate environments and criteria for enrichment.) I then ask, 'Do you really believe that there would be no takers?' I continue: 'You know quite well what would happen. All of a sudden ideology would be left behind and people would be burning the midnight oil coming up with interesting new methods to study animal mind and feeling. And much of what they came up with would be interesting and useful, as in the few examples I have given.' Interestingly enough, few people quarrel with my thought experiment. 'By the same token,' I continue, 'they would not respond in the same way if someone came up with vast sums of money to study the Heavenly Land or angels. So clearly the study of animal consciousness is not as inaccessible or mysterious as your common sense of science leads you to believe.'

Conclusion

We have thus seen that the factors involved in the waxing, waning, and re-waxing of the scientific legitimacy of studying animal consciousness have not followed a rational path. Contrary to the ideology of science, they have been dominated by value judgements of all sorts, including moral judgements; by sloganeering and propaganda and who seizes the microphone; by the medium as much as by the message; by social concerns or lack thereof; by legislation; by the way scientists are trained; and by a host of factors never admitted by the common sense of science to have any relevance whatever to scientific change. We have seen that a major factor involved in easing (if not forcing) the return to talking about animal mentation in science in a variety of forms has been rising social, philosophical, cultural, and political concern with the treatment of animals. Thus we have seen that social ethics has a major impact on the development of science.

What is too often ignored, especially by those most zealous in prosecuting the welfare and rights of animals, is that this relationship is reciprocal—that science has (or can have) an equally profound effect on ethics. A science that 'proves' that animals don't feel pain, or that we can't know if they do, can unquestionably retard the progress of animal welfare, especially in a society that treats the pronouncements of science as infallible. On the other hand, a science that acknowledges animal mentation and studies its many modes across various species can be an invaluable resource in helping society set just guide-lines and rules for the treatment and protection of animals. And it is quite obvious that the most basic thing we need to know in setting out such rules is what an animal's life or *telos* is like, both physically and mentally, but especially mentally, because physical needs and their non-satisfaction or satisfaction result in pain and suffering or happiness and other modes of awareness. Until we know this, we cannot accurately, precisely, or realistically discuss the all-important details of our moral obligations to animals.

As we extend the umbrella of moral concern over the rest of nature, we are obliged to do so in a way which recognizes the uniqueness of interests and needs built into each creature's *telos*. We do not behave rationally and morally by treating each animal as a little human, any more than the United States has been morally or pragmatically correct in treating all underdeveloped countries as potential Americas. That we have moral obligations to animals can be established a priori, as I have tried to do in my book on animal rights.[31] What these obligations are in any particular case requires detailed knowledge of the animal, both biologically and in terms of the new science we have discussed.

As ethical awareness of this issue continues to grow in the public mind, so the necessity for and legitimacy of talking about animal subjective awareness will continue to increase, as will the demand for and funding of its scientific study, and this will inevitably lead to a demand that researchers articulate and justify their moral position *vis-à-vis* the research animals they use. This will create a deep, difficult problem for researchers into animal consciousness. For the very nature of and social pressures for their research must awaken in them an awareness of the moral status of their subjects. At the same time, built into that same research is the potential for major infringement upon the interests and natures of the creatures being studied.

The study of the most morally relevant modalities of mentation—pain, fear, anxiety, sorrow, boredom, and aggression—all have the potential for dramatic infringement on the subject animal's well-being and happiness. Thus, contrary to value-free ideology, research into animal consciousness must become a moral science, constrained in its experimental design by the moral status of the creatures it studies. Methodologies must be devised which maximize the respect for individual animals studied in the research process, while at the same time acknowledging that without research into animal awareness, moral concern for animals in society must be limited in both scope and detail, and resulting social policy must inevitably be ill-founded.

In this regard, it is worth citing the work of Jane Goodall as exemplary. Goodall, harking back to the tradition of Romanes, Darwin, and Köhler, has spent twenty-nine years living in Africa and observing and describing the behaviour of chimpanzees under natural conditions. Drawn by curiosity rather than careerism, Goodall did much of her work before becoming academically credentialled and thus her approach to animal behaviour was informed by ordinary common sense, rather than thoroughly dominated by the common sense of science. Though she subsequently received her formal training and doctorate from Cambridge in ethology, her work retained a freshness and sparkle often lacking in standard scientific writing, and thus her work spoke as much to ordinary people as to specialists. Her 1986 *magnum opus*, *The Chimpanzees of Gombe*, published by Harvard University Press, is both rigorous and highly readable, and earned her increased credibility with the academic scientific community even while continuing to speak eloquently to non-scientists.

In this book, in many passages reminiscent of Köhler, Goodall does not cavil at imputing consciousness to animals, even anthropomorphically:

All those who have worked long and closely with chimpanzees have no hesitation in asserting that chimpanzees have emotions similar to those which in ourselves we label pleasure, joy, sorrow, boredom and so on. . . . Some of the emotional states of the chimpanzee are so obviously similar to ours that even an inexperienced observer can interpret the behavior.[32]

Thus does the good common sense of Hume and Darwin pervade and illuminate her work. Like the ape language studies mentioned earlier, Goodall's writings as well as the extensive television coverage she has enjoyed are unquestionably a major factor helping not only to restore the legitimacy of talking about consciousness in animals, but also to reclaim and solidify the traditional tendency to take it for granted.

But Goodall's work has been equally profound in shaking what we have seen to be another mainstay of scientific ideology, the notion that science is value-free and has no truck with ethics. Goodall was, from childhood, motivated by empathy and moral concern for animals as much as by curiosity about them, and this moral concern suffuses and ennobles her life's work. At a lecture she gave which I attended, she mentioned her attempts to treat chimpanzee disease whenever she could at Gombe. This confession was greeted with a peevish question from a scientist in the audience, solidly ensconced in scientific common sense: 'Surely, Dr. Goodall,' he snapped, 'you don't interfere with the population you are studying?' 'No,' she replied, 'we help.'

Even more profoundly, Goodall has begun to question many of the uses of chimpanzees (and other animals) in biomedical research, though she is still convinced that some animal research is necessary. None the less, she feels that husbandry of these animals should fit their natures, as I have argued at length elsewhere,[33] and is using her unparalleled knowledge of primate behaviour to develop systems of housing and management congenial to the animals, as well as to criticize traditional, austere, environments developed for researchers' convenience, with total disregard for many of the animals' needs.

In the conclusion to her book, Goodall writes the following moving passage, which is very appropriate also to conclude our discussion.

Let us hope that this new comprehension of the chimpanzees' place in nature will bring some relief to the hundreds who presently live out their lives as

> prisoners in our laboratories and zoos. Let us hope that, even as our greed and shameless destruction of the natural world gradually take from yet more chimpanzees their forests, their freedom, and often their lives, our knowledge of their capacity for affection and enjoyment and fun, for fear and suffering and sadness, will lead us to treat them with at least the compassion we would accord fellow humans. Let us hope that if we continue to use chimpanzees for painful or psychologically distressing experiments, we shall have the honesty to label our actions what they are—the infliction of torture on innocent victims.
>
> If only those responsible for the care of captive chimpanzees (and other animals too) could experience with me some of the more intimate moments at Gombe. For so often they generate overwhelming shame for the behavior of our own species—our arrogant assumption that **our** needs, **our** pleasure, **our** wishes must inevitably come first.[34]

In the case of a scientist like Jane Goodall, then, science is not divorced from ethics, but they are inextricably bound together, each influencing the progress of the other.

Recently I received a letter from a friend at Cambridge, containing a clipping from an English newspaper announcing the appointment of a scientist to the world's first endowed professorship in animal welfare. In announcing the appointment, the spokesman for the university stated that invasive methods of research would be minimized. According to the article 'no electrodes or implants [would be] inserted into the animals' brains. Conclusions would be reached by other types of assessment, including observation, sample analysis, and the gathering of general scientific data.' The article concluded that although the primary focus of research by the chair-holder would be intensive agriculture, there would be 'a more esoteric side [of research] into the ethics of using animals for man's purpose'. Thus the ideology of science dies, but it also dies hard.

11 *The Unheeded Cry* revisited

I

During the decade following the first publication of this book, the power, pervasiveness, and tenacity of entrenched ideology have been demonstrated repeatedly in a variety of venues. From China and the Soviet Union we have learned that societies do not easily shed totalitarian ideologies, even following revolutionary political change. Similarly, the experiences of Eastern Europe and Africa make manifest that ideologically based hatreds, whose origins have been obscured by the passage of time, may, like anthrax spores, re-emerge as virulent and lethal as ever, unweakened by years of dormancy. Most strikingly, perhaps, the work of historian Daniel Goldhagen has demonstrated the enormous power of ideology to overwhelm and obscure both common sense and common decency, even among the most civilized of people.

In his monumental work, *Hitler's Willing Executioners* (New York: Knopf, 1996), Goldhagen has shown that ordinary Germans willingly and voluntarily engaged in genocidal activities, even when it was patently open to them to refuse to do so without fear of recriminations. The killers studied by Goldhagen were neither sadists and psychopaths of the sort attracted to the S.S., nor the sort of street brawlers and bullies that composed the ranks of Ernst Röhm's S.A. Rather, they were normal, largely non-violent family men, who operated neither out of fear of punishment for disobedience (one standard explanation), nor out of the blind obedience suggested by Stanley Milgrim and often invoked to explain Nazi killing. According to Goldhagen, neighbors became killers because of their immersion in two centuries of ideological dogma depicting Jews as pathogens in the body politic, rendering that body ill and infirm and demanding radical excision of the disease-causing organisms. As absurd as this seems to those of us unsteeped in similar ideology, it was common sense to Goldhagen's Germans, and a straightforward justification for actions they would recoil from in non-ideological contexts.

Scientific ideology, then, of the sort constituting what I have called "scientific common sense," licensed the denial of pain and mentation to animals (and human infants) even while the same scientists denied the relevance of ethics (or any value judgments) to science. The fact that scientists were highly educated people—as skeptics often said to me when I explained ideological denial of animal awareness and moral thinking in science—in no way touches our point, for their education was in good part a total and non-dialectical immersion in an ideology insulated from questioning.

I have continued to monitor scientific ideology in hope that its hold on scientists had been somewhat mitigated. What I found did not nourish unequivocal hope for surcease of that hold. I was particularly disheartened when an account of a speech delivered at Michigan State University by the then head of the National Institutes of Health, James Wyngaarden, crossed my desk. Wyngaarden, arguably America's top biomedical scientist by virtue of his position, and certainly as mainstream a scientist as one could be, affirmed while discussing issues facing biomedical research that, while "ethical issues such as gene sequencing are always controversial, research should not be hampered by moral considerations."[1]

No less shocking was a letter by an eminent surgeon to the *Hastings Center Report,* after the *Report* had published a special supplement in May of 1990 entitled "Animals, Science, and Ethics." (This in itself, by the way, is revealing. Though the use of animals in biomedical research had been a significant social issue at least throughout the 1980s, resulting in major legislation, it was not discussed by the Hastings Center, which, along with the Kennedy Center, is the medical establishment's "approved" ethical think tank, until 1990.) In his letter, Robert White of Case Western Reserve complained that he was "extremely disappointed in this particular series of articles, which, quite frankly, has no right to be published as part of the *Report.*" White continued: "Animal usage is not a moral or ethical issue, and elevating the problem of animal rights to such a place is a disservice to medical research and the farm and dairy industry."[2]

Accounts of abuse of human subjects continued to proliferate in the 1990s, most notably perhaps being the reports of the radiation experiments conducted in earlier years by the Department of Energy without informed consent. By and large, scientific meetings and journals continued to eschew discussions of ethical issues occasioned by science, and my 1995 *The Frankenstein Syndrome: Ethical and Social Issues in the Genetic Engineering of Animals* (New York: Cambridge University Press) remains one of only two or three books on the subject, despite the fact that of all twentieth-century technological advances, biotechnology is the one most likely to dramatically alter our lives. And again, the research community did essentially nothing to address the ethical issues associated with mammalian cloning instantiated by the cloned sheep Dolly, thereby leaving the discussion and regulation of biotechnology to an uneducated public, with the ethical lacuna filled by doomsayers, opportunists, and fearmongers.

In a similar vein, the second component of scientific ideology, viz., the denial of mentation and feeling to animals, continued to remain respectable if not dominant in science. This is evidenced, for example, by the continued skeptical writings about animal minds by scientists like Gordon Gallup (whose work was discussed in Chapter 9). Neo-Cartesians, such as Peter Carruthers in *The Animals Question* (New York: Cambridge University Press, 1990), attempted to weave a philosophical cloak to cover the nakedness of claims denying consciousness to animals. (Gallup, incidentally, wrote so scurrilous a review of *The Unheeded Cry* that it prompted a long letter from ethologist Marc Bekoff, which in turn led to a significant and sharp exchange in the journal *Animal Behavior.*)[3]

In a more serious vein, a 1995 article in the *Canadian Veterinary Journal* surveyed practicing veterinarians regarding their knowledge of animal analgesia.[4] Seventy-seven percent indicated that their knowledge of the field was inadequate and, even worse, what they did know had not been taught in veterinary school. Post-surgical analgesia is still very much under-utilized in veterinary medicine, and many of the old shibboleths concerning its use (for example, that if pain is controlled animals will injure themselves) are still operative, so much so that a recent textbook of veterinary anesthesia feels compelled to refute them.

Equally of concern is the fact that many physicians still view felt pain, even in humans, positivistically and as outside their purview, as subjective and thus not of direct concern to scientific medicine. In grim attestation to this claim, a 1991 paper by Ferrell and Rhiner argued that although pain can be controlled effectively in 90 percent of cancer patients, it is in fact not controlled in 80 percent of such patients.[5] Many physicians continue to oppose the medical use of marijuana and narcotics even for terminally ill patients on the grounds that such people might become addicted! And a 1994 paper in the *New England Journal of Medicine* demonstrated that children and infants (who are, of course, powerless and, in the case of infants, non-linguistic beings analogous to animals) receive less analgesia for the same procedures than do adults undergoing those procedures.[6] On the brighter side, by the early 1990s the pediatric community had abandoned the odious practice of doing open-heart and other neonatal surgery using paralytic, curariform drugs (see Chapter 5), and had been forced to reappropriate ordinary common sense, largely as the result of social censure as the practice became known. So, by 1992, a paper by Dr. K. J. S. Anand appeared extolling the medical virtues of proper anesthesia and analgesia for neonatal surgery.[7] The paper demonstrated that when infants undergoing open-heart surgery were deeply anesthetized with high doses of sufentanil and also given high doses of opiates for twenty-four hours post-operatively, they had significantly better recovery and significantly fewer post-operative deaths than a group receiving a lighter anesthetic regimen (halothane and morphine) followed post-operatively by intermittent morphine and diazepam for analgesia. The group that received deep anesthesia and profound analgesia "had a decreased incidence of sepsis, metabolic acidosis, and disseminated intravascular coagulation and fewer post-operative deaths (none of the 30 given sufentanil versus 4 of 15 given halothane plus morphine)."

A particularly egregious example of the persistence of academic scientists ignoring pain in veterinary medicine is worth relating to close our discussion. Not long ago, I was invited by a swine extension veterinarian to lecture to a group of pork producers. Very concerned with the ethics of animal use, they sought counsel on a number of issues. In one case, a farm manager showed me some freshly castrated two-year-old boars while I was touring his facility. The animals were very sore, and I asked him and the extension veterinarian whether they had used anesthesia and post-surgical analgesia, pointing out that the animals would heal more quickly and suffer fewer complications. They replied that they had not, since neither knew of such a regimen, but both were willing, indeed eager, to try it. When I left, they told me they would contact the nearby veterinary school to obtain state-of-the-art information regarding swine anesthesia and analgesia.

Some weeks later, I received a phone call from the extension veterinarian, who was extremely agitated. He informed me that he had contacted experts in two different departments at the school, and all the veterinarians had refused to "get involved." They did not see the issue as important, professed ignorance of swine anesthesia and analgesia, utilizable under farm conditions, and feared liability and compromising food safety. The veterinarian then asked me if I could find a proper regimen, which I was quickly able to do by contacting some of the veterinarians I knew to be sympathetic to pain control. As I write this chapter, the trial of these regimens is being run. What is worth noting in this story is that farmers, possessed of ordinary common sense, were far more open to admitting the existence of pain in animals and trying to control it than were academics steeped in scientific common sense.

II

Given all that we have argued in the book about the stranglehold of scientific common sense on scientific thinking during much of the twentieth century, it is not surprising that one can continue to find evidence of its continuing presence. What is much more interesting is the question of whether one can find signs of its hold abating, and the methodological question of what would count as evidence for such a claim.

According to the account developed in the text, scientific practice and scientific ideology will develop along their own pathways until such time as societal pressure emerging from changes in social thought forces them to change direction—for example, changes forcing them into closer harmony with ordinary common sense, leading to what I call the "reappropriation of ordinary common sense." Is there evidence, during the decade following the original publication of this book, of the requisite social pressures being exerted on the scientific community to force a return on the part of scientists to ordinary common sense and a moral concern with pain and suffering? And equally important, is there evidence of that pressure having a genuine effect?

Engaging the latter question first, I would respond strongly in the affirmative. If one looks at some of the examples provided in the previous section, one can see that they are, as it were, iridescent. While attesting to the continued hold of scientific ideology, many of them also attest to its attrition. In other words, there is relevance here of the old cliché about the pessimist seeing the glass half empty, while the optimist sees it as half full.

Consider the human neonatal surgery case. While it is indeed true that it took the pediatric community some five years and numerous studies to reappropriate common sense in response to social concern, the bottom line was that it occurred, and that there were scientists such as Dr. Anand undertaking research to validate the benefits of anesthesia and analgesia. Similarly in the swine castration case—however solidly ensconced the academics at the veterinary school in question were in scientific ideology, and in dismissing the moral and medical relevance of pain—I was easily able to find others who took quite the opposite view.

That would not have been the case ten or fifteen years earlier. Though there are plenty of Gallups ready to trivialize or localize animal consciousness to a few select animals, there are increasing numbers of other scientists like Bekoff who are working valiantly (and brilliantly) to erode both components of this scientific ideology. See, for example, Bekoff and Jamieson's superb two-volume work, *Interpretation and Explanation in the Study of Animal Behavior* (Boulder, Colorado: Westview Press, 1990).

And to more than offset the letter from Robert White or the statement by Wyngaarden, it is patent that the Animal Care and Use Committees, mandated by the federal laws we discussed in Chapter 7, are wrestling seriously and responsibly with ethical issues occasioned by animal research, most particularly pain and suffering. Thereby they are of necessity weakening the hold of scientific ideology, at least among member of such committees, and to a lesser extent among the researchers whose protocols they review.

The reader will recall the description of the legislation in Chapter 7. Those of us who wrote those laws had placed great hope in their functioning as a vehicle for undercutting scientific ideology and reappropriating ordinary common sense about animal consciousness, animal pain, and the moral questions occasioned by invasive animal use. If federal law assumed that animals felt pain and required its control, scientists, we believed, would give pain and pain control serious attention, thereby building up knowledge of pain recognition and alleviation, until the old ideological skepticism was eclipsed. Similarly, deliberation in committees about control of pain and suffering, proper euthanasia, multiple animal use, alternatives to invasive animal use, and the like, we believed, must inevitably increase researcher sensitivity to ethical questions inherent in animal research.

As a person who has served on a protocol-reviewing Animal Care and Use Committee for almost twenty years, and who has sat in on and consulted for numerous others, I have seen our faith vindicated. To be sure, the changes in attitude, thought, and gestalt are not immediate, and different committees reflect different cultures. For some (usually senior) researchers who have labored under scientific ideology for most of their careers, it took some time to assimilate this new way of thinking. On occasion, the change necessitated what I have elsewhere called "breakthrough experiences" leading to a change in gestalt. For example, I recall one major researcher on the committee who tended to see time spent with committee work as a waste of time done to satisfy bureaucratic requirements. He maintained this cynical attitude until the committee confronted a steel-jawed trap protocol. This pushed his buttons. "Goddammit," he shouted. "We're supposed to be an Animal Care Committee? If that is the case, why the hell are we even discussing this protocol. The steel-jawed trap is a barbaric instrument banned in many countries—we should just throw this damn protocol out!" We were off to the races, and he never took a jaded view of our job again.

Many committees have in fact been so infected with a moral perspective that they have exceeded their authority. I have long argued that animal research should, in addressing the moral issues it raises, at least be required to weigh potential benefits of research against cost to animals (in terms of pain and suffering). This is currently a requirement for committees overseeing research on human subjects, though obviously, from a conceptual perspective, such cost-benefit analysis is highly difficult and even problematic. But no similar exercise is legally mandated for animal research. Nonetheless, many Animal Care and Use Committees explicitly undertake cost-benefit analysis (with others doing so implicitly), especially when the research involves significant pain and suffering. Others specifically ask researchers to address the issue on the protocol forms they submit.

The moral perspective is self-expanding.[8] While many researchers serving on committees may have initially approached such service as simply a bureaucratic necessity, most eventually begin to see the socio-ethical issues behind the legislation, and begin to think in ethical terms. For example, many committees have begun to demand pain control for higher invertebrates, such as cephalopods, even though the laws cover only vertebrate animals. "Pain is pain," one such committee member told me. As another example, even though committees do not, by statute, cover food and fiber research, they have invariably extended the mandated requirements for surgical anesthesia and post-surgical analgesia to agricultural research projects utilizing surgery. A third example of such committee (and researcher) extension of ethical concern beyond the letter of the law is provided by the issues of environmental enrichment. According to the law, as we saw in the book, only primates and dogs are covered by the requirement for enrichment; dogs must be given exercise, and primates must be housed in a manner that "promotes their psychological well-being." Nonetheless, many committees have seen that environments meeting the needs dictated by an animal's *telos* or nature are clearly essential to all animals, and have thus provided enrichment for other species—toys and gang-housing for cats; group housing with tunnels and toys for rabbits; playthings for swine. Trade journals, such as

Lab Animal, regularly run valuable articles on enrichment for a variety of species and, perhaps most significantly, the most recent edition (1996) of the NIH *Guide to the Care and Use of Laboratory Animals* has placed great emphasis on environmental enrichment. It is becoming increasingly evident, as Dr. Thomas Wolfle of the Institute of Laboratory Animal Resources of the National Academy of Sciences has argued for years, that laboratory animals probably suffer more in virtue of the impoverished environments inimical to their natures in which we keep them than from the invasive manipulations we perform on them, especially given the new laws mandating control of procedural and post-procedural pain, and early end points for invasive studies.

In any case, concern and discussion in the above areas inevitably undercut scientific ideology regarding both the denial of ethics in science and the denial of consciousness in animals. Obviously talk of enriched and impoverished environments clearly entails animal awareness of those environments, and also raises issues of animal comfort and even, I dare say, animal happiness. Increasing numbers of researchers and laboratory animal veterinarians are using locutions such as "the animals are happier now that we are housing them in social groups," and supporting such locutions by appeal to animals' demeanor and affect, as well as by reference to more "objective" features, such as better animal health and successful reproduction. The talk of modes of awareness in turn potentiates ethical judgments, such as "old ways of keeping chimps in solitary confinement, without stimulation or companionship, were morally wrong." Such moral attention in turn drives greater attention to animal comfort, and so on. Similarly, dramatic results obtained from administration of analgesia in painful research make researchers realize the extent to which animals not receiving analgesia suffer, which in turn drives greater concern with analgesia.

Thus, as we hoped in writing the federal laws, these laws have catalyzed the reappropriation of ordinary common sense and the correlative suppression of the hold of scientific ideology. And once ordinary common sense is in place, it tends to be self-reinforcing and to grow. Were the laws to be repealed tomorrow, I do not believe that, at least for many scientists, the new perspectives on ethics and animal pain would vanish. Recognizing something's moral status when you have not previously done so involves a major shift in gestalt that one cannot easily abandon. For example, those of us who forty years ago, while not being active racists nevertheless took segregation as a given, could probably not resume that perspective, even if all legal constraints preventing segregation disappeared.

My own institution admirably illustrates the extent to which abandonment of scientific ideology perpetuates itself. Although federal law does not specify how long animals must be given post-surgical analgesia, our committee has required that it be for at least three days post-surgically, and has further insisted on pre-operative or intra-operative analgesia since it became known in veterinary medicine that such prophylactic analgesia prevents the pain response from becoming too severe. It is far easier to thereby keep pain under control, since using moderate doses of analgesia to keep the pain response from getting fully "wound up" is far more effective than using massive doses of analgesia to mitigate a full-blown pain reaction.

The three-day requirement was based on veterinary clinical experience, as well as on anthropomorphic extrapolation from human post-surgical patients, in whom most acute pain has abated by three days after surgery. On the other hand, our faculty has recognized that three days was somewhat arbitrary and, on various occasions, has lengthened the period of post-surgical analgesia considerably (up to two weeks). A year ago, the dean appointed a committee of faculty to study long-term acute pain, as when a slow-healing lesion is deliberately created in an animal, as well as chronic pain, i.e., pain that persists even in the absence of a recognizable lesion and is refractory to standard analgesics. Regrettably, there is a paucity of knowledge about chronic pain, virtually none regarding chronic pain in animals, and very little on chronic pain in humans. Indeed, the primary modalities for managing chronic pain in humans are psychological. What little information exists in veterinary medicine suggests that the same is true in animals—psychological modalities such as increasing environmental stimulation seem to mitigate chronic pain. Research in this area is likely to provide increasing evidence of mentation in animals, much as did Mason's work, discussed earlier, on cognitive modulation of the stress response in animals.

In sum, I believe that the federal laws have certainly moved forward the erosion of scientific ideology vis à vis animal pain and, to a lesser extent, animal "distress"—the federal law's word for suffering—including such notions as fear, anxiety, loneliness, and boredom. Real progress, I believe, has been made in the biomedical research community, the main targets of the laws. One personal sidelight is worth relating. A year or so after the laws passed, I was approached by one of the editors at CRC Press and asked to edit a volume covering those areas that researchers would be responsible for under the law. I wrote back to CRC indicating my appreciation for their offer, but pointing out that such a book would be more credibly undertaken by a laboratory animal veterinarian. I included a list of such eminent and respected veterinarians. Much to my amazement, CRC responded by informing me that the very same veterinarians I had recommended had recommended me, since I had been a major player in thinking through the ethics underlying the laws and because I was acquainted with the relevant science.

With the help of a colleague in laboratory animal medicine, Dr. M. Lynne Kesel, who functioned as my co-editor, I agreed to do the book. The project quickly expanded to two volumes containing almost fifty chapters. Volume I dealt with key concepts that would be instrumental to the control of pain and suffering in animal research—ethics, regulations, aseptic surgery, experimental design, anesthesia, analgesia, environmental stress, etc.—and included six chapters on pain and suffering. Thanks to the visionary authors I was able to recruit, the chapters superbly melded ethics and science, and many are now classic. The second volume was a species-by-species survey of animals used in biomedical research, with accounts of their basic biology and needs, diseases, how to minimize pain and suffering, and, above all, what an optimal enriched environment would be like for these animals. Again, most of my authors rose to the challenge and again transcended scientific ideology. The two volumes of *The Experimental Animal in Biomedical Research* (1989 and 1995) have become standard references, and they have provided an indelible lesson for me in how splendidly scientists can rise above the old restrictive ideology when given the occasion to do so.

Forces other than the law have contributed to the reappropriation of common sense about felt pain, at least in veterinary medicine. In the first place, clients have become more and more concerned about what their pets experience as our social ethic for animals grows and expands and as pet animals occupy ever more important roles in peoples' lives, for example, as companions for the divorced and single, playmates for children with working or single parents, family and protection for the aged and infirm. As the president of one major state veterinary association told me, "Pain control is the one area where clients don't mind paying a veterinarian, especially since the relief analgesia brings is so immediate and dramatic. At the same time, it is very easy money for the veterinarian." He went on to predict that the historical ignorance and disregard of pain management in veterinary medicine would soon disappear, driven both by client demand and the lucrative implications of that demand for veterinarians. Obviously, all this will drive veterinarians to become more educated about animal pain, and veterinary schools to feature pain recognition and pain control in their curricula. In May of 1997, the highly respected veterinary journal, *Seminars in Veterinary Medicine and Surgery (Small Animal),* published an entire issue on pain and pain control, edited by pain management pioneer Bernie Hansen.

A second force militating in favor of pain control in veterinary medicine has been the pharmaceutical industry or, more accurately, one pharmaceutical company, Pfizer. During the middle 1990s, the company ascertained that a non-steroidal anti-inflammatory drug, carprofen (trade name Rimadyl®), originally intended for human use, had highly dramatic effects in improving the pain of osteoarthritis in older dogs. Through a series of clever television advertisements, the company reminded pet owners of their moral obligation to control pain in their animals. At the same time, the company took out advertisements in veterinary journals pressing for the use of carprofen to control pain, reminding veterinarians that in the end this drug helps them do that which originally motivated them to go into veterinary medicine–relieve suffering. And Pfizer is assiduously working to educate both pet owners and veterinarians to recognize the symptoms of pain in animals. This has indeed been an opportunity for the company to do good while doing well.

III

While all of the factors discussed above have certainly contributed to the reappropriation of common sense regarding pain and its control in the scientific community, they are not the most significant vectors. In my view, what has most driven scientific concern is the change in social ethics regarding animal treatment that has occurred during roughly the last twenty-five years, a change that is reflected in and responsible for the aforementioned federal laws assuring control of pain and distress in animals. So long as society continues to extend ever-increasing levels of moral concern to animals, it will not be possible for those using animals to ignore their pain and suffering.

In 1990, I was hired by the USDA to explain the nature and roots of the new social ethic for animals, so that the agency could understand this new pressure on agriculture. They accepted my report and, for the first time in their history, made farm animal welfare a legitimate research topic for inclusion in their competitive grant awards. The report was published in book form in 1995 by Iowa State University Press as *Farm Animal Welfare: Social, Bioethical, and Research Issues.* The following rational reconstruction of the new social ethic for animals is drawn from that work.

Although society has indeed had a social consensus ethic for animal treatment during most of human history, that ethic has been extremely minimal, and was directed against overt, sadistic, willful, purposeless, and unnecessary infliction of cruelty upon animals, or else against outrageous neglect, as for example when one fails to feed and water one's animals. This ethic may be found in the Bible, in some ancient philosophers, in medieval philosophy, and is expressed today in all Western societies as anti-cruelty laws. The sources of this ethic are twofold. In the rabbinical tradition, it is based in empathetic identification with animal suffering. In the Thomistic tradition, where animals enjoy no moral status, it is based in the realization that if people are allowed to behave cruelly towards animals, they will "graduate" to abusing weaker humans, an insight confirmed by modern research.

Though minimalistic, this ethic sufficed to regulate animal treatment until the middle of the twentieth century. The reason for this lies in the nature of traditional animal use. For essentially all of human history, the overwhelming reason for domesticating and keeping animals has been agricultural. The raison d'être for agriculture has been the use of animals and their products for food, fuel, or fiber, or as a source of power for locomotion, plowing, grinding, carrying, and so on. In terms of numbers, agriculture still uses far and away the largest number of animals of any human pursuit. Other areas of animal use, such as research and testing, were statistically negligible until well into the twentieth century. So it is fair to equate the preponderance of human animal use with agriculture through virtually all of human civilization.

By the same token, certain constitutive features of animal agriculture were essentially constant across space and time until the mid-twentieth century. Most significant for our purposes was the symbiotic nature of animal agriculture, what I call the "social contract" between humans and animals. Humans provided food, forage, protection against extremes of weather and predation, and, in essence, the opportunity for the animals to live lives for which they were maximally adapted–better lives than they would live if left to fend for themselves. The animals in turn provided food, toil, fiber, and power for humans. The situation was thus a win/win one, with both animals and humans better off in the relationship than they would have been outside it.

The key feature, then, of traditional agriculture was good husbandry. And the essence of good husbandry was keeping the animals under conditions to which their natures were biologically adapted, and augmenting those natural abilities by providing additional food, protection, care, or shelter from extremes of climate, predators, disease, drought, and so on, and by selectively breeding better survival traits into their natures. (The very term *husbandry* is highly significant; etymologically, it means "bonded to the house.") Thus traditional animal agriculture involved piggybacking, as it were, on the natures and abilities of animals. If the animals thrived, the producers thrived. The animals' interests were the producers' interests. Ethics and prudence were closely intertwined: the biblical injunction to rest the animals on the Sabbath expressed both concern for animals and prudence; exhausted animals would not reproduce, produce, or work as well as rested ones. Ethics and self-interest were organically united. To this day, extensive cattle ranchers in the American West, those people whose activities most closely approximate traditional agriculture, affirm that "we take care of the animals—and the animals take care of us," and it is axiomatic to ranch children that the animals are fed and watered before they themselves eat or rest.

In this amalgam of human interest and animal interest, ethics and prudence, the strongest possible sanction existed against harming the animals—harm to the animals was harm to the farmer. Any prolonged suffering inflicted on animals by a producer, any systematic attempt to violate or work against the animals' natures would ultimately work just as much against the producers' interest as against the interests of the animals. Traditional agriculture, therefore, was primarily about putting square pegs in square holes, round pegs in round holes, with as little friction as possible. Any affronts to the animals' interests, such as knife castration or hot-iron branding, were of necessity short-lived, and systematic infliction of anything that routinely abraded the animals' biological natures was inconceivable.

This is not to say that traditional agriculture was a bed of roses for the animals. Since human control of nature was severely limited, animals suffered from famine, drought, disease, extremes of climate, and so on. But animal suffering *at human hands* was minimized by the symbiosis described above. The coincidence of ethics and self-interest was nearly perfect—anyone who deliberately hurt an animal, except for the most exigent reasons, was necessarily deviant. To be sure, this idealized picture was sometimes distorted by custom or misconception, as occurred (and still occurs) in rough handling of cattle or brutal training of horses, but wise husbandrymen knew that "gentling" was best. To this day, ranchers are reluctant to hire rodeo competitors as ranch hands, for they are too often guilty of "cowboying" the animals, being athletes interested in developing or exhibiting a skill, not husbandrymen.

If a nineteenth-century agriculturalist, let alone an ancient one, had dreamed of keeping thousands of chickens in one building, such a scheme would have been soon corrected by nature. It would be a rapid path to ruin, bringing quick spread of animal disease, death, and financial disaster. Producers did well if, and only if, animals did well, and–this is critical– "did well" for animals meant playing out their biological natures in an environment for which those powers had been selected by both natural and artificial selection.

Society, therefore, did not need laws mandating good husbandry for animals–that was dictated by self-interest and reinforced by the ancient ethic of care. If a person did not care about self-interest, he or she was unlikely to be persuaded by laws. Punishing bad husbandrymen was redundant, for they were effectively self-punishing. This, in turn, explains why the traditional social consensus ethic for the treatment of animals–the anti-cruelty ethic and, later, the laws expressing it–could be so minimal yet socially adequate. Normal people cared for their animals. Failure to care for animals; failure to provide food, water, and shelter; or eagerness to inflict pain and suffering bespoke sadism or pathology that was irrational, deviant, and needed to be socially punished. To this day, a powerful aversion to animal cruelty–that is, willfully and uselessly harming an animal, or harming an animal for frivolous reasons–is ingrained in virtually all agriculturalists. Indeed, by extracting this powerful and universal ethical maxim from ranchers, one can also quickly extract moral self-examination about rodeo, which, though culturally sanctified among ranchers, is nonetheless a source of moral discomfort to them when they are led to measure it by their own ethic.[9]

In sum, society has had a very minimal consensus ethic for the treatment of animals–the prohibition of deliberate cruelty–largely because animal abuse was fundamentally inimical to what agriculturalists, the primary animal users, were trying to accomplish. Given the almost perfect coincidence of moral concern for animals and producer and social self-interest, it was redundant to write that ethic large, in Plato's phrase, in social morality, for it was already presupposed. Only deviant, sadistic, senseless abuse represented a moral affront, and thus only such behavior became the focal point of concern for the social ethic about animal treatment.

All of this changed, however, in the mid-twentieth century with the advent of significant changes in animal use. In the first place, intensive, industrialized animal agriculture was developed in order to feed a burgeoning population with a shrinking farm labor force and less farmland.

Industry replaced husbandry. One could now put square pegs in round holes, and round pegs in square holes, because one had "technological sanders" to make them fit—antibiotics, vaccines, hormones, bacterins, etc. So one could now raise thousands of chickens in one building without their succumbing to disease and make a profit, even though the animals' biologically and psychologically based needs flowing from their natures were not being met. Thus, the ancient and fair contract was broken.

At roughly the same time, significant amounts of biomedical research and toxicity testing on animals were initiated. This too violated the ancient husbandry contract. We began to create diseases, trauma, toxicity, burns, pain, fear, etc., in these animals. Though we and other animals benefited from this research, the research animals themselves received no benefit, only significant suffering. Once again, the ancient and fair contract built into husbandry agriculture was broken.

Neither confinement agriculture nor animal research are ethically, conceptually, or legally captured by the anti-cruelty ethic. Researchers are not sadists—they are trying to help people and animals. Confinement agriculturalists are not trying to hurt animals; they are trying to supply cheap and plentiful food. But both of these activities cause enormous amounts of animal suffering on a scale hitherto unimaginable, with millions of animals being used in research and billions being produced in confinement agriculture.

Most people easily recognize the urgent need for a new ethic dealing with animal treatment that grows out of these new animal uses and the conceptual inadequacy of the traditional anti-cruelty ethic. Whenever I speak to any audience about these issues, something I have done hundreds of times, I ask the same question: If I were to draw a pie chart representing all the suffering animals currently experience at human hands, how much of that suffering would be the result of deliberate, sadistic cruelty? All of my audiences, be they rodeo cowboys in Montana or animal rights activists in San Francisco, say the same thing: "Under 1 percent, a tiny slice."

This realization, linked with urbanization, the emergence of the companion animal as the paradigm for animals in the urban mind, the rise of extensive media coverage of animal issues, the appearance of powerful arguments advocating higher moral status for animals, and the sensitivity to oppression and injustice that has emerged in society during the last forty or so years, all converge to sharpen public demand for a new ethic and new laws protecting animals. Whereas forty years ago one would have found virtually no legislation pending or being passed to protect animals, today one finds scores of such laws and bills in state, local, and federal legislatures. These laws range from the laboratory animal laws in the United States and the United Kingdom we discussed in Chapter 7, to the 1988 Swedish law abolishing confinement agriculture, to state laws banning steel-jawed traps, to municipal laws prohibiting rodeos. All of these laws go well beyond the prohibition of cruelty.

There is no need to explain the details of the new ethic that has emerged in response to social concern for animal treatment; this is done in detail in my *Farm Animal Welfare* (Ames: Iowa State University Press, 1995) or, in a more philosophical way, is rationally reconstructed in my *Animal Rights and Human Morality,* 2d ed. (Buffalo, N.Y.: Prometheus Books, 1993). All we need stress here is that there is a new ethic and that its concern is the control of pain and suffering of all sorts in animals, not the result of cruelty, and that it aims at restoring the fairness and respect for animal nature that was epitomized in traditional husbandry agriculture.

In the face of such an ethic, especially one that continues to grow, it is manifest that society will not let the scientific community ignore either the ethical issues in scientific research or the pain and suffering associated with animal use in science. Recall that society passed the 1985 laboratory animal laws despite predictions from the biomedical research community that such laws would jeopardize advances in human health care. (To the credit of that community—or at least large portions of it—it now seems to have accepted the laws as not only protecting animals, but as actually making better science.)

Thus, I am not worried about researchers backsliding in such a way as to ignore animal pain and pain control. And, as I mentioned, Animal Care Committees will inevitably keep awareness of ethics alive in animal researchers. So significant blows have been dealt to the component of scientific ideology affirming the value-free nature of science, and to the component denying the reality of felt pain and the need to control it in animals.

Nascent scientists, both undergraduate and graduate students, now seem far less willing to buy into scientific ideology than members of my generation did during the ascendance of positivism. This is perhaps because these young people have been brought up with ethical issues in science a constant matter of media and public discussion. Graduate students today would, in their formative years, have been bombarded with ethical concerns in science—animal use, improper use of human subjects, data falsification, medical research done solely on males, genetic engineering, cloning, etc. And NIH, seeming at last to realize that one cannot teach that science is value-free and ethics-free and then express surprise when people falsify data, steal results, cheat on grants, etc., has mandated courses in science and ethics for any institution receiving training grants from them. I have taught such a course since the advent of the mandate and find, frustratingly but gratifyingly, that these students find it hard to believe that anyone ever took scientific ideology seriously!

IV

It remains to consider briefly the last aspect of scientific ideology—the denial of consciousness or awareness beyond pain and suffering to animals. Although it is conceptually difficult to see how one can accept pain yet deny consciousness or awareness to animals, since pain *is* a morally relevant mode of awareness or consciousness, it appears that some psychologists and ethologists continue to do so. Instead of compartmentalizing ordinary common sense versus scientific common sense, they seem to have added a third compartment—what mentation we must admit as real as scientists because of legislation and regulation. Thus, though they may concern themselves about pain and distress when submitting protocols to the Animal Care and Use Committees, they remain agnostic about animal consciousness in their scientific work, where morally based social concern has no impact and pain is not an issue.

This basis for agnosticism about animal consciousness remains pretty much as we described it in this book. Scientists continue to eschew mentalistic language in their empirical work, for fear of being perceived as "unscientific" or "fringe," a fate that has befallen Donald Griffin in virtue of his spirited defense of animal awareness as scientifically legitimate. This agnosticism has probably been fed by cognitive science, with its emphasis on a computer model of the mind and denial of awareness as a private introspective state. Much cognitive science historically was based in an artificial intelligence model of consciousness, which was agnostic about awareness as a subjectively introspectable state. Instead, it viewed consciousness as certain sorts of behavior, creating, in essence, a new behaviorism (see the quote from Steven Harnad in Chapter 9.). Though this is changing at the cutting edge of theorizing about cognitive science, it continues to thrive in the trenches of science.

This is not, of course, to suggest that all psychologists and ethologists remain overt, militant agnostics or atheists about animal awareness. Much excellent work on anthropomorphism, deception, primate behavior, and other areas has been done with implicit or explicit acknowledgment of awareness in animals. And certainly it is more respectable to delve into these issues than it was a decade ago.

What has not occurred is an overwhelming community acceptance among psychologists and ethologists of the legitimacy of talking about animal awareness. Nor are they initiating research in the area, or philosophically supplanting positivistic, ideological doubts about its reality or knowability with a sounder philosophy—the sort of thing that did occur in the biomedical community vis à vis animal felt pain. And, in my view, this is the case because there is nothing in social morality to drive the reappropriation of ordinary common sense about animal awareness in general, the way social ethical concern about animal pain and suffering, expressed in the law, has driven the reappropriation of common sense about such pain and suffering.

One possible response to my view would point to the legal requirement of environmental enrichment for laboratory animals, and to growing social concern about the austerity of environments for farm animals raised in industrialized confinement systems, particularly hens in battery cages and sows in gestation crates. Both of these socio-ethical concerns have driven some research that essentially—at least implicitly—focuses on what animals experience and are aware of. But the bulk of research done on enrichment seems to come from the laboratory animal science community or the animal science community, or from ethologists associated with these communities rather than with ethology or psychology per se. (One exception to this is research done on environmental enrichment in zoos, which is, however, a small field.) So mainstream ethology and psychology remain basically untouched by these concerns. Thus there is no strong societal pressure on ethology and psychology to reappropriate common sense about animal awareness, or what has been called "folk psychology." In other words, to put it baldly, society doesn't care what scientists think in these areas; even as we saw in this book they didn't care what behaviorism believed.

There is no question, however, that the public possesses an endless fascination with animal awareness—this is evidenced by countless newsmagazine, newspaper, and television coverage of animal awareness and animal "mind," and by the proliferation of popular books about dog and cat and other animal thinking. As funding for research continues to shrink (especially in areas like ethology, perceived by the public as esoteric), there will inevitably be a thrust to make research seem more relevant to public concerns. Reciprocally, as concern for animal treatment continues to mount, more and more questions about animal awareness will be raised–for example, questions about boredom, loneliness, fear, happiness, etc., modalities that go beyond the more obvious aspects of pain and suffering currently being looked at. If confinement agriculture becomes as abhorrent to the U.S. public as it has become to the Swedish public, questions about what really matters to an animal will be thrown into prominence. For example, do swine need large pastures or is the quality of the space provided as important, less important, or more important than the quantity? In other words, increasingly sophisticated social-ethical concerns about animal treatment could well demand more sophisticated answers from scientists about animal awareness. And if, all of a sudden, millions of dollars in research money is available to study animal thought and feeling, ideological cavils will rapidly disappear.

V

I am guardedly optimistic about the ultimate downfall of scientific ideology regarding animal awareness in general, and I am quite optimistic regarding the demise of its denial of pain in animals and ethics in science.

As I was writing this chapter, a friend of mine, Dr. Steve Davis, an animal scientist at Oregon State University, sent me some preliminary results of a survey he had done asking a variety of groups, including both life scientists and non-scientists, whether animals "had minds and the ability to think." Though zoologists tended to be skeptical of the question as undefined (faculty more so than graduate students), 38 percent said yes while 25 percent said no. Sixty-six percent of animal scientists said yes, and fully 94 percent of veterinarians also said yes. Among ordinary people, 90 to 100 percent said yes. The majority of ordinary people also affirmed that the presence of a mind means not only that we should not be cruel to these animals, but that "management and production systems should be designed to allow the animals to exhibit their species-specific needs and/or behaviors." The latter was specifically identified by the authors as "Rollin's new social ethic."

This survey seems to accord well with our discussion. Leaving aside one methodological question, namely did scientists and veterinarians respond as scientists (i.e., cloaked in scientific common sense) or as ordinary people (wearing ordinary common sense), the results are quite plausible. The scientists' responses, as distinct from those of non-scientists and undergraduates in all majors, showed the hold of scientific ideology on credentialed scientists (except for veterinarians). On the other hand, that hold, in my view, is considerably less than it would have been 10 to 20 years earlier. The fact that the hold of ideology is less on veterinarians and animal scientists than on zoologists may be a function of the fact that the former groups have felt the pressure of social ethical concern far more significantly than zoologists have. The majority of nonscientists clearly accept the new social ethic (61–75 percent), while a smaller percentage (25–40 percent) of scientists and veterinarians do. In my view, this indicates that while scientists have not yet caught up to the social ethic, it has nonetheless established a significant beachhead in that community.

In sum, Davis's results accord well with my own experience. The new social ethic is undercutting scientific ideology, although it has not done so completely. The more that ethic impinges on the arena in which a scientist works, the more rapidly that ideology will be undercut. Skepticism about animal awareness in general will probably remain the last bastion of scientific ideology among ethologists, psychologists, zoologists, and others whose work escapes public scrutiny more than biomedical research, veterinary medicine, or animal agriculture does. But, the bottom line is, things have gotten better in the past decade, and have done so surprisingly quickly given the degree to which scientific ideology seemed to be entrenched.

NOTES

Preface

1 See B.E. Rollin, *Animal Rights and Human Morality*; id., *The Teaching of Responsibility*.
2 See D. Britt, 'Ethics, ethical committees and animal experimentation'.

Chapter 1

1 D. Hume, *An Enquiry Concerning Human Understanding*, p. 9.
2 G. Allport, 'The psychologist's frame of reference', pp. 382–3.
3 A. Eddington, *The Nature of the Physical World*, p. 342.
4 Allport, 'Psychologist's frame of reference', p. 390.
5 For typical examples of such discussions in recent textbooks, see S. Mader, *Biology*, pp. 11–15, and W. Keaton and J. Gould, *Biological Science*, pp. 2–6.
6 M.B. Visscher, Review of *Animal Rights and Human Morality*, by B.E. Rollin, p. 1303.
7 P. Feyerabend, *Science in a Free Society*.
8 B.E. Rollin, 'On the nature of illness'; id., *Animal Rights and Human Morality*.
9 In T. Beauchamp and L. Walters (eds.), *Contemporary Issues in Bioethics*, p. 89.
10 Rollin, 'Nature of illness'.
11 Rollin, 'Nature, convention, and the medical approach to the dying elderly'; *id.*, 'The right to die'.
12 See R. Munson, *Intervention and Reflection*.
13 Thomas Szasz, *The Myth of Mental Illness*; id., *Ideology and Insanity*.
14 N. Chomsky, Review of *Verbal Behavior*, by B.F. Skinner.
15 Hume, *A Treatise of Human Nature*, p. 272.
16 Ibid., p. 176.

Chapter 2

1 M.F. Seabrook, 'The psychological relationship between dairy cows and dairy cowmen and its implications for animal welfare'.
2 S. Walker, *Animal Thinking*.

3 C. Darwin, *The Descent of Man and Selection in Relation to Sex*. p. 448.
4 G. Romanes, *Animal Intelligence*, p. xi.
5 Ibid., p. vi.
6 Ibid.; emphasis added.
7 Ibid.
8 Ibid., p. vii.
9 Ibid., pp. viii–ix.
10 Ibid., p. 6.
11 Ft. Collins *Colorodoan*, 30 Aug. 1987, p. A5.
12 R. Rosenthal, *Experimental Effects in Behavioral Research*.
13 This is beautifully described in Russell McCormmach's recent, scholarly, fictional reconstruction *Night Thoughts of a Classical Physicist*.
14 Nicholas Wade and William Broad, *Betrayers of the Truth: Fraud and Deceit in the Halls of Science*, p. 19.
15 W.W. Stewart and N. Feder, 'The integrity of the scientific literature'.
16 Romanes, *Animal Intelligence*, p. 2.
17 Ibid., pp. 2–3.
18 Ibid., p. 4.
19 Ibid., p. 5.
20 Ibid., p. 399.
21 Ibid., p. 363; Romanes, 'Tails of rats and mice'.
22 Romanes, 'Experiments on the sense of smell in dogs'.
23 M.F. Washburn, *The Animal Mind: A Textbook of Comparative Psychology*, p. 104.
24 Darwin, *The Formation of Vegetable Mould Through the Action of Worms, with Observations on their Habits*, p. 95.

Chapter 3

1 T.S. Kuhn, *The Structure of Scientific Revolutions*.
2 G. Boas, 'The Mona Lisa in the history of taste'.
3 See the discussion of Linus Pauling's approach to DNA recounted in James Watson, *The Double Helix*, pp. 30, 47, 102 ff., and *passim*.
4 P. Feyerabend, *Science in a Free Society*.
5 Kuhn, *Scientific Revolutions*; E.A. Burtt, *The Metaphysical Foundations of Modern Science*; P.K. Feyerabend, *Against Method*; and A. Koyré, *From the Closed World to the Infinite Universe*; and id., *Metaphysics and Measurement*.
6 P. Forman, 'Weimar culture, causality and quantum theory, 1918–1927: adaptation by German physicists and mathematicians to a hostile intellectual environment'.
7 H.I. Brown, *Perception, Theory, and Commitment*, p. 96.
8 See Kuhn, *Scientific Revolutions*; and I. Lakatos, 'Falsification and the methodology of scientific research programmes'.
9 F. Suppe (ed.), *The Structure of Scientific Theories*, p. 7.

10 Quoted in M.H. Mandelbaum, *History, Man, and Reason: A Study in Nineteenth Century Thought*, p. 234.
11 L. Kolakowski, *The Alienation of Reason*, p. 133.
12 Quoted in A. Janik and S. Toulman, *Wittgenstein's Vienna*, p. 96.
13 R. Carnap, 'The elimination of metaphysics through logical analysis of language'.
14 For example, J.B. Watson, 'Psychology as the behaviorist views it'.
15 Thomas Reid, *Inquiry into the Human Mind on the Principles of Common Sense*.
16 In *Nature*, 20 (1879), a book entitled *The Rights of an Animal: A New Essay in Ethics*, which defended the rights of animals, was reviewed anonymously. Later correspondence in *Nature* between author and reviewer revealed the reviewer to be Romanes. In his review and in subsequent letters, Romanes indicated that we have the right to eat animals, since they eat each other, and since eating them benefits us. 'Man', he tells us, 'as an intellectual, moral, and social being has rights additional to those of a mere sentient being' (p. 427). As an example of the fact that animals do have 'some rights', he states that 'man may not kill or torture them needlessly without incurring some moral blame' (p. 288). The 'rights' accorded to animals by Romanes are thus identical to the standard prohibitions of 'unnecessary' killing, cruelty, beating, and so forth found in anti-cruelty laws. This shows that Romanes's work and belief in animal mentation did not lead to a moral position beyond that of any ordinary civilized Englishman not sharing his knowledge of phylogenetic continuity of consciousness. In an 1883 review in *Nature* (Vol. 28) of a book entitled *Physiological Cruelty*, a pro-vivisection tract, Romanes reveals himself to be a solid opponent of antivivisectionism.
17 E.P. Evans, *Evolutional Ethics and Animal Psychology*.
18 Quoted in J.F. Smithcors, 'History of veterinary anesthesia', p. 20.
19 Quoted in ibid., p. 19; emphasis added.
20 Quoted in ibid., p. 20.
21 I have discussed this pernicious process at length in my book *Animal Rights and Human Morality*.

Chapter 4

1 E.G. Boring, *A History of Experimental Psychology*.
2 D. Robinson, Preface to Volume II of Significant Contributions to the History of Psychology 1750–1920.
3 M. Marx and W. Hillix, *Systems and Theories in Psychology*, p. 168.
4 C.L. Morgan, *An Introduction to Comparative Psychology*, p. 53.
5 Ibid., p. 323.
6 Ibid., p. x.
7 Morgan, 'The limits of animal intelligence', p. 230.
8 Ibid., p. 231.
9 Morgan, *Comparative Psychology*, p. 36; emphasis added.

10 Morgan, *Comparative Psychology*, p. 50.
11 Morgan, 'Limits', p. 229.
12 Ibid., p. 239.
13 Ibid.
14 Robinson, Preface, p. xxxv.
15 R. Lowry, *The Evolution of Psychological Theory: A Critical History of Concepts and Presuppositions*, p. 157.
16 J. Loeb, *The Mechanistic Conception of Life*, p. xvi.
17 Ibid., pp. 33–64.
18 Loeb, *Comparative Physiology of the Brain and Comparative Psychology*, p. 6.
19 Ibid., p. 3.
20 Loeb, *Mechanistic Conception*, pp. 40–1.
21 Loeb, *Comparative Physiology*, p. 11.
22 Ibid., p. 12; emphasis original.
23 H.S. Jennings, *Behavior of the Lower Organisms*, p. vii.
24 Ibid., p. v.
25 Ibid.
26 Ibid., p. vii.
27 Ibid., p. 179.
28 Ibid., p. 299.
29 Ibid., pp. 332–3; emphasis original.
30 Ibid., p. 335.
31 Ibid., pp. 336–7; emphasis original.
32 J.B. Watson, Review of *Behavior of the Lower Organisms*, by H.S. Jennings.
33 E.B. Titchener, *Experimental Psychology: A Manual of Laboratory Practice*, vol. 2, pt. 1, p. xix.
34 Ibid., p. 3; emphasis original.
35 Titchener, *A Text Book of Psychology*, p. 27.
36 Ibid., pp. 31–2.
37 E.L. Thorndike, *Animal Intelligence*, p. 4.
38 Ibid., p. 6.
39 Ibid., p. 8.
40 Ibid., p. 9.
41 Ibid., p. 11.
42 Ibid., p. 12.
43 Ibid., p. 13; emphasis original.
44 See B.E. Rollin, 'Thomas Reid and the semiotics of perception'.
45 Thorndike, *Animal Intelligence*, p. 15.
46 Ibid.
47 Watson, 'Psychology as the behaviorist views it', p. 459.
48 Ibid., p. 461.
49 Ibid., p. 462.
50 Ibid., p. 463.

51 Ibid.
52 Ibid., p. 466.
53 Rollin, 'Animal consciousness and scientific change', p. 142.
54 K. Lashley, 'The behavioristic interpretation of consciousness', p. 238.
55 Ibid.
56 Ibid., p. 240.
57 Watson, 'Psychology', pp. 459–60.
58 See B.F. Skinner, 'Behaviorism at fifty!', and comments thereon.
59 D.O. Hebb, 'The American revolution', p. 736.
60 C. Burt, 'The concept of consciousness', p. 230.
61 G. Allport, 'The psychologist's frame of reference', p. 380.
62 Watson, 'Psychology', p. 465.
63 Ibid.
64 Ibid., pp. 466, 470.
65 Titchener, 'On psychology as the behaviorist views it', p. 14.
66 Cf. R.B. Joynson, *Psychology and Common Sense*. See also chaps. 8 and 9 below.
67 Allport, 'Psychologist's frame of reference', p. 383.
68 Ibid., p. 382.

Chapter 5

1 See J. Mitford, *Kind and Unusual Punishment*; P.A. Freund (ed.), *Experimentation with Human Subjects*; and R. Munson, *Intervention and Reflection*.
2 See J. Katz, 'The regulation of human experimentation in the United States'.
3 Ibid., p. 3.
4 Ibid., p. 1.
5 See, e.g. American Veterinary Medical Association, *Principles of Veterinary Medical Ethics*.
6 R. Lindsey, 'Differences in Moral Sensitivity Based Upon Varying Levels of Counseling Training and Experience'.
7 M.B. Visscher, Review of *Animal Rights and Human Morality*, by B.E. Rollin.
8 B.E. Rollin, 'The moral status of research animals in psychology'.
9 H. Lansdell, 'Rollin's article may condone immorality'.
10 D. Rosenhan and M. Seligman, *Abnormal Psychology*, p. 154.
11 Rollin, 'Seven pillars of folly: barriers to reason in laboratory animal welfare'.
12 Rollin, 'Animal pain'.
13 E.M. Wright and K.L. Marcella, 'Animal pain: evaluation and control', p. 21.
14 N. Miller, 'Value and ethics of research on animals'.
15 L. Davis, 'Species differences in drug disposition as factors in alleviation of pain', p. 175.

16 P.A. Flecknell, 'The relief of pain in laboratory animals', p. 147.
17 G. Moran, 'Severe food deprivation: some thoughts regarding its exclusive use'.
18 H. Rachlin, 'Pain and behavior'.
19 H. Selye, *The Stress of Life*.
20 W.B. Cannon, 'Stresses and strains of homeostasis'.
21 Rollin, 'Euthanasia and moral stress'.
22 W. Isaac, 'Causes and consequences of stress in the rat and mouse'.
23 R.M. Nerein *et al.*, 'Social environment as a factor in diet-induced atherosclerosis'.
24 D. Gärtner *et al.*, 'Stress response of rats to handling and experimental procedures', p. 267.
25 Rollin, 'Moral, social, and scientific perspectives on the use of swine in biomedical research'.
26 R. Kilgour, 'The application of animal behavior and the humane care of farm animals'.
27 M. Pernick, *A Calculus of Suffering: Pain, Professionalism and Anesthesia in Nineteenth Century America*, pp. 44, 46, 49, and 52 respectively.
28 Ibid., pp. 47–8.
29 Ibid., p. 58.
30 Ibid., pp. 151, 149, and 172 respectively.
31 Ibid., p. 3.
32 Ibid., p. 172.
33 Ibid., p. 154.
34 Ibid., p. 153.
35 Quoted in ibid., p. 318.
36 Ibid., pp. 155–6.
37 Quoted in ibid., p. 156.
38 See ibid., p. 183.
39 M. Shearer, 'Surgery on the paralyzed, unanesthetized newborn', p. 79.
40 Ibid.
41 R. Poland *et al.*, 'Neonatal anesthesia'.

Chapter 6

1 L. Soma, personal communication.
2 I. Kant, *Critique of Pure Reason*.
3 E. Muir, *Collected Poems: 1921–1958*.
4 See D. Davidson, 'Thought and talk'; J. Bennett, *Rationality*; R.G. Frey, *Interests and Rights: The Case Against Animals*; but also N. Malcolm, 'Thoughtless brutes', for an interesting exception.
5 L. Wittgenstein, *Philosophical Investigations*, pp. 223, 174, 90, 153, 166; id., *Zettel*, pp. 70, 91.
6 See B.E. Rollin, 'There is only one categorical imperative'.
7 See Rollin, *Natural and Conventional Meaning: An Examination of the Distinction*.

8 T. Reid, *Inquiry into the Human Mind on the Principles of Common Sense*, chap. 5.
9 Kant, *Foundations of the Metaphysics of Morals*, pp. 5–6.
10 F.J.J. Buytendijk, 'Toucher et être touché'.
11 See H.H. Price, *Thinking and Experience*.
12 D.O. Hebb, 'Emotion in man and animal'.
13 C.L. Morgan, *An Introduction to Comparative Psychology*, pp. 41–2.
14 B. Spinoza, *Ethics*, pts 3–5.
15 R.L. Kitchell, personal communication.
16 See Rollin, *Animal Rights and Human Morality*, pp. 38 ff.
17 R.L. Kitchell and R.D. Johnson, 'Assessment of pain in animals', p. 3; C.J. Vierck *et al.*, 'Evaluation of electrocutaneous pain'.
18 R. Sternbach, *Pain: A Psychophysiological Analysis*, p. 21.
19 B. Wolff and S. Langley, 'Cultural factors and the response to pain: a review', p. 145.
20 Ibid., pp. 146–7.
21 B. Tursky and R. Sternbach, 'Further physiological correlates of ethnic differences in responses to shock'.
22 H.K. Beecher, 'Relation of significance of wound to pain experienced'.
23 M.W. Fox, 'Observations of paw raising and sympathy lameness in the dog'.
24 R. Melzack and T.H. Scott, 'The effects of early experience on the response to pain'.
25 Stephen Walker's recent book *Animal Thought* elegantly documents the neurophysiological similarity between humans and animals.
26 M. Fettman *et al.*, 'Naloxone therapy in awake endotoxemic Yucatan minipigs'.
27 J.A. Gray, *The Neuropsychology of Anxiety*.
28 B. Russell, *The Problems of Philosophy*.
29 Logical positivism, in some of its later forms, transcended some of the positions detailed here; but it seems to have had little influence on the common sense of science.
30 M.P. Evans, *Criminal Prosecution and Capital Punishment of Animals*.
31 See Rollin, *Animal Rights*, pt. 2.
32 M.P. Evans, *Evolutional Ethics and Animal Psychology*; H.S. Salt, *Animals' Rights Considered in Relation to Social Progress*.

Chapter 7

1 See B.E. Rollin, 'Environmental ethics and international justice'.
2 See Rollin, *Animal Rights and Human Morality*.
3 Ibid., pp. 47 ff.
4 S. Sapontzis, *Morals, Reason and Animals*, p. 175.
5 For a summary of the Animal Welfare Act and other laws and regulations relevant to animal research, see Rollin, 'Laws relevant to animal research

in the United States'; and U.S. Congress, Office of Technology Assessment, *Alternatives to Animal Use in Research, Testing, and Education.*

6 See Rollin, 'Laws', p. 327.

7 Animal Legal Defense Fund, Submission on pain and anesthesia with reference to the improved standards for Laboratory Animals Act of 1985.

8 See A. Rowan and B.E. Rollin, 'Animal research—for and against: a philosophical, social, and historical perspective'; U.S. Congress, Office of Technology Assessment, *Alternatives.*

9 See Rollin, *Animal Rights.*

10 See U.S. Congress, Office of Technology Assessment, *Alternatives*, pp. 310–12.

11 U.S. Department of Health and Human Services, *Guide to the Care and Use of Laboratory Animals.*

12 See Rollin, 'Laws', p. 330.

13 Ibid., pp. 329–30.

14 Ibid., p. 332.

15 *Federal Register*, 31 Mar. 1987, p. 10314.

16 R. Rissler, speech given at the American Association of Laboratory Animal Science annual meeting, Chicago, 9 Oct. 1986.

17 U.S. Congress, Office of Technological Assessment, *Alternatives*, p. 341.

18 U.S. Department of Health and Human Services, Guide Supplement for Grants and Contracts, p. 8.

19 See Rollin, *Animal Rights*, pp. 92–5.

20 S. Littlewood, *Report of the Departmental Committee on Experiments with Animals*, p. 15.

21 See J. Turner, *Reckoning with the Beast*, p. 144, nn. 25, 26.

22 See R.D. French, *Antivivisection and Medical Science in Victorian Society.*

23 Littlewood, *Report*, p. 14.

24 Quoted in French, *Antivivisection*, p. 104.

25 See Littlewood, *Report.*

26 See J. Hampson, *Pain and Suffering in Experimental Animals in the United Kingdom.*

27 See C. Hollands, 'The Animals (Scientific Procedures) Act 1986'.

28 See Rollin, *Animal Rights*, pp. 95–104; Hampson, 'Laws relating to animal experimentation', p. 27; Rollin, 'Animal ethics and toxicology'.

29 See Rollin, *Animal Rights, passim.*

30 See D. Britt, 'Ethics, ethical committees and animal experimentations'; D. Porter, personal communication.

31 See Rollin, *Animal Rights*, pt. 2.

32 J. Barnes, 'Offences Against Animals'.

33 See Rollin, *Animal Rights*, pt. 2.

34 Barnes, 'Offences', pp. 35–6.

35 Committee for Research and Ethical Issues of the International Association for the Study of Pain, 'Ethical standards for investigations of experimental pain in animals', pp. 141–3.

36 Committee for Research and Ethical Issues, 'Ethical guidelines for investigations of experimental pain in conscious animals', pp. 109–10.
37 Committee for Research and Ethical Issues, 'Ethical standards', p. 141.
38 Ibid.; emphasis added.
39 Ibid.
40 Ibid., pp. 141–2.
41 See F. Wemelsfelder, 'Animal boredom: is a scientific study of the subjective experiences of animals possible?'
42 A. Guyton, *Textbook of Medical Physiology*, p. 703; emphasis added.
43 D.D. Kelly, 'Central representations of pain and analgesia', p. 199; emphasis added.
44 See Rollin, *Animal Rights*, pp. 107–17.
45 Even the 1978 Universities Federation for Animal Welfare *Handbook on the Care and Management of Laboratory Animals*, the most exhaustive book extant on laboratory animal care and welfare, written by experts for a major sophisticated animal welfare organization, had only one brief reference to analgesia in its 619 pages.
46 P.A. Flecknell, 'The relief of pain in laboratory animals', p. 147.
47 E.M. Wright and K.L. Marcella, 'Animal pain: evaluation and control', p. 20.
48 Ibid.; emphasis added.
49 Ibid., p. 21.
50 Ibid., p. 22.
51 Ibid., p. 33; emphasis added.
52 P. Taylor, 'Analgesia in the dog and cat', p. 5.
53 D.B. Morton and P.H.M. Griffiths, 'Guidelines on the recognition of pain, distress and discomfort in experimental animals and an hypothesis for assessment', p. 431.
54 Ibid.
55 Ibid.
56 Hampson, 'Laws', p. 29.
57 Quoted in J.W. Mason, 'A re-evaluation of the concept of "non-specificity" in stress theory', p. 325; emphasis added.
58 Quoted in ibid.
59 Ibid.
60 Ibid., p. 326.
61 Ibid., pp. 326–7.
62 Rollin, 'Moral, social, and scientific perspectives on the use of swine in biomedical research'.
63 J.C. Russell and D.C. Secord, 'Holy dogs and the laboratory: some Canadian experiences with animal research', p. 379.
64 See M. Kiley-Worthington, *Behavioral Problems of Farm Animals*; and P. Hemsworth, 'The social environment and the sexual behavior of the domestic boar', p. 306.
65 See M.W. Fox, *Farm Animals: Husbandry, Behavior, and Veterinary Practice*,

pp. 46–50; and M. Schmidt and H.C. Adler, 'Danish studies on behavior of early weaned piglets: preliminary results', pp. 211 ff.

66 D.L. Ingram, 'Physiology and behavior of young pigs in relation to the environment', p. 40.

67 R. Dantzer and P. Mormède, 'Can physiological criteria be used to assess welfare of pigs?', pp. 53–74.

68 R. Kilgour, 'The application of animal behavior and the humane care of farm animals', pp. 1483–4.

69 H. Bogner, 'Animal welfare in agriculture—a challenge for ethology', p. 48.

70 M.F. Seabrook, 'The psychological relationship between dairy cows and dairy cowmen and its implications for animal welfare', pp. 295–8.

71 J.L. Barnett *et al.*, 'The welfare of confined sows: psychological, behavioral, and production responses to contrasting housing systems and handler attitudes', pp. 217 ff.

72 Hemsworth, 'The behavioral response of sows to the presence of human beings and its relation to productivity', pp. 67 ff.

73 L.M. Panepinto, 'A comfortable minimum-stress method of restraint for Yucatan miniature swine'.

74 See Rollin, *Animal Rights*, pp. 38–43.

75 A. Stolba, 'Discussion', p. 49.

76 J. Weiss, 'Psychological factors in stress and disease', pp. 101 ff.

Chapter 8

1 B.E. Broadbent, *Behaviour*, p. 35.

2 See E.G. Boring, *A History of Experimental Psychology*, p. 653.

3 C.L. Hull, *Principles of Behavior*, p. 24.

4 Ibid., p. 27.

5 B.F. Skinner, 'Behaviorism at fifty!'

6 Skinner, *Beyond Freedom and Dignity*, p. 160.

7 Skinner's *Verbal Behavior*, e.g. tried to explain language in terms of the same sort of conditioned stimulus and response by which he explained maze-learning in rats.

8 G. Allport, 'The psychologist's frame of reference', p. 378.

9 Ibid., p. 381.

10 Ibid., pp. 383–4.

11 e.g. T.J. Kalikow, 'Konrad Lorenz's ethological theory: explanation and ideology, 1938–1943'.

12 K.S. Lashley, Introduction to C.H. Schiller (ed.), *Instinctive Behavior.*

13 Ibid., p. x.

14 N. Tinbergen, Preface to C.H. Schiller (ed.), *Instinctive Behavior*, p. xv.

15 K. Lorenz, *Studies in Animal and Human Behavior*, pp. xiv–xv.

16 Tinbergen, Preface, p. xvi.

17 Ibid., p. xvii.
18 Ibid., p. xv.
19 A notable exception is the British ethologist W.H. Thorpe, who seems to have had no problems with consciousness, perhaps in part as a result of being steeped in the rich British tradition which attributed consciousness to animals, stretching from Hume to Darwin, Romanes, Lloyd Morgan, and McDougall. See his *Animal Nature and Human Nature*, chap. 9.
20 Lorenz, *Studies*, p. xiii.
21 Ibid., p. 268.
22 Ibid.
23 Ibid., p. 324; emphasis added.
24 Ibid., p. 267.
25 Ibid., pp. 266–7.
26 Ibid., pp. 336–7.
27 Ibid., pp. 326–7.
28 Ibid., p. 327.
29 Ibid., p. 334.
30 Ibid.
31 Ibid., p. 272.
32 D. Krech, Introduction to 1967 edition of E.C. Tolman, *Purposive Behavior in Animals and Men*, p. xiii.
33 E.C. Tolman, *Purposive Behavior*, pp. 206–7; emphasis original.
34 Ibid., p. 573.
35 Ibid., p. 423.
36 Tolman, 'Psychology vs. immediate experience', p. 94.
37 Tolman, *Purposive Behavior*, p. 422.
38 W. McDougall, *Body and Mind: A History and A Defense of Animism*, p. x.
39 McDougall, *Outline of Psychology*, pp. 46–8.
40 Ibid., pp. 48-9.
41 McDougall, *Body and Mind*, pp. 263–4.
42 See B.E. Rollin, *Natural and Conventional Meaning: An Examination of the Distinction*; id., *Animal Rights and Human Morality*.
43 Quoted in W. Köhler, *The Selected Papers of Wolfgang Köhler*, pp. 422–3.
44 Köhler, *The Task of Gestalt Psychology*, p. 45.
45 Köhler, *Selected Papers*, pp. 429–30.
46 Köhler, *Gestalt Psychology*, p. 9.
47 Köhler, *Selected Papers*, p. 431.
48 Köhler, *The Mentality of Apes*, p. 262.
49 Ibid., p. 288.
50 F.J.J. Buytendijk, *Pain: Its Modes and Functions*, p. 72.
51 Ibid., p. 73.
52 Ibid., p. 74.
53 Ibid., p. 79.
54 Ibid., pp. 79-87.

55 T. Beer, A. Bethe, and J. von Uexküll, 'Vorschlage zu einer objectivirender Nomenclatur in der Physiologie der Nervensystems'.
56 J. von Uexküll, 'A walk through the worlds of animals and men', p. 6.
57 Ibid., p. 5.
58 Ibid., p. 34.
59 Ibid., pp. 30–1.
60 R.B. Joynson, *Psychology and Common Sense.*
61 Skinner, *Beyond Freedom*, p. 160.
62 Lorenz, *Civilized Man's Eight Deadly Sins*, p. 86.
63 L. Graham, *Between Science and Values*, p. 185.
64 See ibid. and Kalikow, 'Lorenz's ethological theory'.

Chapter 9

1 See S. Walker, *Animal Thought*, p. 73.
2 See also G. Allport, 'The psychologist's frame of reference'.
3 See B.F. Skinner, *The Behavior of Organisms: An Experimental Analysis.*
4 F. Beach, 'The snark was a boojum', p. 121.
5 Ibid.
6 Ibid., p. 122.
7 Ibid., p. 123.
8 Ibid., p. 125.
9 L. Graham, *Between Science and Values*, p. 166.
10 K. Breland and M. Breland, 'The misbehavior of organisms'.
11 M.E.P. Seligman, 'On the generality of the laws of learning'.
12 See I. Rock, 'The role of repetition in associative learning'.
13 N. Chomsky, Review of *Verbal Behavior*, by B.F. Skinner, p. 142.
14 See J. Rieux and B.E. Rollin (eds. and trans.), *The Port Royal Grammar.*
15 Chomsky, *Rules and Representations*; *id.*, 'Rules and representations'.
16 C. Burt, 'The concept of consciousness', p. 229.
17 Ibid., p. 233.
18 Ibid.
19 C.L. Hull, *Principles of Behavior.*
20 Burt, 'Concept', p. 234.
21 Ibid., pp. 239–40.
22 R.B. Joynson, *Psychology and Common Sense*, p. 74.
23 Quoted in ibid., p. 79.
24 Ibid.
25 E. Hilgard, R.C. Atkinson, and R.L. Atkinson, *Introduction to Psychology*, 6th edn., p. 159.
26 Rollin, 'Animal pain, scientific ideology, and the reappropriation of common sense'.
27 Hilgard, 'Consciousness in contemporary psychology', p. 8.
28 R.R. Holt, 'Imagery: The return of the ostracized', p. 254.

29 W.K. Estes, *Handbook of Learning and Cognitive Processes*, p. 5.
30 Hilgard, 'Consciousness', p. 10.
31 S. Harnad, 'Aspects of behaviorism', p. 901.
32 R.A. Gardner and B.T. Gardner, 'Teaching sign language to a chimpanzee'; id., 'Early signs of language in child and chimpanzee'; E.S. Savage-Rumbaugh and D.M. Savage-Rumbaugh, 'Symbolic communication between two chimpanzees'; id., 'Symbolization, language, and chimpanzees: a theoretical re-evaluation based on mutual language acquisition processes in four young *Pan troglodytes*'; id. and S. Boyson, 'Linguistically mediated tool use and exchange by chimpanzees (*Pan troglodytes*)'; D. Premack and G. Woodruff, 'Does the chimpanzee have a theory of mind?'; F.G. Patterson and E. Linden, '*The Education of Koko*'.
33 Chomsky, *Rules*; id., 'Rules'.
34 G. Kimble and L. Perlmutter, 'The problem of volition', p. 372.
35 Ibid.
36 R. Shepard and L. Cooper, *Mental Images and Their Transformations*, p. 3.
37 Ibid., p. 2.
38 Walker, *Animal Thought*, p. 1.
39 Ibid., p. xiii.
40 Ibid.
41 G. Burghardt, 'Animal awareness: current perceptions and historical perspective', p. 905.

Chapter 10

1 D. Griffin, *Listening in the Dark*.
2 Griffin, personal communication.
3 B.E. Rollin, *Natural and Conventional Meaning: An Examination of the Distinction*.
4 Griffin, *The Question of Animal Awareness*, pp. 87 ff.
5 N.K. Humphries, Review of *The Question of Animal Awareness*, by D. Griffin, p. 521.
6 Ibid.
7 Ibid.
8 Ibid.
9 Ibid.
10 Rollin, 'The Frankenstein thing'.
11 F.W.R. Brambell, *Report of the Technical Committee to Enquire into the Welfare of Animals Kept under Intensive Livestock Husbandry Systems*, p. 9.
12 D.B. Morton and P.H.M. Griffiths, 'Guidelines on the recognition of pain, distress and discomfort in experimental animals and an hypothesis for assessment'.
13 Council for Agricultural Science and Technology, 'Scientific aspects of the welfare of food animals'.

14 A. Stolba, 'Discussion', p. 49.
15 G. Burghardt, 'Animal awareness: current perceptions and historical perspective'.
16 I.J.H. Duncan, 'Animal rights–animal welfare: a scientist's assessment', p. 491.
17 J. Spinelli and H. Markowitz, 'Prevention of cage-associated distress'.
18 Griffin, *Animal Awareness;* id., *Animal Thinking;* S. Walker, *Animal Thought;* M. Dawkins, *Animal Suffering: The Science of Welfare.*
19 R.J. Herrnstein, D. H. Loveland, and C. Cable, 'Concepts in pigeons'.
20 For example, G. Gallup, 'Do minds exist in species other than our own?'
21 D. Premack and G. Woodruff, 'Does the chimpanzee have a theory of mind?'
22 Gallup, 'Do minds exist?'
23 D. S. Blough, 'Experiments in animal psychophysics'.
24 G. Kanizsa, 'Subjective contours'.
25 D.J. Ingle, 'The goldfish as a retinex animal', p.652.
26 Ibid., p. 653.
27 J. Bennett, *Rationality.*
28 J. L. Gould, 'How bees remember flower shapes', p. 1492.
29 T. Matsuzawa, 'Use of numbers by a chimpanzee'.
30 H. Davis, 'Counting behavior in animals: a critical evaluation'; id., 'Discrimination of the number three by a racoon (*Procyon lotor*)'.
31 Rollin, *Animal Rights and Human Morality.*
32 J. Goodall, *The Chimpanzees of Gombe,* p. 118.
33 Rollin, *Animal Rights and Human Morality.*
34 J. Goodall, *The Chimpanzees of Gombe,* pp. 593–4.

Chapter 11

1 James Wyngaarden, quoted in *Michigan State News,* 'Director addresses health research,' February 27, 1989, p. 8.
2 Robert White, 'Animal Ethics,' *Hastings Center Report,* 20, no. 6 (Nov.-Dec. 1990), p. 43.
3 G.G. Gallup, 'Review of Rollin' (1989), *Animal Behavior* 41, 1990, pp. 200-201; Beckoff and Jamieson, '*The Unheeded Cry* revisited,' *Animal Behavior* 43, 1992, pp. 349-51; B.E. Rollin, '*The Unheeded Cry* unheeded: author's response,' *Animal Behavior* 43, 1992, pp. 352-53; G.G. Gallup, 'Mental states must become matters of fact, not faith,' *Animal Behavior* 43, 1992, pp. 354-55.
4 S.E. Dohoo and I.R. Dohoo, 'Postoperative use of analgesics in dogs and cats by Canadian veterinarians,' *Canadian Veterinary Journal* 37 (September 1996), pp. 546-51.
5 B.R. Ferrell and M. Rhiner, 'High-tech comfort: ethical issues in cancer pain management for the 1990s,' *Journal of Clinical Ethics* 2, 1991, pp. 108-15.

6 G.A. Walco, R.C. Cassidy, and N.L. Schechter, 'Pain, hurt and harm: the ethics of pain control in infants and children,' *New England Journal of Medicine* 331, 1994, pp. 541-44.
7 K.J.S. Anand, and P.R. Hickey, 'Halothane-morphine compared with high-dose sufentanil for anesthesia and postoperative analgesia in neonatal cardiac surgery,' *New England Journal of Medicine* 369, 1992, pp. 1-9.
8 B.E. Rollin, 'Ethics, animal welfare, and ACUCs.' Forthcoming in anthology on Animal Care Committee Issues edited by John Gluck and Barbara Orlans (Purdue University Press).
9 B.E. Rollin, 'Rodeo and recollection: applied ethics and western philosophy,' *Journal of Philosophy of Sport* 23, 1996, pp. 1-10.

BIBLIOGRAPHY

Allport, G., 'The psychologist's frame of reference', presidential address to the American Psychological Association, 1939; repr. in E.R. Hilgard (ed.), *American Psychology in Historical Perspective* (Washington, DC: American Psychological Association, 1978), pp. 371–99.

American Veterinary Medical Association, *Principles of Veterinary Medical Ethics* (Chicago: AVMA, 1973).

Animal Legal Defense Fund, 'Submission on pain and anesthesia with reference to the improved standards for Laboratory Animals Act of 1985 (New York, 1986).

Barnes, J., 'Offences against animals', Substantive Criminal Law Project, Law Reform Commission of Canada (1985).

Barnett, J.L., *et al.* 'The welfare of confined sows: psychological, behavioral, and production responses to contrasting housing systems and handler attitudes', *Annales de Recherches Veterinaires*, 15 (1984), 217–26.

Beach, F., 'The snark was a boojum', *American Psychologist*, 5 (1950), 115–25.

Beauchamp, T., and Walters, L., *Contemporary issues in Bioethics* (Belmont, Calif.: Wadsworth, 1978).

Beecher, H.K. 'Relation of significance of wound to pain experienced', *Journal of the American Medical Association*, 161 (1956), 1609–13.

Beer, T., Bethe, A., and von Uexküll, J., 'Vorschlage zu einer objectivirender Nomenclatur in der Physiologie der Nervensystems', *Biologisches Zentralblatt*, 19 (1899), 517–21.

Bennett, J., *Rationality* (London: Routledge and Kegan Paul, 1964).

Blough, D.S., 'Experiments in animal psychophysics', *Scientific American*, 206 (July, 1961), 113–22.

Boas, G., 'The Mona Lisa in the history of taste', *Journal of the History of Ideas*, 1 (1940), 207–24.

Bogner, H., 'Animal welfare in agriculture—a challenge for ethology', In W. Sybesma (ed.), *The Welfare of Pigs* (The Hague: Martinus Nijhoff, 1981), pp. 46–53.

Boring, E.G., *A History of Experimental Psychology* (New York: Appleton-Century-Crofts, 1957).

Brambell, F.W.R., *Report of the Technical Committee to Enquire into the Welfare of Animals Kept Under Intensive Livestock Husbandry Systems* (London: HMSO, 1965).

Breland, K., and Breland, M., 'The misbehavior of organisms', *American Psychologist*, 16 (1960), 661–4.

Bridgman, P.W., *The Logic of Modern Physics* (New York: Macmillan, 1927).

Britt, D., 'Ethics, ethical committees and animal experimentation', *Nature*, 311 (1985), 503–6.

Broad, W., and Wade, N., *Betrayers of the Truth: Fraud and Deceit in the Halls of Science* (New York: Simon and Schuster, 1982).

Broadbent, B.E., *Behaviour* (London: Eyre and Spottiswoode, 1961).

Brown, H.I., *Perception, Theory, and Commitment* (Chicago: University of Chicago Press, 1977).

Burghardt, G., 'Animal awareness: current perceptions and historical perspective', *American Psychologist*, 40 (1985), 905–19.

Burt, C., 'The concept of consciousness', *British Journal of Psychology*, 53 (1962), 229–42.

Burtt, E.A., *The Metaphysical Foundations of Modern Science* (New York: Doubleday, 1954).

Buytendijk, F.J.J., *The Mind of the Dog* (London: Allen and Unwin, 1935).

—— *Pain: Its Modes and Functions* (Chicago: University of Chicago, 1943; repr. 1961).

—— 'Toucher et être touché', *Archives Neerlandaises de Zoologie*, 10, suppl. 2 (1953), 34–44.

Cannon, W.B., 'Stresses and strains of homeostasis', *American Journal of Medical Science*, 189 (1935), 1.

Carnap, R., 'The elimination of metaphysics through logical analysis of language', (1931; repr. in A.J. Ayer (ed.), *Logical Positivism* (Glencoe, Ill: Open Court, 1959), pp. 60–82).

Chomsky, N., *Cartesian Linguistics* (New York: Harper and Row, 1966).

—— *Problems of Knowledge and Freedom* (New York: Pantheon, 1971).

—— *Reflections on Language* (New York: Pantheon, 1975).

—— Review of *Verbal Behavior*, by B.F. Skinner, *Language*, 35 (1959), 26–58.

—— Review of *Verbal Behavior*, by B.F. Skinner (with additional comments), in L.A. Jakobovits and M.S. Miron (eds.), *Readings in the Psychology of Language* (Englewood Cliffs, NJ: Prentice-Hall, 1967), pp. 142–71.

—— *Rules and Representations* (New York: Columbia University Press, 1980).

—— 'Rules and representations', *Behavioral and Brain Sciences*, 3 (1980), 1–63.

Committee for Research and Ethical Issues of the International Association for the Study of Pain, 'Ethical guidelines for investigations of experimental pain in conscious animals', *Pain*, 16 (1983), 109–10.

—— 'Ethical standards for investigations of experimental pain in animals', *Pain*, 9 (1980), 141–3.

Council for Agricultural Science and Technology, 'Scientific aspects of the welfare of food animals', Report No. 91 (Nov. 1981).

Danto, A., *Nietzsche as Philosopher* (New York: Macmillan, 1965).

Dantzer, R., and Mormède, P., 'Can physiological criteria be used to assess welfare of pigs?', in W. Sybesma (ed.), *The Welfare of Pigs* (The Hague: Martinus Nijhoff, 1981), 53-74.

Darwin, Charles, *The Descent of Man and Selection in Relation to Sex* (1871; repr. New York: Modern Library).

—— *The Expression of the Emotions in Man and Animals*, (1872; repr. New York: Greenwood Press, 1969).

—— *The Formation of Vegetable Mould Through the Action of Worms, With Observations on Their Habits* (New York: D. Appleton and Co., 1886).

Davidson, D., 'Thought and talk', in S. Guttenplan (ed.), *Mind and Language* (Oxford: Oxford University Press, 1974), pp. 7-24.

Davis, H., 'Counting behavior in animals: a critical evaluation', *Psychological Bulletin*, 92 (1982), 547-71.

—— 'Discrimination of the number three by a racoon (*Procyon lotor*)', *Animal Learning and Behavior*, 12 (1984), 409-13.

Davis, L., 'Species differences in drug disposition as factors in alleviation of pain', in R.L. Kitchell and H.H. Erickson (eds.), *Animal Pain: Perception and Alleviation* (Bethesda, Md.: American Physiological Society, 1983), 161-78.

Dawkins, M., *Animal Suffering: The Science of Animal Welfare* (London: Chapman and Hall, 1980).

Descartes, R., *Discourse on the Method of Rightly Conducting the Reason* (1637; repr. in E.S. Haldane and G.R.T. Ross (eds.), *The Philosophical Works of Descartes*, vol. 1 (New York: Dover, 1956), pp. 79-130).

Duncan, I.J.H., 'Animal rights—animal welfare: a scientist's assessment', *Poultry Science*, 60 (1981), 489-99.

Eddington, A., *The Nature of the Physical World* (Cambridge: Cambridge University Press, 1928).

Estes, W.K., *Handbook of Learning and Cognitive Processes* (6 vols; Hillsdale, NJ: Erlbaum, 1975-8).

Evans, E.P., *Criminal Prosecution and Capital Punishment of Animals* (London: William Heinemann, 1904).

—— *Evolutional Ethics and Animal Psychology* (New York: Appleton, 1898).

Federal Register, 31 Mar. 1987, pp. 10291 ff.

Fettman, M., *et al.*, 'Naloxone therapy in awake endotoxemic Yucatan minipigs', *Journal of Surgical Research*, 37 (1984), 208-26.

Feyerabend, Paul, *Science in a Free Society* (London: NLB, 1978).

Feyerabend, P.K., *Against Method* (Atlantic Highlands, NJ: Humanities Press, 1975).

Flecknell, P.A., 'The relief of pain in laboratory animals', *Laboratory Animals*, 18 (1984), 147-60.

Forman, Paul, 'Weimar culture, causality and quantum theory, 1918-1927: adaptation by German physicists and mathematicians to a hostile intellectual environment', in R. McCormmach (ed.), *Historical Studies in the*

Physical Sciences (Philadelphia: University of Pennsylvania Press, 1971), pp. 1-116.

Fox, M.W., *Farm Animals: Husbandry, Behavior, and Veterinary Practice* (Baltimore, Md.: University Park Press, 1984).

—— 'Observations on paw raising and sympathy lameness in the dog', *Veterinary Record*, 74 (1962), 895-6.

French, R.D., *Antivivisection and Medical Science in Victorian Society* (Princeton: Princeton University Press, 1975).

Freund, P.A. (ed.), *Experimentation with Human Subjects* (New York: G. Braziller, 1970).

Frey, R.G., *Interests and Rights: The Case Against Animals* (Oxford: Oxford University Press, 1980).

Frisch, K. von, *The Dance Language and Orientation of Bees* (Cambridge, Mass.: Harvard University Press, 1967).

Gallup, G., 'Do minds exist in species other than our own?', *Neuroscience and Biobehavioral Reviews*, 9 (1985), 631-41.

Gardner, R.A., and Gardner, B.T., 'Early signs of language in child and chimpanzee', *Science*, 187 (1975), 752-3.

—— 'Teaching sign language to a chimpanzee', *Science*, 165 (1969), 664-72.

Gärtner, D., *et al.*, 'Stress response of rats to handling and experimental procedures', *Laboratory Animals*, 14 (1980), 267-74.

Goodall, J., *The Chimpanzees of Gombe* (Cambridge, Mass.: Harvard University Press, 1986).

Gould, J.L., 'How bees remember flower shapes', *Science*, 227 (1985), 1492-4.

Graham, L., *Between Science and Values* (New York: Columbia University Press, 1981).

Gray, J.A., *The Neuropsychology of Anxiety* (Oxford: Oxford University Press, 1982).

Griffin, D., *Animal Thinking* (Cambridge, Mass.: Harvard University Press, 1985).

—— *Listening in the Dark* (New Haven, Conn.: Yale University Press, 1958).

—— *The Question of Animal Awareness* (New York: Rockefeller University Press, 1976).

Guyton, A., *Textbook of Medical Physiology* (Philadelphia: Saunders, 1981).

Hampshire, S., *Spinoza* (London: Faber and Faber, 1956).

Hampson, J., 'Laws relating to animal experimentation', in A. Tuffery (ed.), *Laboratory Animals: An Introduction for New Experimentors* (London: John Wiley, 1987), pp. 21-52.

—— *Pain and Suffering in Experimental Animals in the United Kingdom*. London: Royal Society for the Prevention of Cruelty to Animals, 1983.

Harnad, S., 'Aspects of behaviorism', *Times Literary Supplement*, 16 Aug. 1985, p. 901.

Harrison, R., *Animal Machines* (London: Vincent Stuart, 1964).

Harvard University Office of Government Community Affairs, 'The animal rights movement in the United States: its composition, funding sources, goals, strategies and potential impact on research' (Cambridge, Mass.; repr. by Society for Animal Rights, 1982).

Hebb, D.O., 'The American revolution', *American Psychologist*, 15 (1960), 735-45.

—— 'Emotion in man and animal', *Psychology Review*, 53 (1946), 88-106.

Hemsworth, P., 'The behavioral response of sows to the presence of human beings and its relation to productivity', *Livestock Production Science*, 8 (1981), 67-74.

—— 'The social environment and the sexual behavior of the domestic boar', *Applied Animal Ethology*, 6 (1980), 306.

Herrnstein, R.J., and Loveland, D.H., 'Complex visual concepts in the pigeon', *Science*, 146 (1964), 549-51.

—— —— and Cable, C., 'Concepts in pigeons', *Journal of Experimental Psychology: Animal Behavior Processes*, 2 (1976), 285-302.

Hilgard, E., 'Consciousness in contemporary psychology', *Annual Review of Psychology*, 31 (1980), 1-26.

—— Atkinson, R.C., and Atkinson, R.L., *Introduction to Psychology* (6th edn; New York: Harcourt Brace Jovanovich, 1975).

Hollands, C., 'The Animals (Scientific Procedures) Act 1986', *Lancet*, 5 July 1986, pp. 32-3.

Holt, R.R., 'Imagery: the return of the ostracized', *American Psychologist*, 19 (1964), 254-64.

Hull, C.L., *Principles of Behavior* (New York: Appleton-Century-Crofts, 1943).

Hume, D., *An Enquiry Concerning Human Understanding* (1777; repr. in L.A. Selby-Bigge (ed.), *Enquiries Concerning the Human Understanding and Concerning the Principles of Morals by David Hume* (Oxford: Oxford University Press, 1963)).

—— *A Treatise of Human Nature*, ed. L.A. Selby-Bigge (Oxford: Oxford University Press, 1960).

Humphries, N.K., Review of *The Question of Animal Awareness*, by D. Griffin, *Animal Behavior*, 25 (1977), 521-2.

Iggo, A., *Pain in Animals: C. W. Hume Memorial Lecture* (Hertfordshire: UFAW, 1985).

Ingle, D.J., 'The goldfish as a retinex animal', *Science*, 226 (1985), 651-3.

Ingram, D.L., 'Physiology and behavior of young pigs in relation to the environment', in W. Sybesma (ed.), *The Welfare of Pigs* (The Hague: Martinus Nijhoff, 1981), pp. 33-45.

Isaac, W., 'Causes and consequences of stress in the rat and mouse,' paper read at American Association for Laboratory Animal Science Seminar on Stress (Atlanta, Ga., 1979).

Janik, A., and Toulman, S., *Wittgenstein's Vienna* (New York: Simon and Schuster, 1973).

Jennings, H.S., *Behavior of the Lower Organisms* (1906; repr. with an introduction by D.D. Jensen (Bloomington, Ind.: Indiana University Press, 1962)).

Joynson, R.B., *Psychology and Common Sense* (London: Routledge and Kegan Paul, 1974).

Kalikow, T.J., 'Konrad Lorenz's ethological theory: explanation and ideology, 1938–1943', *Journal of the History of Biology*, 16, no. 1 (1983), 39–73.

Kanizsa, G., 'Subjective contours', *Scientific American*, 234, no. 4 (Apr. 1976), 48–65.

Kant, I., *Critique of Pure Reason* (1787; trans. N. Kemp Smith (London: Macmillan, 1963)).

—— *Foundations of the Metaphysics of Morals* (1785; trans. L. Beck (Indianapolis: Bobbs-Merrill, 1959)).

Katz, J., 'The regulation of human experimentation in the United States—a personal odyssey', *IRB*, 9, no. 1 (1987), 1–6.

Keeton, W.T., and Gould, J.L., *Biological Science* (New York: W.W. Norton, 1986).

Kelly, D.D., 'Central representations of pain and analgesia', in R.R. Kandel and J.H. Schwartz (eds.), *Principles of Neural Science* (New York: Elsevier, 1985), pp. 199–212.

Kiley-Worthington, M., *Behavioral Problems of Farm Animals* (London: Oriel Press, 1977).

Kilgour, R., 'The application of animal behavior and the humane care of farm animals', *Journal of Animal Science*, 46 (1978), 1478.

Kimble, G., and Perlmutter, L., 'The problem of volition', *Psychological Review*, 77 (1970), 361–84.

Kitchell, R.L., and Erickson, H.H. (eds.), *Animal Pain: Perception and Alleviation* (Bethesda, Md.: American Physiological Society, 1983).

—— and Johnson, R.D., 'Assessment of pain in animals', in G. Moberg (ed.), *Animal Stress* (Bethesda, Md.: American Physiological Society, 1985).

Köhler, W., *The Mentality of Apes* (1925; repr. New York: Vintage, 1959).

—— *The Selected Papers of Wolfgang Köhler*, ed. M. Henle (New York: Liveright, 1971).

—— *The Task of Gestalt Psychology* (Princeton, N.J.: Princeton University Press, 1969).

Kolakowski, L., *The Alienation of Reason* (New York: Doubleday, 1968).

Koyré, A., *From the Closed World to the Infinite Universe* (New York: Harper and Row, 1958).

—— *Metaphysics and Measurement* (Cambridge, Mass.: Harvard University Press, 1968).

Krech, D., Introduction to 1967 edition of E.C. Tolman, *Purposive Behavior in Animals and Men* (1932; repr. New York: Appleton-Century-Crofts).

Kuhn, T.S., *The Structure of Scientific Revolutions* (Chicago: University of Chicago Press, 1970).

Lakatos, I., 'Falsification and the methodology of scientific research programmes', in I. Lakatos and A. Musgrave (eds.), *Criticism and the Growth of Knowledge* (Cambridge: Cambridge University Press, 1970), pp. 91–196.

Lansdell, H., 'Rollin's article may condone immorality', *American Psychologist*, 41, no. 8 (1986), 842.

Lashley, K.S., 'The behavioristic interpretation of consciousness', *Psychological Review*, 30 (1923), 237–72, 329-53.

—— Introduction to C.H. Schiller (ed.), *Instinctive Behavior* (New York: International Universities Press, 1957).

—— 'The problem of serial order in behavior', in L.A. Jeffress (ed.), *Cerebral Mechanisms in Behavior* (New York: John Wiley, 1951), pp. 112–46.

Lehner, P., *Handbook of Ethological Methods* (New York: Garland, 1979).

Lindsey, R., 'Differences in Moral Sensitivity Based Upon Varying Levels of Counseling Training and Experience', unpublished Ph.D. diss. (University of Denver).

Littlewood, S., *Report of the Departmental Committee on Experiments on Animals*, Cmnd. 2641 (London: HMSO, 1965).

Loeb, J., *Comparative Physiology of the Brain and Comparative Psychology* (New York: Putnams, 1900).

—— *The Mechanistic Conception of Life* (1912; ed. D. Fleming from the 1912 edn. (Cambridge, Mass.: Harvard University Press, 1964)).

Lorenz, K., *Civilized Man's Eight Deadly Sins* (New York: Harcourt Brace Jovanovich, 1974).

—— *Studies in Animal and Human Behaviour* (2 vols; Cambridge, Mass.: Harvard University Press, 1971).

Lowry, R., *The Evolution of Psychological Theory: A Critical History of Concepts and Presuppositions* (New York: Aldine, 1982).

Mader, S., *Biology: Evolution, Diversity, and the Environment* (Dubuque, lowa: W.M.C. Brown, 1987).

Malcolm, N., 'Thoughtless brutes', *Proceedings and Addresses of the American Philosophical Association*, 46 (1973), 5–20.

Mandelbaum, M.H., *History, Man, and Reason: A Study in Nineteenth Century Thought* (Baltimore: Johns Hopkins University Press, 1971).

Manning, A., *An Introduction to Animal Behavior* (Reading, Mass.: Addison-Wesley, 1979).

Marx, M., and Hillix, W., *Systems and Theories in Psychology* (New York: McGraw-Hill, 1967).

Mason, J.W., 'A re-evaluation of the concept of "non-specificity" in stress theory', *Journal of Psychiatric Research*, 8 (1971), 323–33.

Matsuzawa, T., 'Use of numbers by a chimpanzee', *Nature*, 315 (1985), 57–9.

McCormmach, R., *Night Thoughts of a Classical Physicist* (Cambridge, Mass.: Harvard University Press, 1982).

McDougall, W., *Body and Mind: A History and A Defense of Animism* (New York: Macmillan, 1920).

—— *Outline of Psychology* (New York: Scribner's, 1923).

Melzack, R., and Scott, T.H., 'The effects of early experience on the response to pain', *Journal of Comparative and Physiological Psychology*, 50 (1957), 155–61.

Miller, N., 'Value and ethics of research on animals', *Synapse*, 23 (1984), 1.
Mitford, J., *Kind and Usual Punishment* (New York: Knopf, 1973).
Moran, G., 'Severe food deprivation: some thoughts regarding its exclusive use', *Psychological Bulletin*, 82, no. 4 (1975), 543–57.
Morgan, C. L., *An Introduction to Comparative Psychology* (London: Walter Scott, 1894; facs. repr. in Robinson, pp. 1–382).
—— 'The limits of animal intelligence', *Fortnightly Review*, 54 (1893), 223–39; facs. repr. in Robinson.
Morton, D. B., and Griffiths, P. H. M., 'Guidelines on the recognition of pain, distress and discomfort in experimental animals and an hypothesis for assessment', *Veterinary Record*, 20 Apr. 1985, pp. 431–6.
Muir, E. *Collected Poems 1921–1958* (London: Faber and Faber, 1963).
Munson, R., *Intervention and Reflection* (Belmont, Calif.: Wadsworth, 1979).
Nerein, R. M., *et al.*, 'Social environment as a factor in diet-induced atherosclerosis', *Science*, 208 (1980), 1475–6.
Panepinto, L. M., 'A comfortable minimum-stress method of restraint for Yucatan miniature swine', *Laboratory Animal Science*, 33 (1983), 95.
Patterson, F. G., and Linden, E., *The Education of Koko* (New York: Holt, Rinehart and Winston, 1981).
Pernick, M., *A Calculus of Suffering: Pain, Professionalism and Anesthesia in Nineteenth Century America* (New York: Columbia University Press, 1985).
Poland, R. *et al.*, 'Neonatal Anesthesia', *Pediatrics*, 80, no. 3 (1987), 446.
Premack, D., and Woodruff, G., 'Does the chimpanzee have a theory of mind?', *The Behavioral and Brain Sciences*, 1 (1978), 515–26.
Price, H. H., *Thinking and Experience* (Cambridge, Mass.: Harvard University Press, 1953).
Rachlin, H., 'Pain and behavior', *Behavioral and Brain Sciences*, 8 (1985), 43–83.
Reid, Thomas, *Inquiry into the Human Mind on the Principles of Common Sense* (1764; repr. ed. T. Duggan (Chicago: University of Chicago Press, 1970)).
Rieux, J., and Rollin, B. (eds. and trans.), *The Port-Royal Grammar* (The Hague: Mouton, 1975).
Rissler, R., Speech given at American Association for Laboratory Animal Science annual meeting, Chicago, 9 Oct. 1986.
Robinson, D., Preface to Volume II of Significant Contributions to the History of Psychology 1750–1920, series D, *Comparative Psychology*, C. L. Morgan (Washington, DC: University Publications of America, 1977).
Rock, I., 'The role of repetition in associative learning', *American Journal of Psychology*, 70 (1957), 186–93.
Rollin, B. E., 'Animal consciousness and scientific change', *New Ideas in Psychology*, 4, no. 2 (1986), 141–52.
—— 'Animal ethics and toxicology', *Comments on Toxicology*, 5 (1987), 289–302.
—— 'Animal pain', in M. W. Fox and L. Mickley (eds), *Advances in Animal*

Welfare Science 1985–86 (The Hague: Martinus Nijhoff, 1986), pp. 91–106.
—— 'Animal pain, scientific ideology, and the reappropriation of common sense', *Journal of the American Veterinary Medical Association*, 191 (1987), 1222–6.
—— *Animal Rights and Human Morality* (Buffalo, N.Y.: Prometheus, 1981).
—— 'The concept of illness in veterinary medicine', *Journal of the American Veterinary Medical Association*, 182 (1983), 122–5.
—— 'Environmental ethics and international justice', in S. Luper-Foy (ed.), *Ethics and International Justice* (Boulder, Colo.: Westview Press, 1988).
—— 'Euthanasia and moral stress', in R. De Bellis (ed.), *Loss, Grief and Care* (Binghamton, N.Y.: Haworth Press, 1986), pp. 115–26.
—— 'The Frankenstein thing', in J. W. Evans and A. Hollaender (eds.), *Genetic Engineering of Animals: An Agricultural Perspective* (New York: Plenum, 1986), pp. 285–98.
—— 'Laws relevant to animal research in the United States', in A. Tuffery (ed.), *Laboratory Animals* (London: John Wiley, 1987), pp. 323–33.
—— 'Moral, social, and scientific perspectives on the use of swine in biomedical research', in M.E. Tumbleson, *Swine in Biomedical Research* (New York: Allan R. Liss, 1986), pp. 29–38.
—— 'The moral status of research animals in psychology', *American Psychologist*, 40, no. 8 (1985), 920–6.
—— *Natural and Conventional Meaning: An Examination of the Distinction* (The Hague: Mouton, 1976).
—— 'Nature, convention, and the medical approach to the dying elderly', in M. Tallmer *et al.* (eds.), *The Life-Threatened Elderly* (New York: Columbia University Press, 1984), pp. 95–110.
—— 'On the nature of illness', *Man and Medicine*, 4, no. 3 (1979), 157–72.
—— 'The right to die', in A. Kutscher and A. Carr (eds.), *Principles of Thanatology* (New York: Columbia University Press, 1987), pp. 109–31.
—— 'Seven pillars of folly: barriers to reason in laboratory animal welfare', *Alternatives to Laboratory Animals*, 12, no. 4 (1985), 243–7.
—— *The Teaching of Responsibility*, the C.W. Hume Lecture (Hertfordshire: UFAW, 1983).
—— 'There is only one categorical imperative', *Kant-Studien*, 67, no. 1 (1976), 60–72.
—— 'Thomas Reid and the semiotics of perception', *Monist*, 61, no. 2 (1978), 257–71.
Romanes, G., *Animal Intelligence* (London: Kegan Paul, Trench, Trubner and Co., 1898).
—— 'Experiments on the sense of smell in dogs', *Nature*, 36 (1887), 273.
—— *Mental Evolution in Animals* (New York: Appleton, 1884).
—— 'Tails of rats and mice', *Nature*, 12 (1875), 512.
Rosenhan, D., and Seligman, M., *Abnormal Psychology* (New York: Norton, 1984).

Rosenthal, R., *Experimental Effects in Behavioral Research* (New York: Appleton, 1966).

Rowan, A., and Rollin, B.E., 'Animal research—for and against: A philosophical, social, and historical perspective', *Perspectives in Biology and Medicine*, 27, no. 1 (1983), 1–17.

Russell, B., *The Problems of Philosophy* (Oxford: Oxford University Press, 1951).

Russell, J.C., and Secord, D.C., 'Holy dogs and the laboratory: some Canadian experiences with animal research', *Perspectives in Biology and Medicine*, 28 (1985), 374–81.

Ryle, G., *The Concept of Mind* (London: Hutchinson's University Library, 1949).

—— *Plato's Progress* (Cambridge: Cambridge University Press, 1966).

Salt, H.S., *Animals' Rights Considered in Relation to Social Progress* (London: G. Bell, 1892).

Sapontzis, S., *Morals, Reason and Animals* (Philadelphia: Temple University Press, 1987).

Savage-Rumbaugh, E.S., and Rumbaugh, D.M., 'Symbolic communication between two chimpanzees', *Science*, 201 (1978), 641–4.

—— 'Symbolization, language, and chimpanzees: a theoretical re-evaluation based on mutual language acquisition processes in four young *Pan troglodytes*', *Brain and Language*, 6 (1978), 265–300.

—— —— and Boyson, S., 'Linguistically mediated tool use and exchange by chimpanzees (*Pan troglodytes*)', *Behavioral and Brain Sciences*, 1 (1978), 539–54.

Schiller, C.H. (ed.), *Instinctive Behavior* (New York: International Universities Press, 1957).

Schmidt, M., and Adler, H.C., 'Danish studies on behavior of early weaned piglets: preliminary results', in W. Sybesma (ed.), *The Welfare of Pigs* (The Hague: Martinus Nijhoff, 1981), pp. 211–25.

Scottish Society for the Prevention of Vivisection, *Annual Pictorial Review*, 1987.

Seabrook, M.F., 'The psychological relationship between dairy cows and dairy cowmen and its implications for animal welfare', *International Journal for the Study of Animal Problems*, 1 (1980), 295–8.

Seligman, M.E.P., 'On the generality of the laws of learning', *Psychological Review*, 77 (1970), 406–18.

—— and Hager, J., *Biological Boundaries of Learning* (New York: Appleton-Century-Crofts, 1972).

Selye, H., *The Stress of Life* (New York: McGraw Hill, 1978).

Shearer, M., 'Surgery on the paralyzed, unanesthetized newborn', *Birth*, 13 (1986), 79.

Shepard, R., and Cooper, L., *Mental Images and Their Transformations* (Cambridge, Mass.: MIT Press, 1982).

Singer, P., *Animal Liberation* (New York: New York Review of Books, 1975).
Skinner, B.F., *The Behavior of Organisms: An Experimental Analysis* (New York: Appleton-Century-Crofts, 1938).
—— 'Behaviorism at fifty!', *Behavioral and Brain Sciences*, 7 (1984), 615–67. (See also commentaries thereon in same issue.)
—— *Beyond Freedom and Dignity* (London: Jonathan Cape, 1972).
—— *Verbal Behavior* (New York: Appleton, 1957).
Smithcors, J.F., 'History of veterinary anesthesia', in L.R. Soma (ed.), *Veterinary Anesthesia* (Baltimore: Williams and Wilkins, 1971), pp. 1–23.
Spinelli, J., and Markowitz, H., 'Prevention of Cage-Associated Distress', *Lab Animal*, 14, no. 8 (1985), 19–29.
—— 'The prevention of caging-associated distress in laboratory animals', paper read at 35th Annual Session of the American Association for Laboratory Animal Science, Seminar on the Perception of Pain and Distress in Laboratory Animals, in Cincinnati, Ohio, Nov. 1984.
Spinoza, B., *Ethics* (1677; ed. R. Elwes (New York: Saver, 1955)).
Sternbach, R., *Pain: A Psychophysiological Analysis* (New York: Academic Press, 1968).
—— (ed.), *The Psychology of Pain* (New York: Raven Press, 1978).
Stewart, W.W., and Feder, N., 'The integrity of the scientific literature', *Nature*, 325 (1987), 207–14.
Stolba, A., 'Discussion', in D.G.M., Wood-Gush, M. Dawkins, and R. Ewbank (eds.), *Self-Awareness in Domestic Animals* (Hertfordshire: UFAW, 1981), p. 49.
Suppe, F. (ed.), *The Structure of Scientific Theories* (Urbana, Ill.: University of Illinois Press, 1977).
Szasz, T., *Ideology and Insanity* (New York: Doubleday, 1970).
—— *The Myth of Mental Illness* (New York: Harper and Row, 1961).
Taylor, P., 'Analgesia in the dog and cat', *In Practice*, 7 (Jan. 1985), 5–13.
Thorndike, E.L., *Animal Intelligence* (New York: Macmillan, 1911).
Thorpe, W.H., *Animal Nature and Human Nature* (New York: Doubleday, 1974).
Tinbergen, N., Preface to C.H. Schiller (ed.), *Instinctive Behavior* (New York: International Universities Press, 1957).
Titchener, E.B., *Experimental Psychology: A Manual of Laboratory Practice* (New York: Macmillan, 1901).
—— 'On psychology as the behaviorist views it', *Proceedings of the American Philosophical Society*, 53 (1914), 1–17.
—— *A Text Book of Psychology* (New York: Macmillan, 1896).
Tolman, E.C., 'Psychology vs. immediate experience' (1935); repr. in E.C. Tolman, *Behavior and Psychological Man* (Berkeley, Calif.: University of California Press, 1951), pp. 94–115.
—— *Purposive Behavior in Animals and Men* (New York: Appleton-Century-Crofts, 1932); repr. 1967.

Turner, J., *Reckoning with the Beast* (Baltimore: Johns Hopkins University Press, 1980).

Tursky, B., and Sternbach, R., 'Further physiological correlates of ethnic differences in responses to shock', in M. Weisenberg (ed.), *Pain: Clinical and Experimental Perspectives* (St. Louis: C.V. Mosby, 1967), pp. 152–7.

Uexküll, J. von, 'A walk through the worlds of animals and men: a picture book of invisible worlds', in C.H. Schiller (ed.), *Instinctive Behavior* (New York: International Universities Press, 1957), pp. 5–82.

Umbreit, N., *Chemical Restraint of Reptiles, Amphibians, Fish, Birds, Small Mammals, and Selected Marine Mammals in North America* (Denver, Colo.: U.S. Department of the Interior, 1980).

US Congress, Office of Technology Assessment, *Alternatives to Animal Use in Research, Testing, and Education* (Washington, DC: US Government Printing Office, 1986).

US Department of Health and Human Services, *Guide to the Care and Use of Laboratory Animals* (Washington, DC: US Government Printing Office, 1985).

—— NIH Guide Supplement for Grants and Contracts, special edition, Laboratory Animal Welfare (Washington, DC: NIH, 1985).

Vierck, C.J., *et al.*, 'Evaluation of electrocutaneous pain: (1) comparison of operant reactions by humans and monkeys and (2) magnitude estimation and verbal description by human subjects', *Society of Neurosciences Abstracts*, 6 (1980), 430.

Visscher, M.B., Review of *Animal Rights and Human Morality*, by B.E. Rollin, *New England Journal of Medicine*, 306 (1982), 1303–4.

Walker, S., *Animal Thought* (London: Routledge and Kegan Paul, 1983).

Washburn, M.F., *The Animal Mind: A Textbook of Comparative Psychology*. (New York: Macmillan, 1909).

Watson, J., *The Double Helix* (New York: Atheneum, 1968).

Watson, J.B., 'Psychology as the behaviorist views it', *Psychological Review*, 20 (1913), 158; repr. in W. Dennis (ed.), *Readings in the History of Psychology* (New York: Appleton-Century-Crofts, 1948), pp. 457–71.

—— 'Review of behavior of the lower organisms', *Psychological Bulletin*, 4 (1907), 288.

Weisenberg, M. (ed.), *Pain: Clinical and Experimental Perspectives* (St. Louis: C.V. Mosby, 1975).

Weiss, J., 'Psychological factors in stress and disease', *Scientific American*, 226 (Mar. 1972), 101–13.

Wemelsfelder, F., 'Animal boredom: is a scientific study of the subjective experiences of animals possible?', in M.W. Fox and L.D. Mickley (eds.), *Advances in Animal Welfare Science 1984–1985* (The Hague: Martinus Nijhoff, 1984), pp. 115–54.

Wittgenstein, L., *Philosophical Investigations* (Oxford: Blackwell, 1958).

—— *Zettel* (Berkeley: University of California Press, 1970).

Wolff, B., and Langley, S., 'Cultural factors and the response to pain: a review' (1968; in M. Weisenberg (ed.), *Pain: Clinical and Experimental Perspectives* (St. Louis: C.V. Mosby, 1975)), pp. 144–51.

Wood-Gush, D.G.M., Dawkins, M., and Ewbank, R. (eds.), *Self-Awareness in Domestic Animals* (Hertfordshire: UFAW, 1981).

Wright, E.M., and Marcella, K.L., 'Animal pain: evaluation and control', *Lab Animal*, May–June 1985.

INDEX

abortion, 177
accountability in science, 169, 171
Alienation of Reason, The, 67
Allport, Gordon, 3, 6, 101, 104–5, 207, 208
alternatives to animal use, 183–4, 186
American Academy of Pediatrics, 133
American Medical Association: and obesity, 16; and use of anaesthetics on infants, 132
American Physiological Society, 179
American Psychological Association, 40, 187, 207–8
American Psychologist, 111, 231, 245, 250
American Veterinary Medical Association, 243
anaesthesia: as chemical restraint, 117, 123; development of, 131; and differential use on humans and animals, 131–2; and 'minor' procedures, 132–3; and non-use for infants, 132–3; in pain research, 115; regulations, 183; in veterinary surgery, 72–3, 172
analgesics, 26, 115–20, 151, 187–8, 190, 192, 194; and 'minor procedures', 132–3
analogies between animals and humans, 258; and animals as models of human states, 114–16; biological, 25–6, 195; neurophysiological, 153, 249–50
anecdotal method, 34–7, 39, 48, 50, 210
Angell J. R., 253–4
'Animal awareness: current perceptions and historical perspective', 250
Animal Behaviour, 213, 253
animal cruelty laws, 181; and child abuse, 134
Animal Intelligence, 33, 93, 96
Animal Legal Defense Fund, 176; *see also* Attorneys for Animal Rights
Animal Machines, 255
'Animal Pain: Evaluation and Control' 192
Animal Pain: Perception and Alleviation 123
Animal Procedures Committee (Britain), 184
Animal Rights and Human Morality, 111
'Animal rights—animal welfare: A scientist's assessment', 259
animal rights movement, 167–70, 174, 176, 178, 181
Animals (Scientific Procedures) Act of 1986 (Britain), 183–4
Animal Suffering: The Science of Animal Welfare, 256–9
Animal Thought, 251
Animal Welfare Act (US, 1966), 118, 127, 175: amendments, 175, 179–80, 191; and anaesthetics and analgesics, 175; and conduct of research, 175; and exemption of certain animals, 175, 178, 180, 201; and infractions of, 175
anthropocentrism, 100
anthropomorphism, 23–6, 34, 36, 116, 120, 122, 135, 145–6, 164, 180, 195, 206, 243, 258
anti-vivisection, 165, 181, 183, 185
Aristotle, 2, 3, 11, 12, 13, 58, 60, 61, 63, 157, 159
association, 78, 82, 96
Association for Biomedical Research, 177
Attorneys for Animal Rights, 176: *see also* Animal Legal Defense Fund
Australia, 186

banning of some types of research, 184
Barnett, J. L., 202
Beach, Frank, 231–5
Beecher, H. K., 151
Behavioral and Brain Sciences, The, 246
'The behavioristic interpretation of consciousness', 99

behaviour: adaptability in, 44-6, 50, 86; control of, 96, 102, 207-8, 240; 'hard-wired', 262-4; of lower organisms, 83-9; in natural environments, 209; objective, 89; study of, 209, 214, 234-5; unlearned, 233-4
behaviourism, 3, 6, 26, 53, 59, 67-75, 78, 83, 85, 88-9, 91, 97-106, 112, 116, 120-1, 124, 128, 130, 148, 158, 191, 204, 206, 207, 209-11, 216-17, 220-1, 223, 226-7, 229-30, 238, 245, 253; criticisms of, 207-8, 231-41; and human freedom, 237; and innateness of mind, 209-10; and language, 236; and learning, 208-10, 243; and perfectibility of humans, 209; and reintroduction of mental locutions, 241-3; and research on animals, 208
Behaviour of the Lower Organisms, 83
Bell's theorem, 129
Bennett, Jonathan, 137, 266
Bergson, Henri, 67
Berkeley, George, 4, 96, 200
Betrayers of the Truth, 41
Beyond Freedom and Dignity 237
Binet, Alfred, 83
biology, 11, 12, 129, 130
biomedical science, 14-18, 61, 108, 166
Birth, 133
Blough, D. S., 264
Body and Mind: A History and A Defense of Animism, 218
Bohr, Nils, 221
Boring, E. G., 74
Braithwaite, R. B., 217
Brambell Commission (Britain), 256, 259
Breland, K. and M., 235
Bridgman, P. W., 101
Britain, 167, 170, 181-5, 254-6
British associationist psychology, 96
British Veterinary Association, 188
Broad, William, 41-2
Broadbent, D. E., 206, 238
Brücke, E. W., 79
Burghardt, Gordon, 250, 258
Burt, Cyril, 101, 238-41
Burtt, E. A., 61
Buytendijk, F. J. J., 140, 225-6, 260

Caesarean sections on moose, 146
Canada, 185-6
Canadian Council on Animal Care (CCAC), 185
Canadian Law Reform Commission, 186
Canadian Medical Research Council, 186
Cannon, Walter, 125
Cartesian Linguistics, 237
castration of horses, 146
Catholicism: and view of animals, 130
causal relations, 223
Chemical Restraint of Reptiles, Amphibians, Fish, Birds, Small Mammals, and Selected Marine Mammals in North America, 118
Chimpanzees of Gombe, The, 270-2
Chomsky, Noam, 20, 138, 154, 236-8, 247
Clark, Stephen, 169
Code of Practice (Britain), 184-5
codes of medical and veterinary medical ethics, 110
cognitive psychology, 237, 240-1, 244-7; and animal consciousness, 246-7; and computer metaphor, 245-6; and imagery, 245, 247-8
Colorado State University (CSU), 171
common sense, 1-8, 13-22, 25-33, 37, 43, 65, 71, 73, 100, 104, 108, 112, 120-1, 135, 138-40, 142, 144, 159-61, 163-7, 169, 171, 173, 190, 207, 229, 231, 236, 239, 243, 248-9, 252, 254, 256, 259, 270-1; of science, 6-21, 23-5, 27, 30, 32, 39, 42, 52-4, 56, 59, 64, 88, 103, 105-9, 112-13, 129, 136, 145, 152, 156-7, 165, 171, 188, 192, 194, 198, 212, 221-2, 229-30, 239, 243, 247-9, 252-4, 256, 259, 267; *see also* ideology of science
communication, animal: of bees, 266-7; and intentionality, 252, 260, 263; and mentation, 250; of primates, 247
companion animals in research, 168; *see also* pound seizure
Comparative Physiology of the Brain and Comparative Psychology, 81
comparative psychology, 76, 82, 231
compartmentalization, 2, 6, 14, 23, 28, 32, 73, 103, 147, 165
Comte, Auguste, 67
'The Concept of Consciousness', 238
Concept of Mind, 148
concepts: animals' alleged lack of, 136, 143; a priori, 139; as evidenced in animals, 142-3; in pigeons, 262; and public check for correctness, 140-2; and sense of self, 140; and universalizability, 136; *see also* private-language argument; self-awareness
consciousness, 66, 68-70, 76, 81, 83-4, 88, 90, 97-104, 158, 207, 230-1, 238-46; in animals, 3, 23-38, 43-51,

53, 64-5, 68, 71-2, 75-6, 81-3, 87-9, 91-5, 103, 105, 114-15, 135-43, 148-9, 153-4, 163-5, 189-201, 206-7, 211-20, 225-9, 237, 241, 247-52, 254, 256, 261-3, 267-71; in humans, 23, 32, 36-7, 46, 65, 86, 91-4, 137, 148-9, 160-2, 206, 241; *see also* intelligence (in animals); subjective mental states
consensus ethic about animals, 177, 181, 189
'The Control of Pain in Laboratory Animals', 120
Cooper, L., 248-9
cosmetics, 168, 176
Cratylus, 2
Critique of Pure Reason, 137

dartboard theory, 157-8, 226
Darwin, Charles (Darwinianism), 33-5, 50, 70-1, 80, 100, 153, 165, 181, 209, 254, 259, 270-1
Darwin-Romanes view of animal mind, 54-4, 57, 60, 64, 66, 69, 71, 74-5, 77, 82-3, 85, 89-91, 97, 99, 104-5, 207
Davidson, Donald, 137
Davis, H., 267
Davis, Lloyd, 119, 193
Dawkins, Marian, 256-61, 266
Descartes, R., 4, 12-13, 24, 33, 60-1, 79, 95, 105, 123, 129, 130, 136-8, 153, 161, 163, 181, 200, 220, 237, 252
Descent of Man, The, 33
detection of flaws in science, 52-3
discontinuity between animals and humans, 86
Diseases of the Dog, 72
distress and discomfort in animals, 195; *see also* suffering in animals
'Do animals undergo subjective experience?', 214
domestic and wild animals compared, 257-8
Draize eye-irritancy test, 168
Driesch, E., 158
Du Bois-Reymond, E., 79
Duncan, Ian J. H., 259-61
duplication of experiments, 179

Ebbinghaus, H., 221-2
Eddington, A., 6-7
education of scientists, 5, 10, 135, 191, 196, 212, 243; and philosophy, 107, 192, 221
Einstein (Einsteinian physics), 3, 41, 53, 69, 221; *see also* physics
electrochemical activity in cerebral cortex, 116
Electromagnetics, 101
empiricism, 9, 53, 55, 60, 66, 68, 156-8, 237-8
endogenous opiates, 154
enriched environments for animals, 258, 261, 268, 271; and psychological well-being of primates, 179, 250; and toys, 261
entelechies, 158
environmental disasters, 19
Estes, W. K., 245
ethology, 209-30, 235, 251-2, 254, 256; cognitive, 252
Euclidean geometry, 57
euthanasia, 177; of pigs, 147; *see also* termination condition
Evans, E. P., 71, 165
evolutionary theory of continuity of mentation, 27, 32-6, 43, 53, 71, 75-7, 80, 86-7, 89, 93, 181, 252-4
excess in 19th century, 66-8
exercise of dogs, 179, 258
experience, 64-5, 78
experimental psychology, 90
Experimental Psychology: A Manual of Laboratory Practice, 90
experimentation, controlled, 39, 83
Expression of the Emotions in Man and Animals, The, 33

Fabre, J. H., 207
factory farming, 167; *see also* intensive agriculture
Faraday, Michael, 90
farm animals in research, 168
fashions: in biomedical science, 61-2; in philosophy, 54, 58; in psychology, 243, 245; in research and research funding, 56-7
feminism, 181
Feyerabend, Paul, 14, 59, 61, 64
fight or flight reaction, 125
Flecknell, Paul, 192-4
food deprivation, 122, 199
Forman, Paul, 62
Formation of Vegetable Mould Through the Action of Worms with Observations on their Habits, The, 50
fraud and falsification in science, 40-2
Freud, Sigmund, 163

Frey, R. G., 137
Friedell, Egon, 67
Frisch, K. von, 266
funding, public control of, 177–8

Galileo, 12–13, 60–1, 157, 159
Gallup, Gordon, 262–4, 266
Gärtner, D., 126, 258
genetics, 210
Georges, G., 186
Gestalt psychology, 221–4
Glamour survey, 176
'The goldfish as a retinex animal', 266
Goodall, Jane 12, 270–2
Gould, J.L., 266–7
Gould, S. J., 154
Graham, Loren, 230, 235
Griffin, Donald, 251–4, 259–61, 266
Griffiths, P. H. M., 196, 256
'Guidelines on the recognition of pain, distress, and discomfort in experimental animals and an hypothesis for assessment', 194
Guyton, A., 191

Haeckel, E., 80
Handbook of Ethological Methods, 212
Handbook of Learning and Cognitive Processes, 245
happiness in animals, 191, 203
Harnad, Steven, 246
Harrison, Ruth, 255
Hartley, David, 96
health, 15
Health Research Extension Act (US, 1985), 178–9
Hebb, D. O., 100, 142
Hegel, G., 153
Heidegger, M., 243
Heisenberg, W., 221
Helmholtz, H. L. F. von, 66, 79–80, 129
Hemsworth, P., 261
Heraclitus, 2
Herrnstein, R. J., 261–2
Hilgard, Ernest, 242, 244–5
Hillix, W., 75
history of science, 5, 11, 34, 52, 61, 63, 80, 97
Hobbes, T., 162
Holt, R. R., 245
hope in animals, 143
hopelessness and learned helplessness, 116
'How bees remember flower shapes', 266
Hughes, B. O., 261
Hull, Clark, 239; and stimulus-response psychology, 206, 232
Hume, David, 1, 2, 4, 6–7, 21–2, 65–6, 69, 96, 138–9, 158, 162–3, 181, 226, 236, 271
Humphries, N. K., 253–4
Husserl, E., 67

ideology of science, 117, 121–3, 139, 153, 170, 177, 180, 191, 268; *see also* common sense (of science)
Improved Standards for Laboratory Animals Act (1985), 179–80
incoherences in science, 113–14, 116–17, 122–3, 153; *see also* psychologists' dilemma
instinct, 77, 80, 209, 233–4
Instinctive Behavior, 211
intelligence, 56, 78; in animals, 34–5, 45, 48, 77–8, 86; in cats 47; in chimpanzees, 260, 267; in earthworms, 50; in elephants, 47; in rats, 47, 49; in wasps, 219; *see also* consciousness
intensive agriculture, 167, 191, 201, 254–6; *see also* factory farming
interests of animals, 173, 186, 203, 258, 261
International Association for the Study of Pain: guidelines for research, 187–9
Introduction to Animal Behavior, 212
Introduction to Comparative Psychology, An, 74
Introduction to Psychology, 242
introspection(ism), 77, 83, 90–3, 96, 239–40, 242, 245
invasive animal use, 9, 21, 112, 147, 164–5, 171

James, William, 136, 253
Jennings, H. S., 47, 74, 83–90, 253–4
Journal of Mental Imagery, 245
Joynson, R. B., 229, 241–2
Jung, C. G., 163

Kandel, R. R., 192
Kant, I., 45, 64–5, 69, 136–40, 158–9, 162, 213, 218, 222–3, 226–7
Katz, Jay, 109
Kilgour, R., 202, 261
Kitchell, R., 144, 190
Klein, Emmanuel, 182
Köhler, W., 221–4, 270–1

Kolakowski, L., 67
Koyré, Alexander, 4, 61
Kuhn, Thomas, 34, 61, 63, 97

LD50, 19, 168; *see also* toxicity testing
Lab Animal, 192
laboratory experiments under abnormal conditions, 38
Lakatos, I., 63
language: acquisition of, 20, 138, 264; animals' lack of, 136–8; as evidence of mind, 136, 220; and evolution, 154; as a 'form of life', 138, 142–3; and human thought, 137; and native competence, 138
Lashley, Karl, 99, 210–11, 229, 235
learning, 207–9, 235; single-trial, 234, 236; trial and error, 85, 89
legislation on animal use in research, 111, 118, 167–8, 173, 175, 178–80, 182–4, 186–7; and blocking by research community, 175, 185; and enforcement, 175–6, 184
Lehner, P., 212
Leibniz, G. W., 200, 227
Lewontin, R., 154
licensing of researchers, 184
'The limits of animal intelligence', 76
Lindsey, R., 110–11
Littlewood Commission, 183
local committee review of animal research, 177–9, 181, 185, 187, 189
Locke, John, 96
Loeb, J., 74, 79–85, 89–90, 129, 229
logic, 58
logical behaviourists, 89, 148, 160
logical positivism, 9; *see also* positivism
Logic of Modern Physics, The, 101
Loos, Adolf, 68
Lorenz, K., 39, 209–11, 213–16, 226, 229–30, 234, 260
Lotze, H., 227
Ludwig, C., 79

McDougall, William, 216, 218–20, 229
Mach, E., 66, 68–9
Manning, A., 212
Manual of Operative Veterinary Surgery, 72
Marcella, K. L., 118, 194
Markowitz, H., 261
Marx, M., 75
Mason, J. W., 197–201, 203
Massachusetts, 168
master science, 157
mathematics, 60–1, 157
mathematization of nature, 4, 12, 60–1
measurement of data, 38
medical ethics, human, 176–7
medicalization of evil, 16–18, 61
Medical Physiology, 191
Melzack, R., 144, 152
memory, 141–2
Mendel, G., 41
Mental Images and Their Transformations, 248–9
Mentality of Apes, The, 222
Metaphysics, 58
mind-body problem, 208, 213
Mind of the Dog, The, 225
Minor Surgery, 133
Mitchell, Silas, 132
moral concern for animals, 24, 167–9, 170–1, 174–9, 183–94, 200, 247, 250–4, 260–1, 268–70; and atrocities, 178; effect of science on, 269; and environmental awareness, 169; history of, 181–3; and intensive agriculture, 254–6; and killing of animals, 173; and moral philosophers, 169; and the oppressed, 169, 181; and overt cruelty, 255; in science, 171–2, 174, 180–1; *see also* welfare of animals
moral disregard for animals, 71–3, 103, 105, 111–13, 116, 118–19, 121–3, 129, 147, 163–5, 167; and philosophy, 164; and the price of morality, 164; and theology, 164
moral disregard for humans: and anaesthesia use, 131–4; as subjects of research, 166; and women and minorities, 131–2
moral philosophers, 169
'The moral status of research animals in psychology', 250
Morals, Reason and Animals, 173
Morgan, C. L., 35, 45, 74–80, 83, 143, 213, 259
Morgan's Canon, 34, 74, 75–8
Morton, D., 196, 256
Muir, Edwin, 136

National Institutes of Health, 109, 111; agreement with USDA, 180; and enforcement of guide-lines, 175–6, 178–9; guide-lines for care and use of laboratory animals, 175, 181
National Society for Medical Research, 8

Nature, 42
Nazism, 10, 210, 230
Nerein, R. M., 126
nervous system of animals, 91–2
New England Journal of Medicine, 10, 111
Newton (Newtonian physics), 3–4, 12, 41, 53, 60–1, 63, 69; *see also* physics

Occam's razor, 75, 79
Offences Against Animals, 186
operationalism, 101–2
O'Rahilly, A., 101
Origin of Species, 34
Ornament and Crime, 68

pain, 113–23, 140, 153, 165, 225–6, 243; anticipation and memory of, 143–5; control of, 118–21, 179–80, 182, 188–90, 243; definition of, lack of, 194; and evolutionary theory, 154; as experience, 192–3; identification, 194–5; and individual differences in behaviour, 30, 150–3; as insignificant, 136, 144; as momentary, 135; physicalization of, 123–8, 130; presupposition of, 114–16, 155; reduction of, 187; and similar mechanisms in humans and animals, 153–5; simulation of, 143, 151–2; and unique behaviour, 135, 145–6, 193–4
Pain, 187–8
Pain: Its Modes and Functions, 225
pain research on animals, 26, 28–9, 38, 114–17, 188
paralytic drugs: regulation of, 179, 183, 187; succinyl choline chloride, 146–7
Pavlov, Ivan Petrovitsch, 215, 237
perception: and 'boxing', 159; of objects, 139; privacy of, 238; and smell in dogs, 49; and snails, 228; study of, 229; and ticks, 228; and unity of experience, 228
Pernick, M., 131, 133
phenomenalism, 68; and solipsism, 156
phenomenalistic positivism, 156, 158–9
phenomenology, 225
philosophical and valuational commitments and assumptions, 57–8, 62, 65, 104–5, 108, 128–9, 131–4, 156, 158–9, 187, 209; and scientific change, 54, 57, 59–64, 66, 97–8, 157, 170–2, 190–1, 194, 196, 230–1, 248–9, 251–2, 269; *see also* values in science (alleged value-free nature of science)
philosophy: alternative traditions in, 213, 225; applied, 55, 63, 244; history of, 244; idealism, 227; ordinary language, 55, 95, 142, 163; twentieth-century orthodoxy, 137
phylogenetic scale, 27, 35, 42, 46, 81; and degree of consciousness, 225; and distribution of consciousness, 264; and value of lower animals, 190
physics, 3–5, 11–12, 66; *see also* Einstein (Einsteinian physics); Newton (Newtonian physics)
physiological research on animals, 182
Place of Value in a World of Facts, The, 221
Plato, 13, 58, 60–1
Popular Lectures, 66
Port-Royalists, 238
positivism, 27, 53–5, 66–9, 79, 88, 101, 112, 121, 130, 147, 155, 157–9, 225–6; *see also* logical positivism
Poultry Science, 259
pound seizure, 168; *see also* companion animals in research
Pratt, Carol, 222
preference testing of animals, 258–9
preservation of species or ecosystems, 169
Principles of Behavior, 207
Principles of Neural Science, 192
Principles of Veterinary Surgery, 72
private-language argument, 137; *see also* concepts
privileged access to mental facts, 94–5, 149, 218
problem of other minds, 137, 147–8
'The Problem of Volition', 247
Problems of Knowledge and Freedom, 237
productivity in farm animals, 256–7
Psychic Life of Organisms, 83
psychological aptitude and personality testing, 19
psychological profiles of domestic farm animals, 31
psychological research on animals, 26, 38, 48
Psychological Review, 247
Psychologie der niedersten Tiere, 83
psychologists' dilemma, 116; *see also* incoherences in science
psychology, 116, 162–3, 206–9, 229, 234, 238–44
Psychology and Common Sense, 241
Ptolemy, 12
public disenchantment with science, 169–70
Purposive Behavior in Animals and Men, 217
Purposive Behavior of Organisms, 232

purposive behaviour, 217–21
Pythagoreans, 164

quantitative ethology, 212
quantum leaps in evolution, 154
quantum theory, development of, 62, 129
Question of Animal Awareness: Evolutionary Continuity of Mental Experience, The 251–2, 260; review of, 253

Randall, John Herman, 226
reason: in animals, 139; in lower animals, 77–8; in science, 113
reductionism, 12, 15–16, 29, 66–7, 69, 79–84, 101, 128–30, 227, 240: and aesthetic theory, 130–1; and convenience to scientists, 147; and molecular biology, 129
Reflections on Language, 237
reflex action, 43–4
reflexes, 226
reflex republic, 228
Regan, Tom, 169
Reid, Thomas, 2, 71, 138
reinforcement, positive and negative, 100
replication of scientific results, 39
research on human subjects: ethics of, 108–11, 132, 166, 170; regulations on, 109
Robertshaw, David, 126
Robinson, Daniel, 74, 78
Romanes, George, 33–9, 42–50, 71, 75–7, 82, 211, 237, 252, 259, 270
Rosenhan, David, 112
Rosenthal effect, 41
Royal Commission, 181–3
Russell, B., 156
Ryle, G., 89, 95, 148, 160, 218

Sachs, Julius, 80
Sapontzis, S., 173
scepticism, 1, 24, 28, 37, 48, 66, 268
Schiller, Claire, 211
Schrödinger, E., 221
Schwartz, J. H., 192
scientific method, 3, 11
scientific revolution, 12, 63, 156
scientism, 207, 208, 234
Scott, T. H., 152
Scottish Common Sense philosophy, 96, 138, 148, 161
seal hunt, Newfoundland, 185
self-awareness, 260, 262–4; *see also*, concepts
Seligman, Martin, 112, 116, 235–6
Selye, Hans, 124, 197
Shepard, R., 248–9
'The Significance of Tropisms for Psychology', 80
Singer, Peter, 169
Skinner, B. F., 20, 102, 207, 229, 236–7
Smith, Sydney, 77
'The snark was a boojum', 231
Snow, C. P., 64
social ignorance of science, 64, 167
Soma, Lawrence, 135
Sommerhof, G., 217
Spinelli, J., 261
Spinoza, B., 144, 200, 213
Spira, Henry, 168
Stalin, J., 209
standards of care in research, 175, 178–84; for domestic farm animals, 114, 127, 201–3
stereotypical behaviour, 258
stress: and caretakers' effect on, 126, 202; and General Adaptation Syndrome (GAS), 124, 197; and pituitary-adrenal axis, 125, 197; psychological, 128, 197–201, 204; in research animals, 18, 124–9, 195, 197, 201–2, 256
stress-induced behaviours, 129
Studies in Animal and Human Behavior, 213
subjective contours experiment, 264–6
subjective mental states, 160, 211–18, 223, 227–9, 249, 253, 259–60: and behaviour, 148–9, 161; of chimpanzees, 223–4, 270–2; denial of, 155, 159–60, 207; of dogs, 224; and individual differences, 226; and meaning, 220; and moral concern, 162–4, 259; and pleasure and displeasure, 215–16; privacy of, 238; and replication in words, 239; science's presumption of, 161–3, 196–201, 204, 239; study of, 104, 238, 256, 261–6; *see also* consciousness
suffering in animals, 190, 203, 256, 259; and anxiety, 155, 190; assessment of, 256–8; and boredom, 190–1; research on; 259–61, 270; *see also* distress and discomfort in animals
Surgical Anesthesia, 132
survival surgery, multiple and single, 171–2, 176, 179, 181
swine in research, 201–3
Switzerland, 167
Systems and Theories in Psychology, 75
Szasz, Thomas, 18

Task of Gestalt Psychology, The, 222
Taub case (1982), 178
Taylor, J. G., 242
telos, 146, 203, 257–8, 269
termination condition, 172–3, 184; *see also* euthanasia
Tertullian, 113
Text Book of Psychology, 91
Thorndike, Edward, 92–6, 149, 160, 218
Tinbergen, N., 39, 209, 211–12, 229, 234
Titchener, E. B., 90–2, 102
Tolman, Edward, 206, 217–18, 232
'Toucher et être touché, 225, 260
toxicity testing, 25; *see also* LD50
training of animal-users, 179
Treaties of Human Nature, 22, 226
tropisms, plant and animal, 80–1, 85
20/20, 167

Uexküll, Jakob von, 211, 226–9, 251
Umwelt, 138, 142, 227–8
Unitas, Johnny, 150
United States, 168
United States Department of Agriculture, 176: agreement with NIH, 180
United States Department of Defense, 167–8, 179
University of Pennsylvania head-injury laboratory, 178

value of animal life, 172–3
values in science (alleged value-free nature of science), 8–15, 17, 20–1, 24, 96, 107, 112–14, 117, 147, 177, 189, 210, 272; *see also* philosophical and valuational commitments and assumptions
Verbal Behavior, 20; review of, 236
verification, principle of, 54
veterinary care of lab animals, 179
Veterinary Record, 194
vices, 261
vocalizations, 103
vocationalism in higher education, 55

Wade, Nicholas, 41–2
Walker, Stephen, 249–52, 261
'A walk through the worlds of animals and men: A picture book of invisible worlds', 227
Washburn, M., 50
Watson, J. B., 13, 35, 39, 70, 78, 89, 97–104, 116, 206–7, 217, 238–40, 267
Weiss, J., 204
welfare of animals, 259; *see also* moral concern for animals
Wemelsfelder, F., 261
Williams, Ron, 130–1
'Will I be all right, Doctor?', 177
Wittgenstein, L., 89, 95, 137–8, 140–3, 148
World Health Organization, 15
Wright, E. M., 118, 194, 204
Wundt, Wilhelm, 90